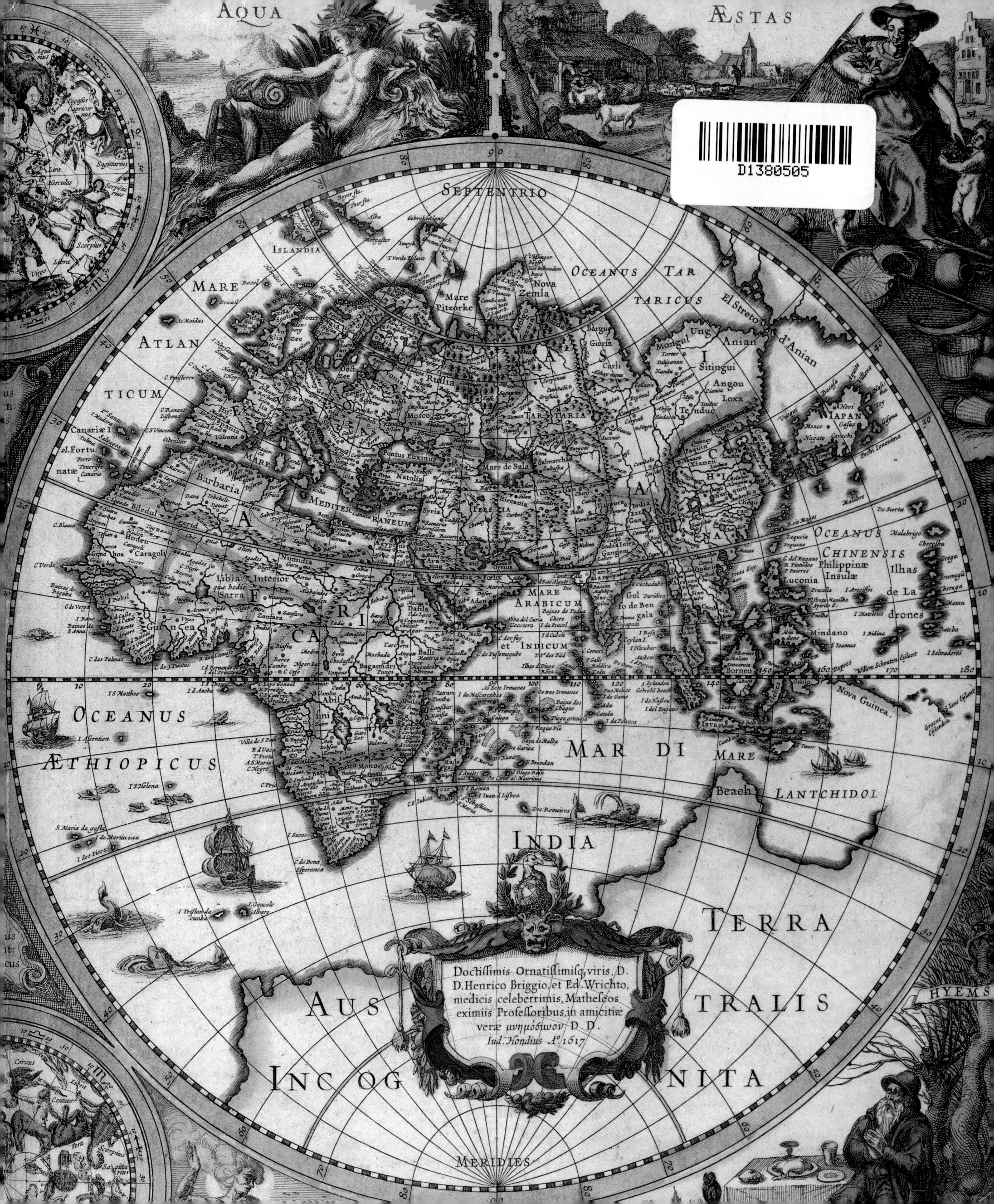
AQUA
ÆSTAS
SEPTENTRIO
ISLANDIA
MARE
ATLAN
TICUM
OCEANUS TAR
TARICUS
Nova Zemla
Mare Pitzorke
El Streto d'Anian
Mongul
Anian
Ung
Carli
Siting
Angou
Loxa
Tenduc
IAPAN
Russia
Moscovia
TARTARIA
Mare de Sala
Pontus Euxinus
Natolia
MEDITER RANEUM
Barbaria
Biledul
gerid
Numidia
Libia Interior
Sarra
PERSIA
India intra Gangem
OCEANUS CHINENSIS
Philippinæ Insulæ
Luconia
Ilhas de La drones
Mindano
MARE ARABICUM
et INDICUM
Borneo
Nova Guinea
OCEANUS
ÆTHIOPICUS
MAR DI
MARE
Beach
LANTCHIDOL
INDIA
TERRA
AUS
TRALIS
INC OG
NITA
MERIDIES
HYEMS
Doctissimis Ornatissimisq; viris D.
D. Henrico Briggio, et Ed. Wrichto,
medicis celeberrimis, Matheseos
eximiis Professoribus, in amicitiæ
veræ μνημόσυνον D. D.
Iud. Hondius A°. 1617

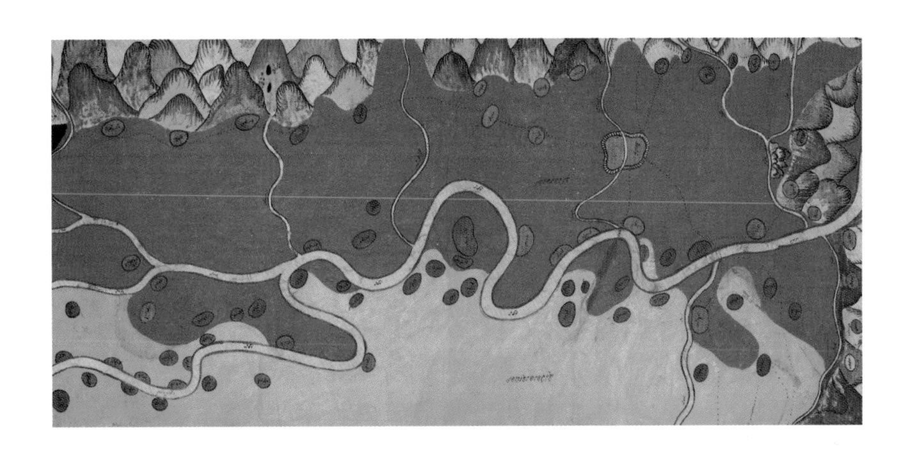

THE MAP BOOK

SMETHK
ROWLEY
R
S
WARTLEIHAL
HALESOWEN
HARB
SALOPIA
ELEY
PAR
NOR
S·KELLVMS
S
FRANCKLEY
ELLINNE

THE MAP BOOK

Edited by

PETER BARBER

WEIDENFELD & NICOLSON

Contents

Introduction

This book is a celebration of the map in its myriad forms over time. It attempts to penetrate beneath the sometimes glossy, sometimes plain surface to look at why they came into being, who their creators were, what purposes they were intended to serve and what their relationship was to the society in which they were created and whose values they inevitably represented. This may seem excessive for an object which for many people is just an ephemeral item of passing need, though there seem always to have been an equal number for whom maps have been an endless source of fascination on scientific, aesthetic, historical, cultural, emotional and symbolic grounds.

But, to begin with the question at the core of the discussion of the last item in this book (2005), what is a map? Over the past twenty years most people working on the academic study of maps have come to accept the definition formulated in 1987 by the late Brian Harley and David Woodward at the beginning of the first volume of their monumental and still incomplete *History of Cartography* that

Maps are graphic representations that facilitate a spatial understanding of things, concepts, conditions, processes, or events in the human world.

This is the definition that I have borne in mind in this book. In 1987, though, it caused some consternation among those who took a more restrictive view of the map as a primarily utilitarian geographical and scientific paper object, usually in the European tradition, whose quality was to be assessed primarily in terms of its geometrical accuracy. This led to a somewhat deterministic and simplistic view of the historical development of mapping propagated primarily by geographers, some of whom had only a passing acquaintance with historical method, with occasional contributions by historical geographers, librarians and map dealers who were anxious to increase popular interest in their area. Historians and pseudo-historians of discovery also used (and have continued to use) old maps to illustrate their texts, sometimes doing research into their creation and publication histories, but this led to a lasting confusion between the history of exploration and the history of cartography which has proved over-restrictive to both and beneficial to neither.

All researchers tended to view the evolution of mapping in terms of the gradual victory of objective 'truth' and scientific method over scientific and geographical ignorance. The achievements of national mapping agencies, such as Ordnance Survey in Britain, the United States Geological Survey or Swiss Topographical Office, were lauded as the pinnacles of cartographic achievement. With this approach came the unwritten assumptions that the only aspect of map history worth studying was its mathematical precision and that with the prospect of the creation of totally geometrically accurate maps made possible by global positioning and remote sensing via satellites and computers, the history of cartography would imminently reach its triumphant conclusion. This view devalued and sometimes totally disregarded the wider cultural context but is still influential in some circles, particularly in the German-speaking world.

Non-European mapping was dismissed in a few pages in the older histories of cartography (in the new *History of Cartography*, by contrast, it has so far had well over 2,000 tightly packed pages devoted to it). Other forms of modern maps, such as the commercial and ephemeral cartography – the types of map with which the vast majority of people are most familiar – were dismissed as debased, derivative and not worthy of study. European mapping from before 1800 and particularly from before 1500 was treated with condescension by most academic geographers, its creators being condemned as ignorant or at least befuddled and many of its most appealing aspects dismissed as 'mere decoration'.

Meanwhile mainstream historians assumed that maps were the province of academic geographers. On the whole they hesitated to utilize them as historical sources in the way that they used written documents. In so far as they took any interest, it was in the form of modern historical mapping that clarified their interpretation of the past (as readers will see if they look at 750 and 1646, historical mapping has a long tradition behind it) and as pretty illustrations or dust jackets for the books intended for a general readership.

Old maps were left to map collectors, who were generally also interested only in their decorative appeal, and their detailed study was relegated to antiquaries and local historians, many of whom were collectors, and to librarians. The typical products of their efforts were detailed listings of maps and the different states in which they were to be found. Much of this work was fascinating in its own right and was essential for any further progress. But it was marked by diffidence and quite often a lack of intellectual rigour. It was as if the authors did not have the collective courage to say out loud what their emotions told them loud and clear: that scientific measurement was not the be all and end all of mapping.

It has only been over the past twenty-five years that this has radically changed, particularly in the Anglo-Saxon world. In large part this has paradoxically been due to the disappearance of cartography and its history as a major element in geography as taught in English-speaking universities, and to the false assumption that digital Geographical Information Systems (GIS) have replaced cartography rather than providing a new medium for it. Having been lost to Geography, maps became available to other disciplines, a development encouraged by the growing popularity of interdisciplinary studies and by the increasing awareness and appreciation of the importance of the visual, which may be a consequence of the spread of television and, more recently, of the internet and the ease with which images can be created and manipulated in a digital environment. Academic historians of all types – social, political, diplomatic and fine art – and literature specialists began to take an interest in maps and to find that they sometimes offered perspectives on their subjects that were not possible from other sources. To these numbers have more recently been added the family historians and the lifelong learners who have become prominent features of early twenty-first-century life.

All have contributed to a re-evaluation of the subject. It is accepted that for some purposes, such as terrestrial and maritime navigation, scientific discovery, resource discovery and exploitation and for war and administration, mathematical accuracy did and still does play a major and even sometimes a paramount role in cartography. In other contexts, such as maps of underground railway systems or propaganda, it was not only irrelevant but inimical. Conversely the very aspects that tended traditionally to be condemned or disregarded, such as the distortions and the decoration become of enormous significance since they can give particularly precious insights into the mentalities of past ages and the views and lives of their creators as well as being packed with more general cultural information such as the receptiveness to artistic fashions.

For many map enthusiasts the fascination of maps ironically stems from their necessary lack of truth. They can be regarded as the most successful pieces of fiction ever to be created because most of their users instinctively suspend disbelief until the unpleasant reality is forced on them when they find that the enclosure map they were using does not give truthful outlines for the buildings or that the military map does not accurately depict the shapes of the fields that they were interested in. Yet it has to be that way. Given the impossibility of representing the total reality, with all its complexity, on a flat surface – be it of paper, parchment, gold or tapestry – hard decisions have to be taken as to what features to select for accurate representation or indeed for representation at all. For most of the time this process of selection is almost instinctive. The mapmaker knows the purpose that he wants his map to serve, and beyond that he is unwittingly guided by the values and assumptions of the time in which he lives – unless these are in conflict with his own value systems, as was the case with Fra Mauro in 1450, Nicholas Philpot Leader in 1823 (p. 258) or Frank Horrabin in 1926.

In order to meet that purpose, the information that is represented will be prioritized according to importance as perceived by the mapmaker – and not necessarily in accordance with actual geographical size. Even on modern national topographic mapping such features as motorways will be shown far larger than they actually are because they are important to drivers and users will expect to see them without difficulty. Conversely large features that are considered unimportant might be completely ignored or reduced in size, like parks and other public spaces in A to Z town maps. Often maps will show things that cannot actually be seen in the real world, such as relative financial wealth, as in the mapping of London instigated and financed by Charles Booth in the late nineteenth century (1891), or the geology far beneath the ground, as in the mapping of the land around Bath in 1823.

Sometimes the purpose of the map is even simpler and has nothing to do with geography. The Hereford World Map proclaims the insignificance of man and his achievements in face of the divine and the eternal. A German propaganda poster of 1938 glorifies Hitler, with the map in the background simply acting as a barometer of his achievements to that date, just as the plan of Ostia harbour of AD 64 primarily serves as a demonstration of the Emperor Nero's benevolence. Sometimes, as in the various depictions of Utopia which can be seen in this book, physical reality is totally absent or so

distorted as to be geographically meaningless. Instead the map serves as a commentary on the gap between the aspirations and feeble achievements of mankind. The quality of a map cannot be judged simply by its scientific precision but by its ability to serve its purpose and in that context aesthetic and design considerations are every bit as important as the mathematical, and often more so.

Viewed in this light, it is plain that to interpret maps as having followed a deterministic path of ever-increasing scientific perfection over time is to miss the main point. In fact they have responded to the mentalities and met the requirements of the societies in which they have been created – and will forever continue to do so. In ancient Greece and Babylon, and in eighteenth- and twentieth-century Europe, the preoccupation with precision and the scientific has indeed predominated (and see 1719 for an exaggerated manifestation). In ancient Rome, early modern China and nineteenth-century Europe the administrative use of mapping came to the fore. By contrast, for long periods of time and in many civilizations the major preoccupation was to define and to depict man's place in relationship to his surroundings, to the universe and to his God. These preoccupations were particularly evident in medieval Europe, in Aztec Mexico and at times in ancient Babylon, in China and in India. They brought forth astounding images which, despite the enormous differences in the time and place where they were created, have certain striking similarities.

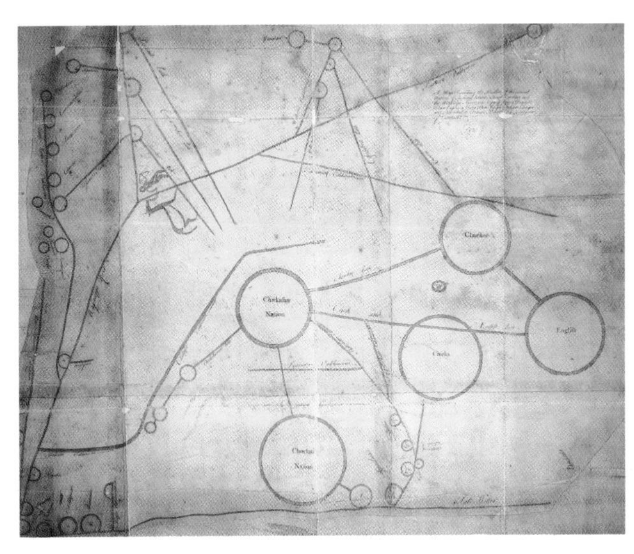

Copy of a map by a Chickasaw Indian showing the location of several nations of Indians between South Carolina and the Mississippi, c.1723.

In this book we have tried to convey something of the enormous variety and interest of maps though there are areas that could have been expanded on, such as the close relationship between man and his land which, while touched on in the context of prehistoric European mapping, could with more time and space have been more fully explored in the context of the indigenous mapping of North America and Australia (see map above). For similar reasons we have not been able to include any of the magnificent – and terrifying – depictions of the Aztec universe though the reader will get a taste of them in the depiction of the headless body that forms the centre of the 1524 map of Tenochtitlan. Nevertheless the book conveys something of the enormous range of types of map and of the media on which maps are to be found. You will see imposing images of the world and its parts (e.g. 1290, 1580) cheek by jowl with private jottings intended to give an approximate idea of thirteenth-century Palestine or the twentieth-century Antarctic (1250, 1914 p. 310), while the materials used range from scraps of paper to the plaster of magnificent painted galleries, by way of copper coins, marble, woollen tapestries, clay tablets, rock and silk.

Rock art – a record of the prayers of non-literate peoples where shamans interceded with the gods.

Fossilized Prayers

THE CHANGE FROM hunting-gathering to sedentary agricultural economies in later prehistoric times brought an unprecedented sense of insecurity as regards the community's food supply. It was no longer possible just to go out and hunt an animal or collect fruit when hungry. Crops take months to develop from seed to cereal, during which time they remain in fields exposed to all sorts of hazards, and are no less vulnerable when stored after harvest. At risk is both the daily meal and, more critically, next year's seed grain. No wonder, then, that when danger threatens the help of the gods is invoked and prayers are said to protect fields, barns, crops and animals.

Rock art is a record of the prayers of non-literate peoples. Throughout the world, in high places such as mountain passes, below a particular peak, or in a remote valley, concentrations of figures hammered (petroglyphs) into or painted (pictographs) on to rock indicate that these were once sacred places where the shaman interceded with the gods on behalf of the local people. In the Alps, at the foot of Monte Bego (Tende) and in Val Camonica (Brescia) are thousands of figures of animals, weapons, carts, people or gods, ploughs with yoked oxen, and other markings, some of which have been interpreted as plans of homesteads and fields. At Bedolina, Val Camonica, an unusually large, and slightly atypical, assemblage (4.6 by 2.3 metres/15 by 7½ feet) is spread over the glacially smoothed surface of a rock high on the hillside. What we see today is a palimpsest, in which different episodes of prayer, possibly separated by hundreds of years, have left a succession of rectangles, irregular lines and dots and, on top of all, outlines of timber-framed houses. The earliest markings are thought to have been made in the Bronze Age (*c.*1200 BC), the latest (the houses) in the Iron Age (*c.*900–700 BC). It is extremely unlikely that the fields, springs and interconnecting paths thus represented ever 'mapped' an actual landscape (there is no archaeological evidence for any matching settlement). Instead, we should see the petroglyphic markings as representing the community's cultivated plots and as an expression of acute anxiety and insecurity. Smaller maps, representing even more clearly a couple of buildings surrounded by two or three enclosed plots, are relatively common not only elsewhere in the Alps, but also in distant regions, such as the upper Yenisei River, Russia.

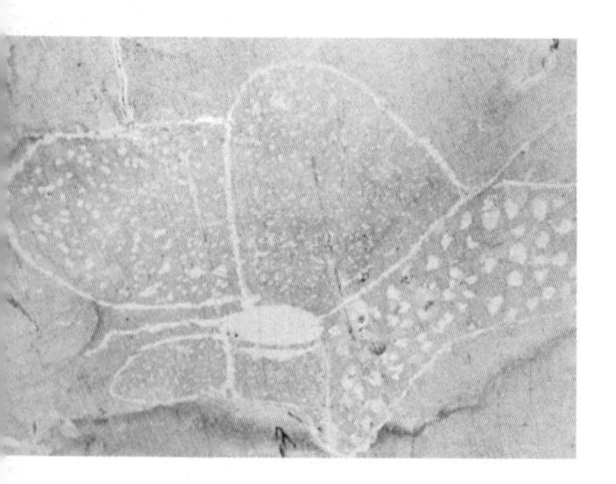

Right Redrawing of the oldest known plan of an inhabited site, Bedolina, Val Camonica, Italy.
Above Photograph of a map of a hut and fields, Monte Bego, Italy.

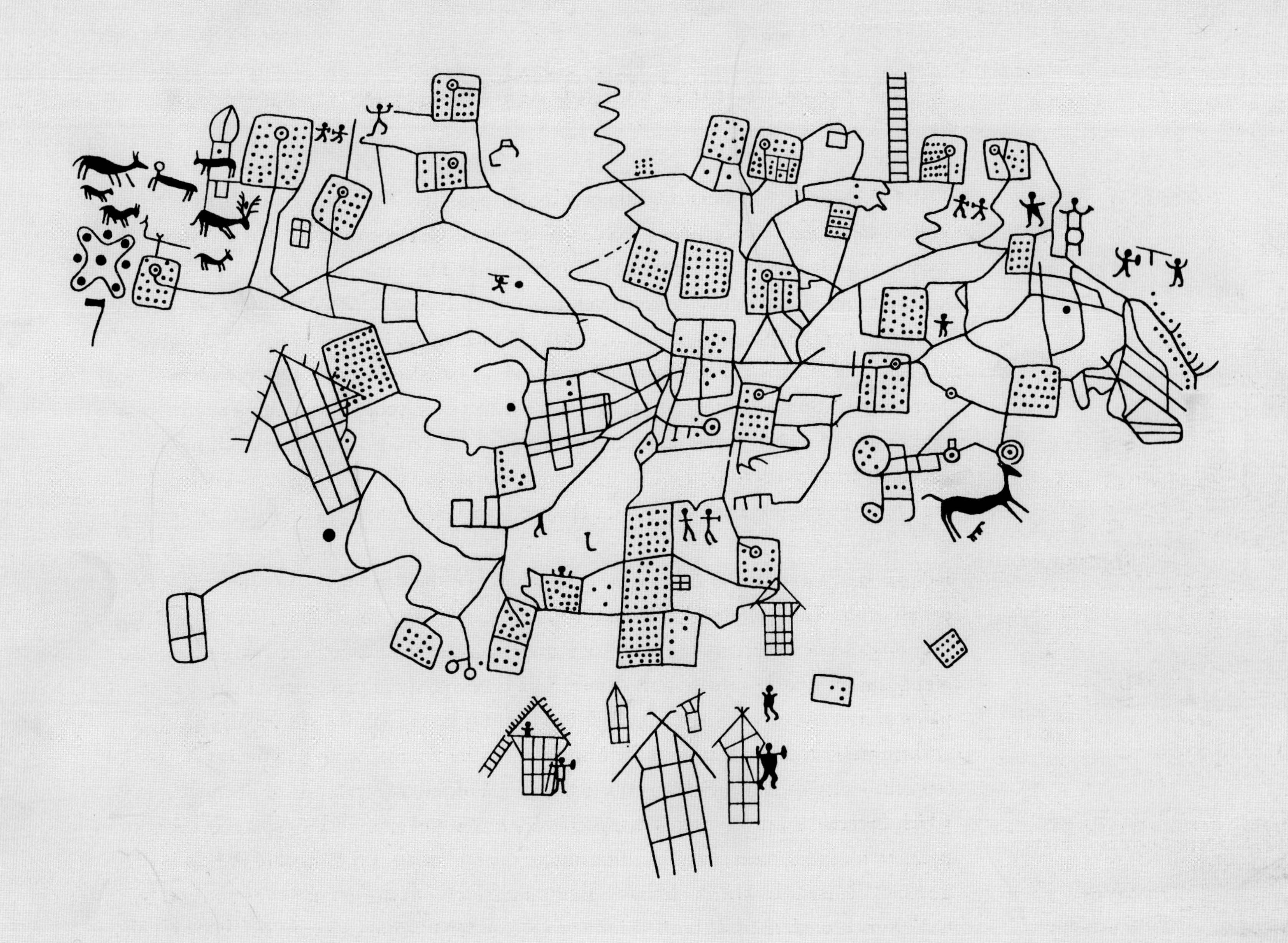

The fortifications, principal buildings and major topographical features of the holy city of Nippur, in what is now modern-day Iraq.

The Earliest Town Plan

Right Plan of Nippur, c.1500 BC.
See also 200, 565.

THE BABYLONIANS FIRST developed some of mankind's key mathematical concepts such as the 360 degrees dividing a circle and, hundreds of years before the birth of Pythagoras, what is now known as his theorem. They also created the first maps to incorporate two key concepts behind much modern mapping: explicitly stated orientation and an explicit scale.

Very few of their maps, which were generally drawn onto small tablets, have survived. They included plans of single buildings, maps of agricultural land, which were created in the context of sales and legal disputes, and, as will be seen, world maps which served educational purposes.

This map dates from about 1500 BC, some seven hundred years later than the earliest known Babylonian plan (which shows the outline of a building and is to be seen on the lap of a headless statue in the Louvre in Paris). It is one of the most important. It is the earliest known, though incomplete, town plan. About a quarter of the city of Nippur, a holy city of the Sumerians and later the Babylonians on the Euphrates between Baghdad and Basra, in what is now Iraq, is shown. In a way that anticipates many European town plans of over 3,000 years later, the map, which may be drawn to a consistent scale, concentrates on the essentials: the town walls, the waterways, what may be two large storehouses by the river and the principal features of the town, identified in cuneiform script, that needed protection. These are two temples and a large open space at the bottom left. The most important temple, dedicated to the Sumerian god Enlil, had double walls and two courtyards. The Euphrates can be seen curving to the left and a canal runs vertically through the middle of the town. Seven town gates are shown and named. For clarity's sake, other buildings are omitted. The map may have been intended, like its later European counterparts, to assist in overhauling the town's defences.

The map is an amazing achievement, testifying to the sophistication of Babylonian civilization. It was as much to commemorate their civilization as because of its place in the Bible that the city of Babylon continued to be shown on European world maps through the medieval period.

Ancient Egyptian mapping – a rough plan of a burial chamber used by workmen refashioning the tomb of Rameses IV.

Fit for a Pharaoh

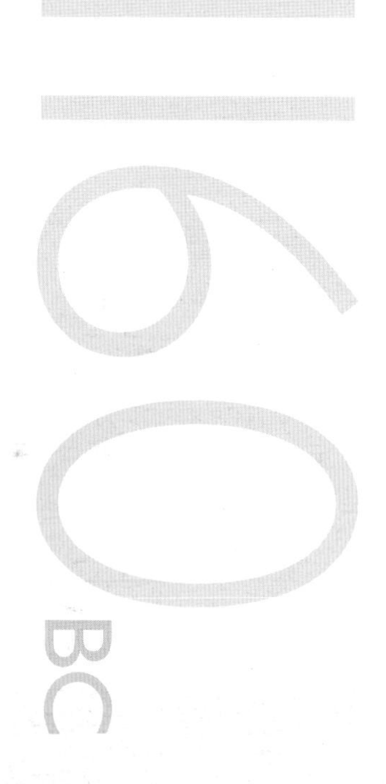

RAMESES IV'S TOMB was unfinished when he unexpectedly died. The series of rooms needed to be refigured so that the large hall which would normally precede the burial chamber could be adapted to house the sarcophagus. Drawn plans seem to have been used in the process, of which two, one on a broken pottery shard and this one, on papyrus, have survived.

The plan, which is incomplete and damaged at the bottom, sets the tomb in context and seems to show its final arrangement. The red-outlined area with red and black dotted lines on a brown background surrounding the plan represents the desert. The plan itself is drawn with double lines as though to represent a free-standing building rather than one hewn out of rock. The sequence of the rooms corresponds to that of Rameses's tomb but they are not to scale or even in all cases to shape. The user of the plan would have been dependent for precise information about size on the written legends, which also give information about the progress of the works. The plan then, for all the care taken in its execution, could not have stood on its own as an expression of the reality that it helped to illustrate.

In this respect it was less sophisticated than the maps and plans being produced in the same years in the Babylonian Empire. This seems to have been characteristic of mapmaking as a whole in ancient Egypt. Ancient Egyptians apparently created most of their pyramids and temples without drawing plans and, with rare exceptions, such as a surviving pictorial plan showing gold mines in Nubia, they successfully administered their empire without recourse to maps or plans, though the very patchy nature of the evidence means one cannot be certain. They did create pictorial plans on their coffins of journeys through the fictional land of the dead that were to be undertaken by the deceased, and depictions of idealized lands to be farmed by them. They similarly created beautiful carved and painted pictorial cosmological depictions of the earth and the heavens. These, however, while fulfilling their intended purposes, were not based on measurement of any kind. Sophisticated, mathematically derived mapping of lands in Egypt only begins to be found much later during the reigns of the dynasty founded by Ptolemy, one of Alexander the Great's generals, who introduced numerous innovations derived from Greek practices.

Right Plan on papyrus of Rameses IV's tomb, not drawn to scale.

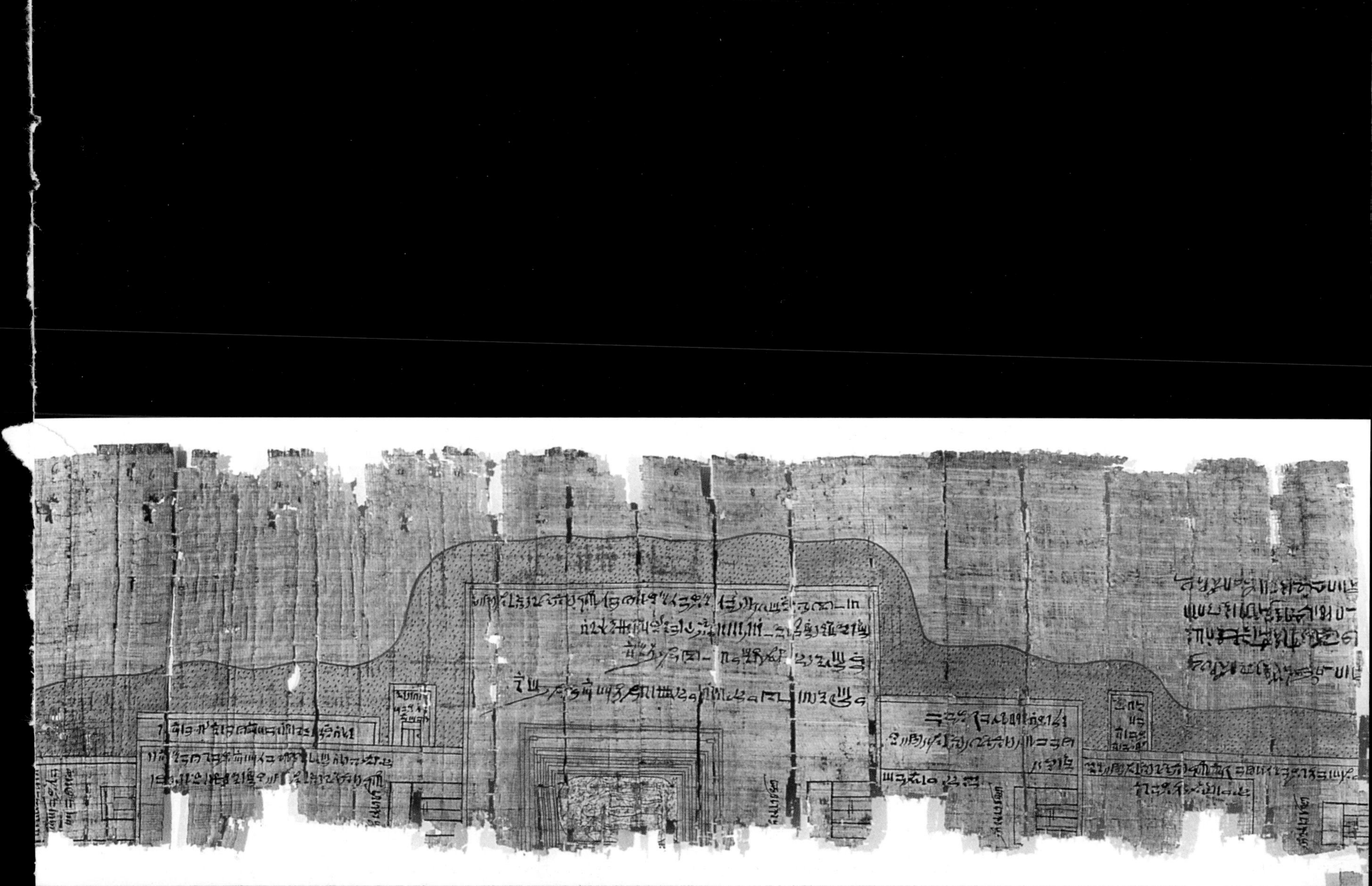

The four corners of the earth with the great Babylonian Empire at its centre depicted on a very early clay tablet.

Babylonia

Right A Babylonian map of the world.

A BAKED CLAY TABLET now in the British Museum (Gallery 55, Case 15) contains a text accompanied by a map. Although believed to have been found in Sippar (or possibly Borsippa), Iraq, the tablet is likely to have been made in Babylon in the seventh or sixth century BC. A note inscribed on the tablet, however, indicates that it was copied from an older tablet made (it would seem from internal evidence) in the ninth century. The tablet, somewhat damaged around the edges, measures 122 by 82 millimetres (5 by 3 inches). About a third of one side (and all the reverse) is filled with text and the map occupies the remaining two-thirds.

The tablet was made at a time when the Babylonian and Assyrian empires had reached their maximum extent and the map highlights the relationship of the most distant regions of the earth to the Babylonian heartland. Its features are clearly delineated and most are labelled. North is at the top of the map. The earth is shown as a flat disk surrounded by, or floating on, an ocean, beyond which the outer regions are indicated by triangles, originally probably eight in number, on which the distance between each is indicated. The remote lands were inhabited by legendary beasts. In one, it is noted that 'the sun is not seen'; in another the 'great wall' refers to the birthplace of a demonic figure mentioned in Sumerian texts. A circle defines the Babylonian world. Parallel lines stand for the unnamed River Euphrates, which flows into a swamp (labelled as such and marked by a rectangle at the mouth of the river) before arriving at the sea in two outflows, each also labelled. A canal or waterway, *bitqu*, possibly an antecedent of the modern Shatt-al-Arab, is shown. The central place, Babylon, is singled out by having both a name and a prominent oblong place sign, whereas other cities are marked with a circle containing its name or just a dot. Regions, such as Assyria, 'Der', 'Bit-Yakin' (an Aramaean territory around the southern Euphrates), 'Habban' (part of Kassite territory around Kermanshah) and 'Urartu' (an independent kingdom in the Iranian-Turkish-Russian borderland), are named. The ancient author concluded by commenting that his sketch showed, as we would put it today, 'the four corners of the world'.

82
7-14
509
92687

Emperor Nero's new artificial harbour at Ostia, at the mouth of the Tiber, depicted on coins to his greater glory.

Bread and Circuses

64

Right Nero. Bronze sestertius, AD 64–68 showing Ostia harbour.

THE ROMANS WERE among the first to use maps as propaganda symbols on coins. This sestertius commemorates the completion, under Nero, of the artificial harbour at the mouth of the Tiber, about 2 miles (3 kilometres) north of the coastal town of Ostia, that had been commenced by the Emperor Claudius. Until that time, grain, the all-important staple ingredient for bread, had had to be landed at Pozzuoli, a considerable distance to the south, and taken up to Rome in small boats. By the mid-first century the grain was not reaching Rome in sufficient quantities to feed its expanding population. Now it could be landed in bulk just a few miles west of the city.

The depiction of the harbour coincides with the description given by the historian Suetonius of two circular breakwaters stretching out into the sea. Showing a fine disregard for consistency, the designer employed a variety of techniques to convey the features that were important to Nero. The shape of the harbour, and what seem to be the wharves on the right side, are depicted in plan to emphasize their size. To the left, the arcade leading to a small temple, possibly akin to a chapel, at the end of the jetty, with a priest sacrificing at an altar in front of it, are depicted flattened but in profile. The large sceptered statue, possibly of the Emperor, the numerous, accurately depicted large and small ships and galleys, the small reclining statue at the mouth of the harbour and the large figure of Neptune with his rudder and dolphin in the foreground are, however, seen in elevation.

The combination of styles makes it possible to highlight the engineering achievement, the size of the harbour, the number of ships that could enter the port (and implicitly the amount of grain that could be landed there), the divine sanction and, perhaps through the statue and certainly through the portrait on the other side of the coin, the Emperor's benevolence. Nero doubtless intended that on contemplating the coin, the Roman town-dweller would remember that the Emperor himself was personally responsible for ensuring that his stomach was pleasantly filled. A few decades later, the satirist Juvenal was to remark that the authorities sought to distract the potentially rebellious Romans with bread and circuses. And, as if in confirmation, Roman sestertii depict circuses as well as harbours.

AVGVSTI
S POR OST C

A Byzantine world map mathematically constructed from ancient Greek coordinates invented by Ptolemy.

The World of Classical Antiquity

PTOLEMY (Klaudios Ptolemaios, *c.* AD 100–170) wrote the *Geographia*, or 'Guide to Drawing a Map of the World', towards the end of his prolific scholarly life in Alexandria. He had already laid the mathematical groundwork for the description of the earth in his earlier and more renowned astronomical work, the *Almagest*. In the *Geographia*, relying on and revising the work of Marinus of Tyre, Ptolemy added a system of meridians which he used for the description of the *oekoumene*, the inhabitable part of the earth. He also devised three projections for laying the *oekoumene* on a map. The main bulk of the work is a technical listing of the coordinates of about 8,000 places and natural features, arranged by continents and provinces.

The translation of Ptolemy's *Geographia* into Latin at the beginning of the fifteenth century is a key moment in the history of European geography, even if some historians have exaggerated its revolutionary effect. Renaissance scholars could find in Ptolemy a system of global geography based on mathematical principles, as well as a model for regional mapping. The *Geographia* retained its canonical status well into the seventeenth century.

Renaissance Ptolemaic world maps are frequently reproduced (usually from the beautiful 1482 Ulm print edition). Here is shown a less familiar map, taken from British Library Ms. Burney 111, a Byzantine manuscript from the late fourteenth (or early fifteenth) century. This choice highlights the contribution of Byzantine scholarship to Renaissance geography: it was Maximus Planudes (*c.*1260–1310) who had searched for and found a copy of Ptolemy's neglected work, and Manuel Chrysoloras (*c.*1355–1415) who initiated the translation of the text into Latin. A unique element worth noting is the group of zodiacal signs in the bottom right corner of the map. It reminds us that geography was closely tied to both astronomy and astrology – Ptolemy himself was the author of an influential astrological work, the *Tetrabiblos*. Finally, Ms. Burney 111 has an unusual number of regional maps: sixty-four, each devoted to a single province, rather than the common twenty-six. This variant feature, found only in four other *Geographia* manuscripts, helps demonstrate the complexity of the transmission process from second-century Alexandria to Renaissance Europe, which involved editorial interventions and scribal errors. Thus even if the *Geographia* was written as a detailed guide for creating maps, we cannot determine today what Ptolemy's own maps may have looked like.

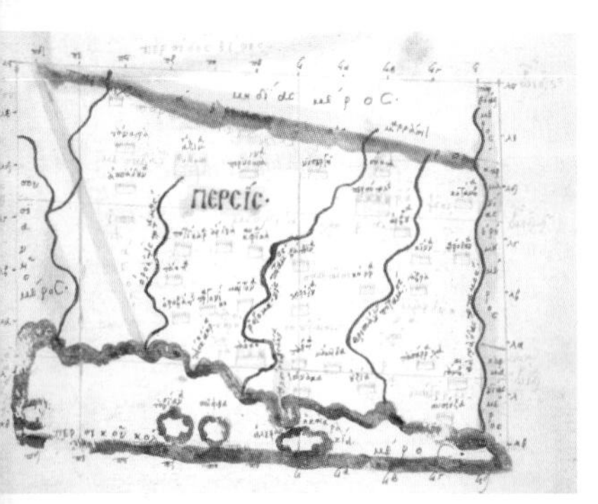

Right World map in Claudius Ptolemy's *Geographia*. **Above** A map of Persia, one of sixty-four regional maps depicted in the Byzantine version in the British Library.

A vast map incised in painstaking detail on 150 marble slabs and intended to inspire awe at the size and splendour of ancient Rome.

Imperial Rome

AS AN ORIGINAL example of a major Roman map, the marble plan is unique (the Peutinger map, by contrast, is known only through a copy). Even so, fragments alone remain, around 1,200 of them, comprising about 10 per cent of the whole, mostly isolated pieces. However, survival of the wall to which the plan was clamped assists reconstruction. The plan covered 150 slabs spanning a colossal 18 metres long by 13 metres high (59 by 43 feet) in a spacious side hall of the Temple of Peace complex that was renovated by the Emperor Septimius Severus after a fire in AD 192. Here the plan could be viewed from as far back as 26 metres (85 feet), but close inspection (visit http://formaurbis.stanford.edu) remained impossible because the bottom lay at least 2 metres ($6\frac{1}{2}$ feet) above floor-level.

The plan's cartographic design is ambitious and apparently original (a forerunner has been suspected, but seems unlikely). Oriented approximately south-east and centred upon the Capitoline Hill, it maintains a consistent scale of about 1:240 to record in 'plan-view' literally every ground-level feature throughout the city (not just rooms, but also stairwells, fountains, columns, steps). Exceptionally, a few major structures like theatres are rendered in 'bird's-eye view' to become more prominent and recognizable; certain temples and other structures are coloured red for the same reason. However, this is the sole use of colour; even the River Tiber is no more than a blank band. Naming of structures is equally sparing.

The orientation and generous scale appear to be those long in use by the professional surveyors charged with maintaining city plans. Thus the need was not to adapt existing data or gather more, but rather to collate, select from, and simplify, the existing record. The plan omits property owners' names, for example, and opts for a single linework style that fails to differentiate a wall and a step. Most puzzling is the rendering of some structures in a form known to be long outdated, while elsewhere account is taken of recent construction.

Today, it may seem absurd that such creativity and painstaking control of detail should permit no practical application. But the fact remains that the gigantic plan can only have served as a proud, impressionistic showpiece of the empire's capital, the Roman world's greatest city. As such, it is a brilliant success: the countless details, individually so trivial, command awe in this monumental form.

Right and Above Fragments of the Forma Urbis Romae, or Severan marble plan of Rome, showing the Subura neighbourhood including the Porticus of Livia. **See also** 1580.

PORT
LIV
IAE

A vividly depicted coastal map on parchment commemorating a region thriving under Roman authority.

Celebrating the Roman Black Sea

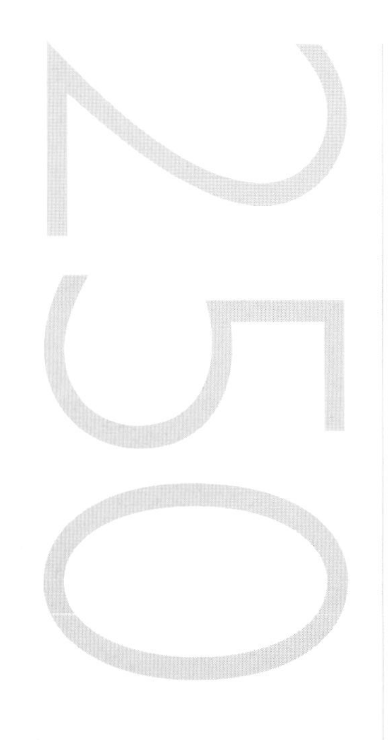

THIS POORLY PRESERVED map fragment was one of several parchments unearthed in the so-called 'Tower of the Archers' at Dura-Europos (modern Sâlihîyah, Syria) by the Belgian scholar Franz Cumont in 1923. The city was a Seleucid Greek foundation dating back to 303 BC, strategically located on the land route along the west bank of the Euphrates overlooking the Mesopotamian plain. It became an important garrison on the Roman Empire's eastern frontier, but fell to the Persians in AD 255/6 and was never reoccupied.

The map could have been the decoration for a shield. Oval shield-boards were among the finds from the Tower, with designs brightly painted either on their wooden planking or on parchment glued to it; in this event, the map may be a souvenir of earlier service by the shield's owner (comparable non-cartographic examples are known). Alternatively, there may be no link to a shield, and the map had some different role. Most of the surviving fragment is dark blue water, on which appear two cargo vessels and a rowboat manned by sailors. To the right, a pale, smooth arc evidently delineates shoreline: into it flow the dark channels of two rivers, and along it are marked vignettes of buildings in pale green stone symbolizing cities. Both rivers are named (Istros, Danoubis), as is each of the vignettes.

Right Detail of the Black Sea Map from Dura-Europos. **Above** The whole of the surviving map.

Since Odessos (modern Varna, Bulgaria) is the name preserved nearest the top, and Arta (perhaps signifying the Straits of Kertch) nearest the bottom, the map's orientation is evidently more or less south. All names, as well as the figures for mileage between successive vignettes (as high as 84), appear in Greek. This is the typical format for a Latin land-route itinerary, although it is also attested for a Greek maritime one (even with distances in Roman miles rather than Greek stades). Here the likelihood is that the mapmaker has coastal voyaging in mind; if there was a corresponding ancient land route from the Danube delta northwards, it is otherwise unknown. We may fairly imagine that the full arc of the map encompassed the Black Sea, but there is no telling whether it extended beyond (and if so, where). Without doubt, the design was vivid, artistic and consciously selective. While reflecting informed knowledge of geography and distances, its primary purpose was surely not to assist navigation, but to celebrate and commemorate the thriving state of the Black Sea region under Roman sway.

ΟΔΕΘ
ΚΑΛ
ΤΟΜΕΑ
ΤΡΟϹΠΟΤ
ΖΑΝΟΥ ΒΙϹΠΟΤ

A map of the Roman world created to affirm faith in what was in fact a battered and embattled empire.

Peutinger's Roman Map

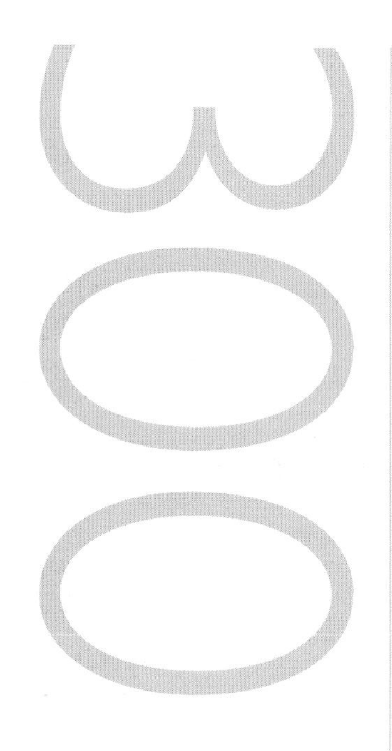

Right Aspects of the empire: eastern England, France, Spain and North Africa on the *Tabula Peutingeriana*, segment 1.
See also 1025, 1210, 1941 (p. 332).

OUR APPRECIATION OF Roman cartography depends heavily upon a parchment roll somehow obtained around 1500 by the humanist Konrad Celtes, who later gave it to his friend Konrad Peutinger. Engravings of the ambitious map it preserves were published in the 1590s, and since 1738 it has belonged to the Austrian National Library. In 1863 the roll was separated into its eleven segments; these had been created for the convenience of the copyist, who probably worked in southern Germany around 1200. As originally designed (around AD 300?), the map was a single seamless creation. Although our unique copy lacks the western end, the complete map must have spanned the limits of Roman claims to rule, from the Atlantic to Sri Lanka, with Rome itself sited at the centre. Even so, the puzzling choice of such a squat, elongated frame (34 by 671 centimetres/1 by 22 feet), with its virtual abandonment of a north–south dimension, calls for selective coverage. The land, and routes across it, are especially featured; in consequence, landmasses like Gaul and Asia Minor are manipulated to fit the frame, and the main expanses of open water (Mediterranean and Black Seas) are reduced to narrow channels. There is no uniform scale; instead, the ingenious mapmaker varies it, with Italy appearing notably large, and Persian territory eastwards small.

The map has been valued primarily for its representation of land routes with approximately 2,700 associated place names and distance figures. Roman itineraries have been seen as its basis, therefore. Only recently, however, has awareness developed that this remarkable exploitation of route data must be underpinned by the basic elements of physical landscape and by accurate placement of principal settlements, marked with eye-catching vignettes. Such extensive geographical grasp, and the confidence with which it is reflected, probably derive from Greek sources.

There is now growing realization, too, that with its indifference to up-to-date information or direct routes, the map would serve land travellers poorly, and that it is better considered a propaganda item for display on a palace wall, say, perhaps even forming part of a larger artistic creation. The map captures the aspirations of the Tetrarchy around AD 300, when four co-emperors struggled to repossess and restabilize the battered empire, as well as to affirm their own fragile harmony; the design and presentation likewise reflect late Roman artistic and cultural tastes.

Pretoriu Agrippine
Lugduno
Albanianis
Nigropullo
Caspingio
Grinnibus
Foro Adriani
Flenio
Tablis
Turnaco
Tervanna
Ratumagus
Brevoduro
Augustoduro
Coriallo
Cosedia
Legedia
Condate
Fanomartis
Reginea
Vorgium
Conbaristum
Iuliomago
CADVRCI
Lemuno
Portunamnetu
Segora
Brigiosum
Cassinomago
Betvriges
Fines
Pretorio
Aginnum
Tolosa
Lactora
Cosa
Biolindum
Carcassione
Iuncaria
Beclana
Insummo pyreneo
Rusuccuru colon
Iomnio Municipio
Tamascani Municipium
Lemelli presidium
Tigisi
Tamannuna Municipium et castel
Adsaua Municipium
Sitifi col
Salinas Nubonenenses
Adoculum Marinum
Vareis
Tubonis
MUSONIORUM
Ad capsu iuliani

The new Roman town of Terracina showing the grants of land made to veteran soldiers.

350

Land for Old Soldiers

Right Terracina from the *Corpus Agrimensorum*.
See also 1597 (p. 136).

IF THE BABYLONIANS and later the ancient Greeks were fascinated by heaven and earth and by theories for their precise measurement, the Romans were concerned with the more humdrum measurement of the land immediately beneath their feet. And necessarily so, since their military expansion was accompanied by the creation of military camps, new towns, and settlements of military veterans in the newly conquered lands, to be followed by drainage, the provision of amenities, the creation of roads and finally the administration of their vast empire. Detailed land surveys, enabling the setting of boundaries and the rectangular subdivision of land known as centuriation (from a division of land into a hundred parts), were an essential preliminary.

From the first century BC until the collapse of the empire nearly five hundred years later, numerous texts, initially derived from Greek sources, were written on surveying. In about AD 350 a collection of the most important was brought together under the title of *Corpus Agrimensorum* or 'body of writings of the surveyors' (literally 'field measurers'). The texts were illustrated by diagrams and by small, vignette-like maps. These were very different from the mathematically precise large-scale maps on bronze and marble created by surveyors following the rules and techniques contained in the texts. They were instead pictorial and intended to give examples of the points being made in the texts. A few sites were identifiable.

None of the original manuscripts – written on papyrus – survive, but the Vatican Library has a copy of about AD 800. It is considered to be quite close in appearance to the original. This drawing contains a bird's-eye view, lacking the internal buildings, of the new town of Terracina ('Colonia A[n]xurnas') founded in 329 BC between mountains, the Pontine Marshes ('paludes') and the Mediterranean on the road from Rome to Naples. It is included to demonstrate how the centuriation of the land to the left of the town is centred on the Appian Way, with grid-like allotments of two *jugera* (approximately one acre) for the three hundred veterans who were the first settlers.

Approximate though the proportions may be, the drawings give one of the best surviving impressions of the appearance of the landscape of the Roman Empire.

finio per demonstrationes & locorum uocabula.

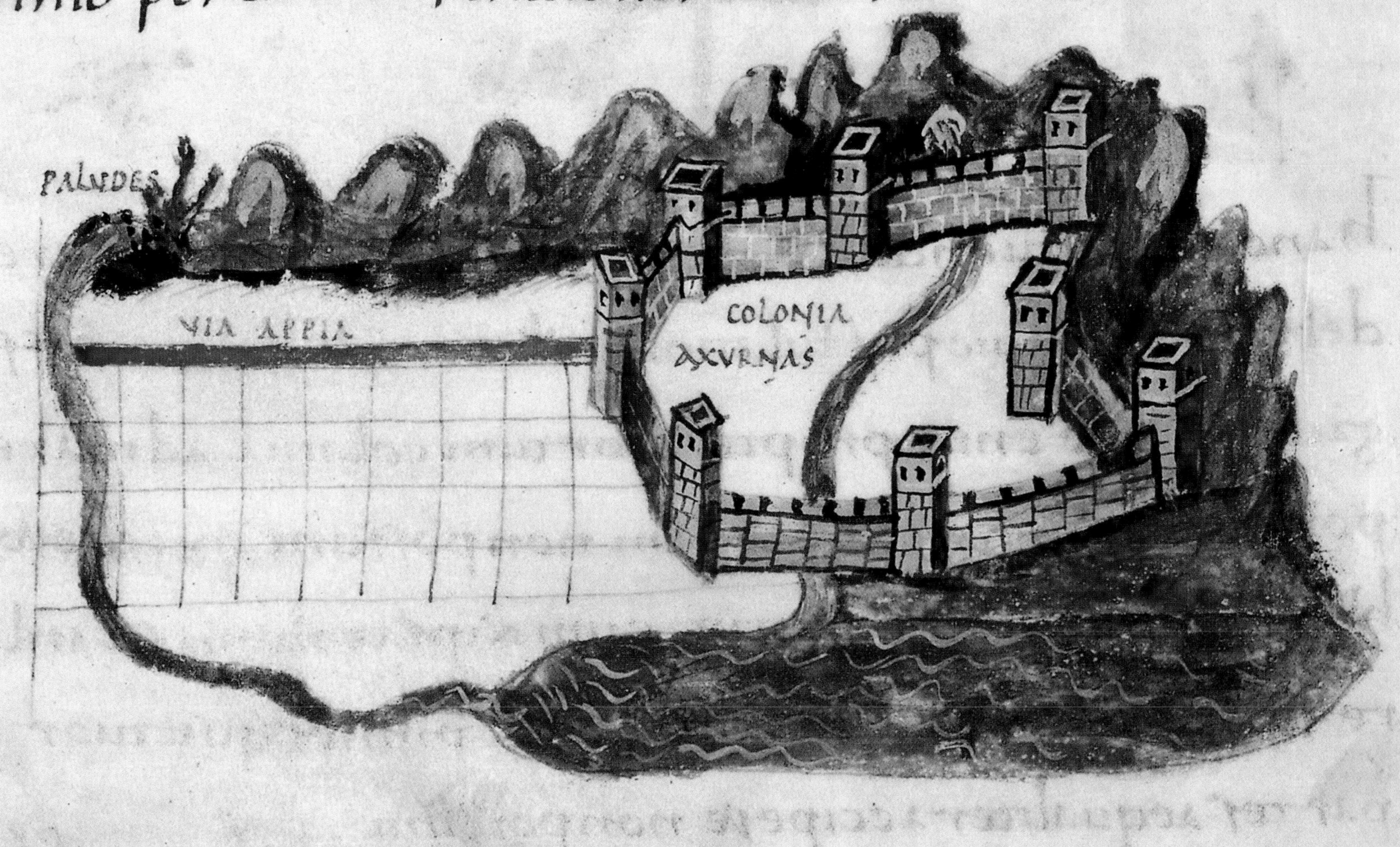

An ancient Roman spherical world map depicting climate zones, frigid, temperate and equatorial, based on the work of an early Greek scientist.

A Scientific World Map

425

HERE THE ENTIRE eastern hemisphere of the earth is shown, divided into five zones: two polar or frigid zones, two temperate, and one equatorial, torrid zone. The concept derives from the Greek scientist, Crates of Mallos, in the second century BC, who hypothesized that there were four landmasses on the earth, each containing a habitable zone. An impassable ocean, swept by tides, divided these lands from one another. Early Christians found this concept difficult to stomach. If each of these lands were inhabited, how did the descendants of Adam get there? And how was the mission of the apostles, to convert the entire world, feasible? Despite these concerns, Macrobius's book and map circulated throughout the Middle Ages in hundreds of manuscripts and was a basic text of medieval science.

Macrobius wrote his *Commentary on the Dream of Scipio* in the early fifth century, basing it on the last part of Cicero's *De Republica*, in which the Roman general, Scipio Africanus the Younger, is transported to the heavens by the spirit of his famous grandfather. From this vantage point he is able to look down upon the earth. Cicero's theme is the transience of human glory and the importance of ruling justly, but Macrobius's lengthy commentary expands on its cosmic vision. One of a group of energetic encyclopedists of the late empire, Macrobius transmitted to future generations some part of classical science when the original works were lost. There is no evidence that Macrobius was a Christian, but the neo-Platonist ideals to which he subscribed were easily comprehensible to his readers.

Macrobian maps have little space for geographical details, as the northern temperate zone is relatively small. Usually only a few place names, marking the extremes of the known world, are shown. Here, for example, we have the Orkney Islands (farthest north-west), the Caspian Sea (north-east), Indian Ocean (south-east), and the Red Sea (south), while Italy indicates the centre. Occasionally a larger and more detailed map was made on this plan, but mostly zonal maps happily co-existed with other forms of world map.

Right A zonal world map by Macrobius, ninth century.

debis caspiu q; mare unde oriatur inuenies. Licet n ignore ee n nullos. q ei de oceano ingressu neget. nec dubie in illa qq aust'rat gnris tepata mare de oceano simlit influere. sed describi hoc nra adtestacione n debuit. cuius situs nob incognitus per seuerat.

Qd aut nram habitacione bile dixit angusta uerticib; laterib; laciore. in eade descripcione poterim aduertere. na quanto longior e tropicus circus septentrionali circo tanta zona uerticib; qua laterib; angustiore e. q summitas eius in artu extremi circuli breuitate contrahit. Deductio aut lateru cu longitudine tropici. ab utraq; parte distendit. Deniq; ueteres omne

Simple 'T-O' map diagrams were used from Roman times onwards on coins and in paintings and texts to symbolize the world.

Maps as Diagrams

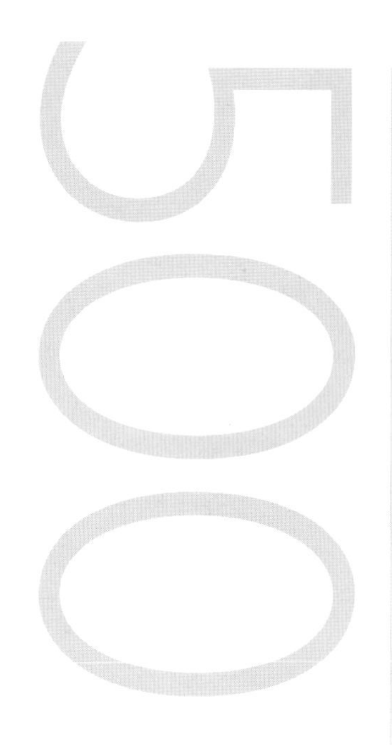

Right Top: Redrawing of an ornately Christianized T-O map diagram for the geographical description of the world in the pagan writer Salust's *Jugurtha*, tenth century. Bottom: Classical T-O diagram illustrating Isidore of Seville's 'The globe is so-called for its roundness… It is divided into three parts, which are called Asia, Europe, Africa.' Ninth or tenth century. **See also** 1460.

THE DIAGRAM IS the commonest form of mapping. Everybody creates a map in diagrammatic style for one reason or another – ruthlessly selecting only the essential topographical detail and reducing direction to straight or boldly curved lines – if only to guide visitors to their destination. Not everyone, though, recognizes that their scribbles are a form of mapping, or that the simple schematic figures they see in books are maps cleverly designed for a specific purpose – that of instant and unambiguous communication between individuals familiar with the subject under discussion.

Diagrams have always been important aids in teaching and explanation in learned treatises as well as in textbooks or on the school blackboard. Abstract notions are grasped and details memorized more easily with the aid of a diagram. In Greek and Roman times, individual countries were already known by their shapes – Sicily and Britain as triangles, Iberia as an ox-hide – and the course of the Nile likened to a Z. Geographical texts describing the globe and its division into the three (Old World) continents were supplied with a T-O diagram. In this, a circle outlined the globe and two straight lines divided the inner space into three portions. East was usually at the top, so the line representing the rivers that separate Asia from Europe and Asia from Africa forms a horizontal axis across the circle (left, i.e. to the north, the Don; right, i.e. to the south, the Nile). The vertical line dividing the lower half of the circle, separating Europe from Africa, stands for the Mediterranean Sea. So familiar was the diagram from Roman times onwards that, portrayed three-dimensionally as a tripartite orb, it was commonly used on coins and in paintings to symbolize the world and to communicate power over it.

T-O diagrams drawn in classical times have not survived except in medieval copies (and some printed editions) of the ancient texts they illustrated. From the eighth century AD, however, copyists began to embellish the diagrams with details irrelevant to the pre-Christian geographical description, but of great contemporary interest, notably biblical features such as the names of Noah's sons, Jerusalem, Mount Sinai, the Red Sea Crossing, the Earthly Paradise. At the same time, compilers of new medieval treatises continued to use the T-O diagram, adapting it to fit their own requirements. Likewise, the orb symbol was widely used throughout the Middle Ages.

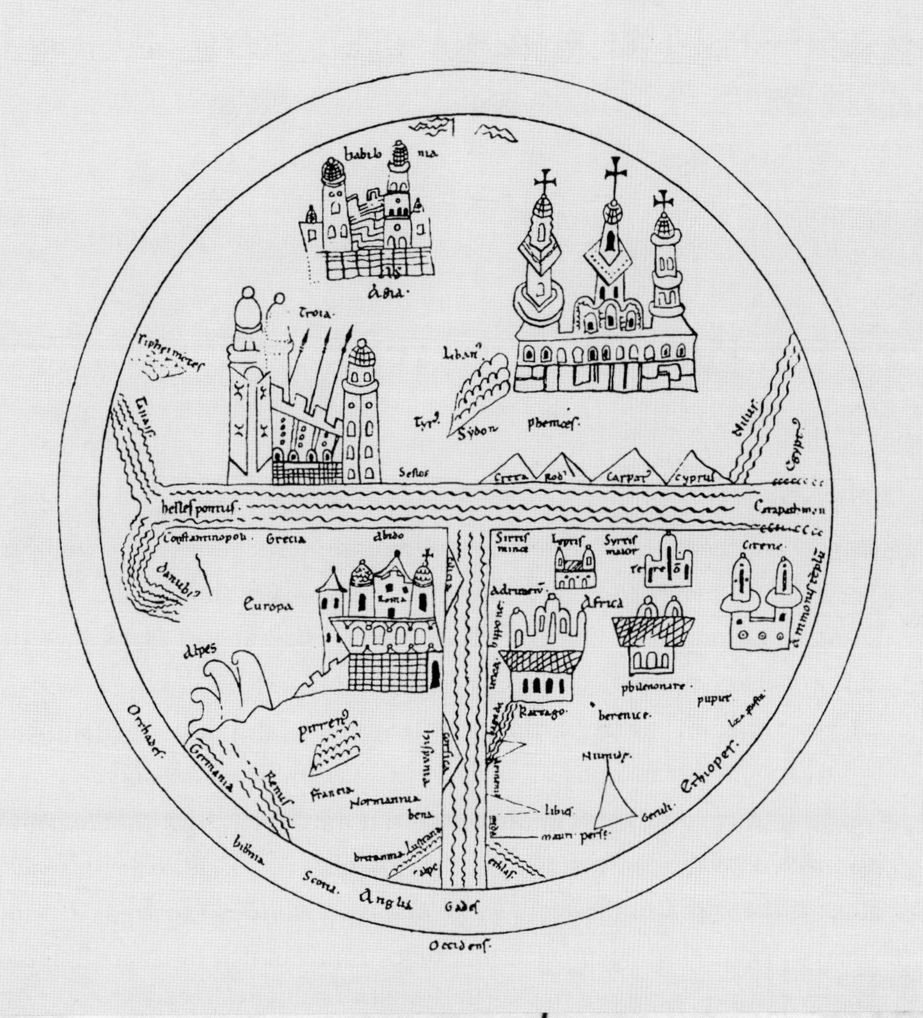

Orbis a rotunditate circuli dicitur quia sicut rota est; unde brevis etiam rotella orbiculus appellatur. Undique enim oceanus circumfluens eius in circulo ambit fines. Divisus est autem trifarie. E quibus una pars Asia, altera Europa, tertia Africa nuncupatur. Quas tres partes orbis veteres non equaliter diviserunt. Nam Asia a meridie per orientem usque ad septentrionem pervenit; Europa vero a septentrione usque ad occidentem. Atque inde Africa ab occidente usque ad meridiem. Unde evidenter orbem dimidium sola Asia. Sed ideo iste due partes facte sunt quia inter utramque ab oceano mare magnum ingreditur quae eas intersecat. Quapropter si in duas partes orientis et occidentis orbem dividas, Asia erit in una; in altera vero Europa et Africa. II

DE ASIA III

ASIA

EUROPA

AFRICA

Asia ex nomine cuiusdam mulieris est appellata, quae apud antiquos imperium tenuit orientis. Haec in tertia orbis parte disposita ab oriente ortu solis, a meridie oceano, ab occiduo nostro mari finitur; a septentrione Meotide lacu et Tanai fluvio terminatur. Habet autem provincias multas et regiones, quarum breviter nomina et situs expediam, sumpto initio a paradiso. Paradisus est locus in orientis partibus constitutus, cuius vocabulum ex greco in latinum vertitur ortus. Porro hebraice eden dicitur, quod in nostra lin-

A map of a flat world surrounded by an impassable rectangular ocean created by a sixth-century Christian monk.

The World is Flat!

THE SIXTH-CENTURY Christian merchant-turned-monk of Alexandria, known as Cosmas Indicopleustes (literally, 'Mr World sailing to India'), knew the world was flat. Like many practical businessmen, he had no time for theorists – in this case those whose works were preserved in the great library of Alexandria. He ridiculed Plato and Aristotle who believed the world was spherical. His experience and reading of the Bible convinced him that the universe was shaped like a tent or tabernacle, with the flat earth at its base.

This he illustrated in his only surviving work, the *Christian Topography*, a polemical work written in Greek in about 540 when the Byzantine Empire was wracked by religious controversies. Cosmas, copying the Greek historian Ephorus (405–330 BC), produced a map of the world to illustrate his words. The world is shown as a rectangle, with the inhabited world or *ekumene* at the centre, surrounded by a rectangular great ocean that humans could not cross. In the east (right), beyond the ocean, in Cosmas's schema, was the Earthly Paradise. From it the four rivers of paradise made their way underground, transforming themselves into the Nile, Tigris, Euphrates and Ganges (or Indus). Four seas flowed from the great ocean into the *ekumene*. Three, the Caspian Sea, the Red Sea and the Persian Gulf, were represented by circles attached to the ocean. The depiction of the fourth, the Mediterranean (left), was more realistic. In line with classical ideas its length from west to east was shown twice that of its length from north to south. The Adriatic, Aegean and Black Seas are shown but not named.

Cosmas's flat-earth theories appealed to the mid-nineteenth-century American writer Washington Irving, the creator of Rip van Winkle and author of an influential biography of Christopher Columbus, who took a dismissive view of the intelligence of medieval man. His views were followed by numerous later writers who should have known better and most western schoolchildren were taught that medieval man believed the earth was flat. Cosmas's theories, however, actually had very little influence on his contemporaries and may indeed have been completely unknown in western Europe. Only three copies of his text survive, all of them in Greek.

Right Map of a flat world by Cosmas Indicopleustes, eleventh-century copy.
Above Byzantine gold coin (Histamenon) of Romanus IV, c.1070, showing emperors holding orbs representing a spherical world.

καὶ ὁ παράδεισος κατὰ ἀνατολὰς κεῖται· διαγράφο
μεν τοίνυν πάλιν τὸ πλάτος τῆς γῆς καὶ τοῦ
ὠκεανοῦ καὶ τῶν κόλπων καὶ τῆς περαιτέρω γῆς· καὶ
τὸν παράδεισον· ἔσω μὲν τῆς γῆς τῆς ἀντ[illegible]·
ὅπως [illegible] γένηται τοῖς ἐντυγχάνουσιν· ἐπειδὴ
τὸ σχῆμα τῆς γῆς πάσης κατὰ ταύτην τὴν ἐπιφά
νειαν καὶ τὸ πλάτος, τοιόνδε :-

A very early mosaic map of the 'holy land' including a very accurate representation of sixth-century Jerusalem.

Mapping in Mosaic

IN 1884, THE GREEK Orthodox Patriarch of Jerusalem was told about a mosaic map of Palestine during the reign of the Emperor Justinian (527–565), newly discovered on the floor of a village church near Amman in Jordan. The largest surviving portion is 10.5 by 5 metres (34 by 16 feet), but the original compositions may have measured 24 by 6 metres (79 by 19½ feet) and covered the entire width of the nave. The map still excites wonder. One reason is that it offers a cartographical portrait of a single region at a time when maps of this kind were rarely made, and another is that it is the first known map of Palestine. The third reason is more thought-provoking, for what we glimpse in the map is the picture of the Holy Land that was forming in the minds of Christians in Byzantium in the middle of the sixth century, scarcely a hundred years after the idea of a Christian holy land was beginning to be accepted by the Church authorities.

When complete, the map would have extended from the Mediterranean Sea to east of the Jordan River, to Syria in the north, and to the Nile delta in Egypt. Features of physical and human geography are wonderfully depicted. Rivers, lakes, deserts, oases, cities and towns are represented pictorially. There are fish, palm trees, coursing animals, and a ferry across the Jordan. Texts describe the events associated with the biblical places, which are drawn from both Old and New Testaments. The signs for the smallest towns are stylized, but Jerusalem is portrayed prominently and so accurately that scholars have been able to identify the principal buildings, streets, gates, towers and the public baths.

The mosaicist took his material from books and maps, notably Eusebius's gazetteer of biblical places (*Onomasticon*, *c*.303), and recent and contemporary pilgrim accounts (such as Theodosius's, written by 518). He left a map of the land as it was in his day, not merely an assemblage of holy places, but a cohesive territory just coming to be known to Christians as the Holy Land.

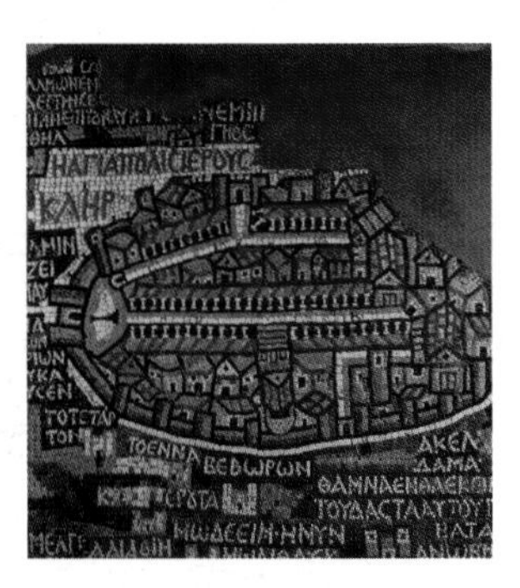

Right Detail of central portion of mosaic, 'Old Church' of Madaba, Jordan, c.565. **Above** Detail showing Jerusalem. **See also** 750, 1250, 1695.

ΘΕΡΜΑ ΚΑΛΛΙΡΟΗΣ
ΑΙΝΩΝ ΕΝΘΑ ΝΥΝ Ο ΣΑΠΣΑΦΑΣ
ΙΕΡΙΧΩ
ΝΙΚΟΠΟΛΙΣ
ΕΦΡΑΘΑ
ΒΗΘΛΕΕΜ
ΣΩΧΩ
ΙΟΥΔΑ
ΣΑΦΙΘΑ
ΑΚΕΛΔΑΜΑ
ΓΑΒΝΗΛΗ Κ ΑΙ ΓΑΜΝΙΑ
ΕΝΕΤΑΒΑ
ΠΡΟΣ ΔΑΝ
ΓΕΘ ΝΥΝ ΓΙΤΤΑ
ΜΟΔΕΕΙΜ Η ΝΥΝ ΜΩΔΙΘΑ ΕΚ ΤΑΥΤΗΣ ΗΣΑΝ ΟΙ ΜΑΚΚΑΒΑΙΟΙ
ΒΕΘΖΑΧΑΡ

A late eighth-century map manipulates the form of the actual world to reflect the Christian belief in the Trinity and in a divinely ordered universe.

A Christian World Map

650

UNTIL RECENTLY IT was an accepted wisdom that because the New Testament had little to say about geography, the coming of Christianity had relatively little effect on mapmaking. This continued to be based on ancient Roman, Greek and Babylonian learning. The distortions on medieval maps were largely caused by successive misunderstandings and corruptions of that learning. Recently the German scholar Brigitte Englisch has challenged these assumptions. Taking this map as a key example, she has suggested that what has traditionally been seen as medieval distortion was in fact a mid-seventh-century attempt to manipulate the classical, and geographically more correct, image of the world to reflect the Christian belief in the Trinity and in a divinely ordered universe.

The map has east at the top and is centred on the Mediterranean, with a triangular Sicily prominently to be seen, a red-speckled Red Sea and a rose-like Earthly Paradise at the east. It was created in 776 in southern France, and Provence ('prouivincia') receives special attention. Though it is now one of the two oldest surviving detailed European world maps, its origins probably stretch back still earlier to the mid-seventh century, just after the Christian patriarchate of Antioch fell to the Muslim Arabs. The surviving patriarchates are indicated with star-like symbols.

Professor Englisch has pointed out how the distances between the four surviving patriarchates have been intentionally distorted so as to create a hidden equilateral triangle, symbolizing the Christian Trinity of Father, Son and Holy Ghost. It is centred on Crete, with Constantinople and Carthage and Alexandria and Jerusalem respectively forming two of the angles and marking the middle of two of the sides. The rest of the map is constructed around three equidistant concentric circles, symbolizing perfection, radiating from Crete with the first linking Jerusalem and Alexandria and the second the historically significant cities of Babylon and Rome, which is shown as a mini-patriarchate. Twenty-four lines at 15 degrees' distance, resembling a clock face and symbolizing the harmony of time and space, also radiate from Crete with the junctions between the lines recalling the Holy Cross.

In addition to their spiritual role, the lines and circles provided the framework for constructing a world map. Professor Englisch suggests that this structure underlay virtually all the medieval European world maps that we know of. The similarities between this and the Evesham World Map, created six hundred years later, are indeed striking.

Right Top: One of the two oldest surviving detailed European world maps, the anonymous, so-called 'Isidore' World Map. Provence, 776. Bottom: Diagram illustrating the structure of the map.
See also 1415.

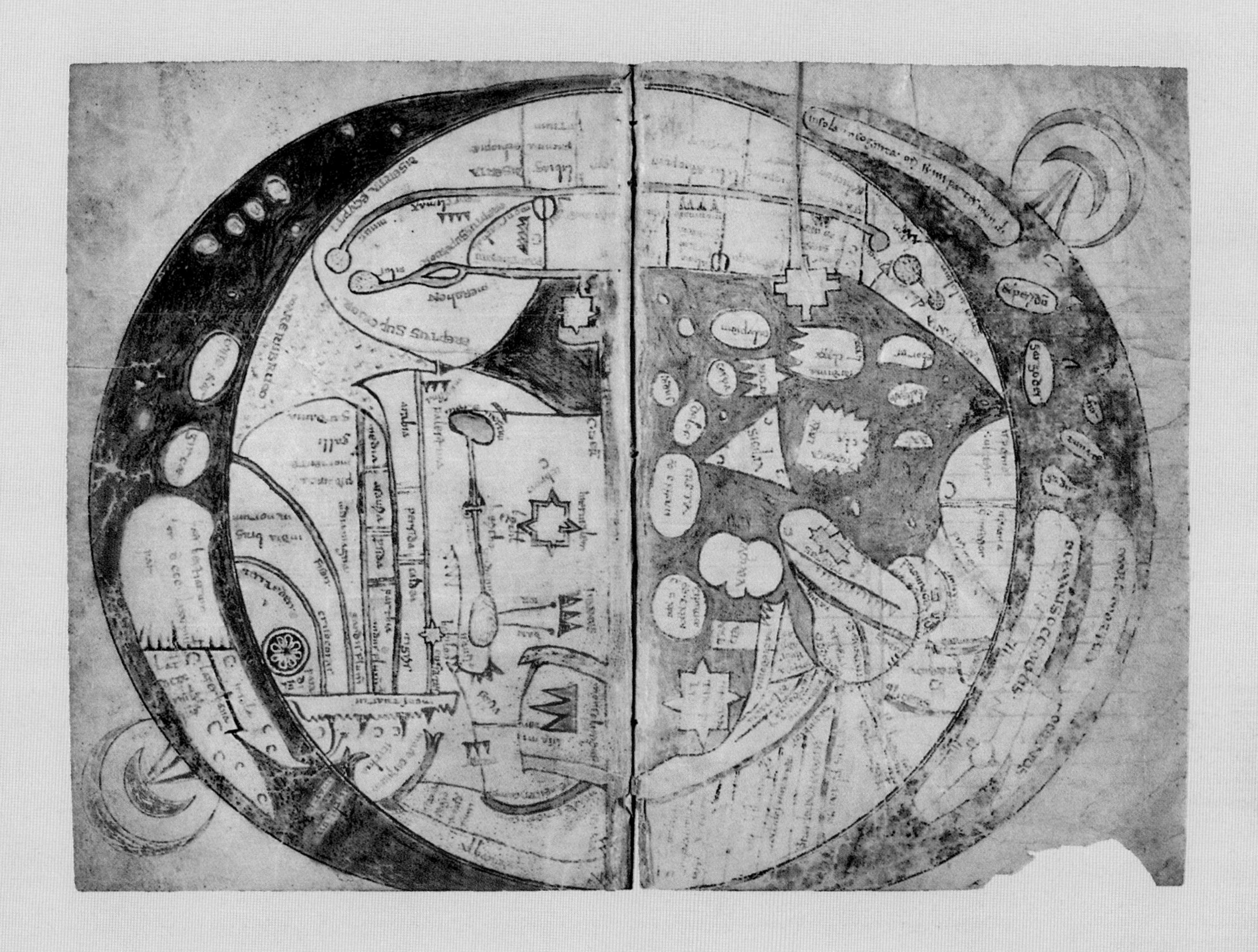

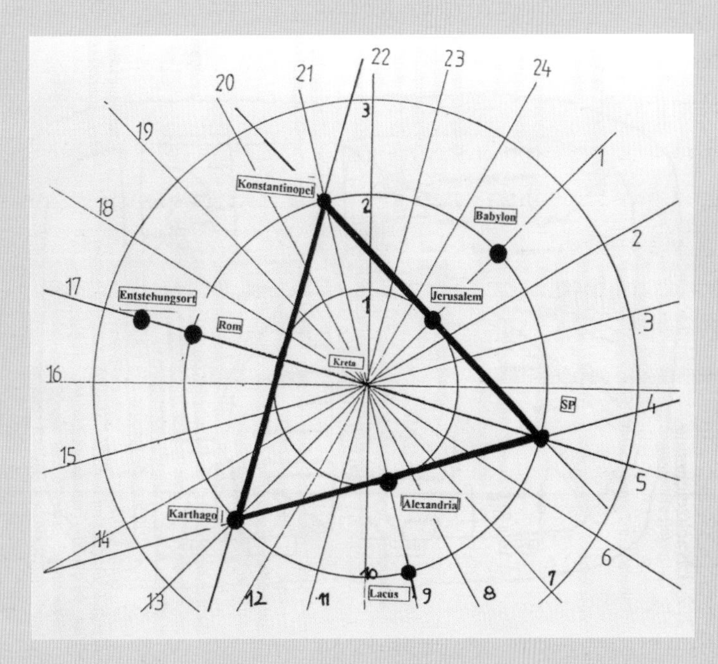

Konstantinopel
Babylon
Jerusalem
Entstehungsort
Rom
Kreta
SP
Alexandria
Karthago
Lacus
1
2
3
4
5
6
7
8
9
10
11
12
13
14
15
16
17
18
19
20
21
22
23
24

One of the first historical maps attempts to place on record the division of land granted to the tribes of Israel after the defeat of the Canaanites.

750

Deciphering a Book with a Map

THIS MAP, ENTITLED *The Shape of the Promised Land*, comes from a commentary on the biblical Book of Joshua originally written in eighth-century Ireland. Medieval Christians viewed the Book of Joshua thus: 'In [it] we read how Joshua led the sons of Israel across the Jordan, their circumcision, their defeating the Canaanites, and then how they received their allotted inheritance in the division of the land between them' (Isidore of Seville). But if it seemed straightforward, Joshua bristles with obscurities and conundrums to be deciphered. The basic problem was that Joshua was written five hundred years after the time it claims to narrate, and intended to portray an ideal past when there was a strong leader, military success, and the whole land was occupied by Israelites. But the described boundaries are contradictory and sometimes impossible, so interpreters resorted to using a map 'to clarify' the situation. Unambiguous cartographic space replaced the textual inconsistencies of an imagined past.

This is the earliest extant map of the tribal divisions and it was not intended as an 'accompanying illustration' – such as is found in modern Bibles – but as an integral part of the commentary: the map is the explanation of over half of the biblical book.

The map's shape is determined by the two key reference points in the biblical text: the Mediterranean, and the River Jordan presented in the Bible as flowing down from the Mountains of Lebanon to the Dead Sea. Hence the map appears to us to have north at the top. The tribal divisions to the east of the Jordan are relatively simple. However, to the west, the commentator has picked on particular verses while ignoring others for the relationship of one block of land to another – and the final result hides all the difficulties. The map silently tells its viewer: 'Yes, the Bible is true and free from error. The confusion you experienced when reading the book was only in your head. My graphic clarity reveals the truth!'

The map is highly focused: it is a genuine historical map as it is dedicated to just one moment in Israel's history (i.e. one text). Hence Jericho is shown as destroyed and other biblical places – even those of great interest to Christian pilgrims – are ignored. It is thus not only a forerunner of modern biblical maps, but of historical maps in general.

Right Map of the land of the Tribes of Israel.

ITEM FIGURA TERRE REPROMISSIONIS. mons libani.

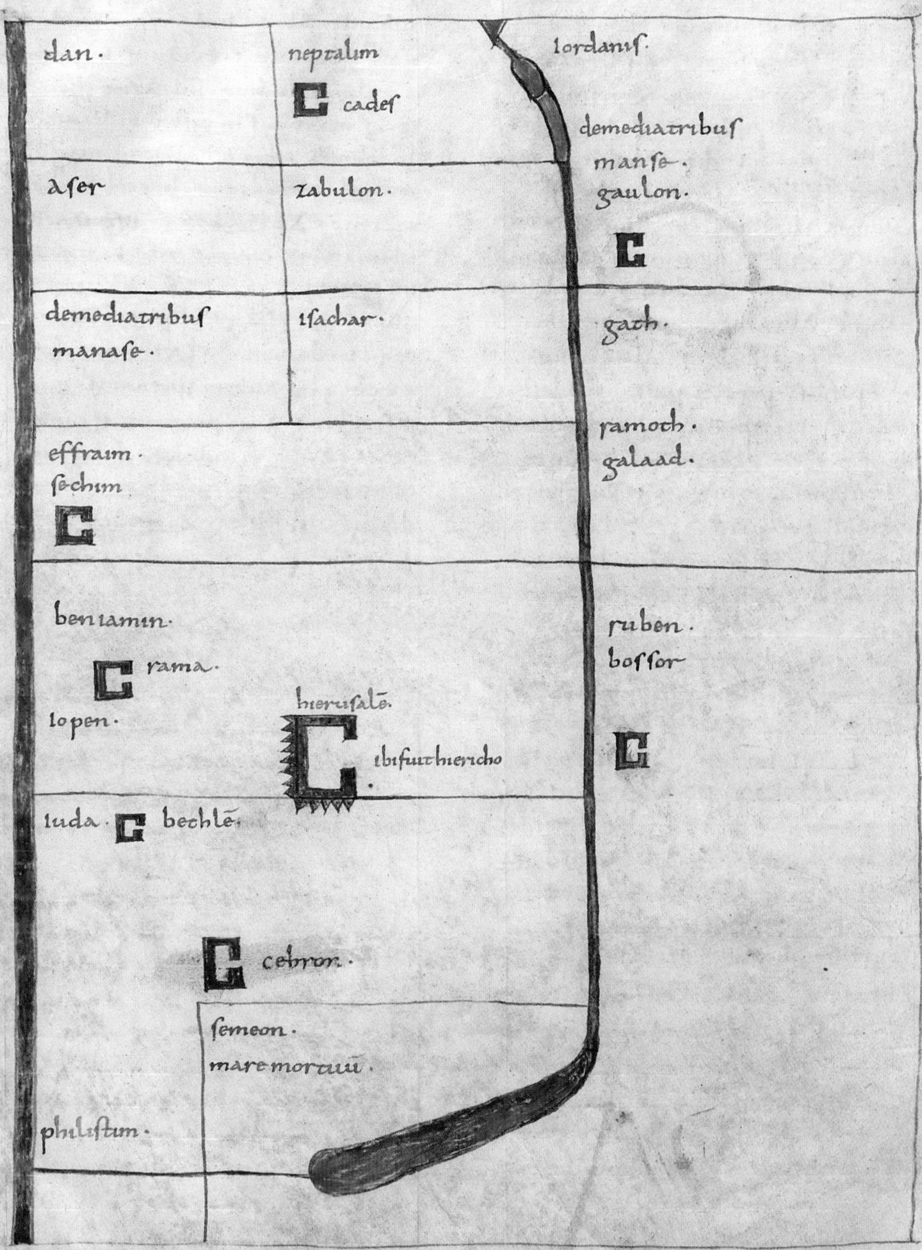

dan.

neptalim

cades

Iordanis.

de media tribus

manse.

gaulon.

aser

zabulon.

de media tribus

manase.

isachar.

gath.

ramoth.

galaad.

effraim.

sechim

beniamin.

rama.

ruben.

bossor

hierusalē

Iopen.

ibi fuit hiericho.

Iuda. bechlē.

cebron.

semeon.

mare mortuū.

philistum.

A Christianized classical world map from the time following the Dark Ages which saw a resurgence of interest in the Roman world.

Saved From the Wreckage

Right The Albi World Map.

AS THE WESTERN Roman Empire descended into political and military anarchy from about 300 onwards, Chistian writers such as Orosius (*fl*.410–20), a pupil of St Augustine of Hippo, Eusebius, Bishop of Caesarea (*c*.260–340), Jerome (*c*.360–420) and much later Bishop Isidore of Seville (*c*.560–636) tried to preserve as much as possible of the ancient learning, albeit with a Christian slant, through their encyclopedic writings. These usually included a section describing the world and its occupants, human, apparently semi-human and animal, and their history. These sections tended to be accompanied by maps. Most were diagrammatic, so-called T-O maps, showing a circle split into three parts, representing the three continents with Asia at the top (see 500). Occasionally, however, there was an effort to depict the world more realistically.

This is one of the two earliest surviving detailed world maps in this tradition and dates from a time following the Dark Ages when there was renewed interest in the Roman world. It comes in a volume of mainly geographical texts, faces a list of the twenty-five seas and twelve winds of the world and is followed by the description of the world by Orosius. The map is likely to have been copied from an updated and Christianized late classical map, perhaps ultimately stemming from a world map on marble created under the direction of Emperor Augustus's son-in-law, Vipsanius Agrippa (63–12 BC) in a portico lining the Via Flaminia near the Capitol in Rome. It has east at the top (the origin of the modern word 'orientation'). It schematically shows the Roman provincial boundaries, though those in northern Europe (left) are unnamed, and towns that were important in classical antiquity like Athens and Carthage. It alludes to the historical succession of world empires, naming Babylon, Persia, Macedonia and Rome. The model must have shown the Earthly Paradise at the top, since the Albi map depicts the four rivers of paradise (though not all are named). The biblical influence can also be seen in the triangular Mount Sinai and the naming of Jerusalem, while the mention of Armenia is probably a reference to Noah's Ark. The prominence given to Ravenna is presumably a reflection of its contemporary importance: it had been the capital of the Byzantine Empire in the west until 752, only a few years before the map was created.

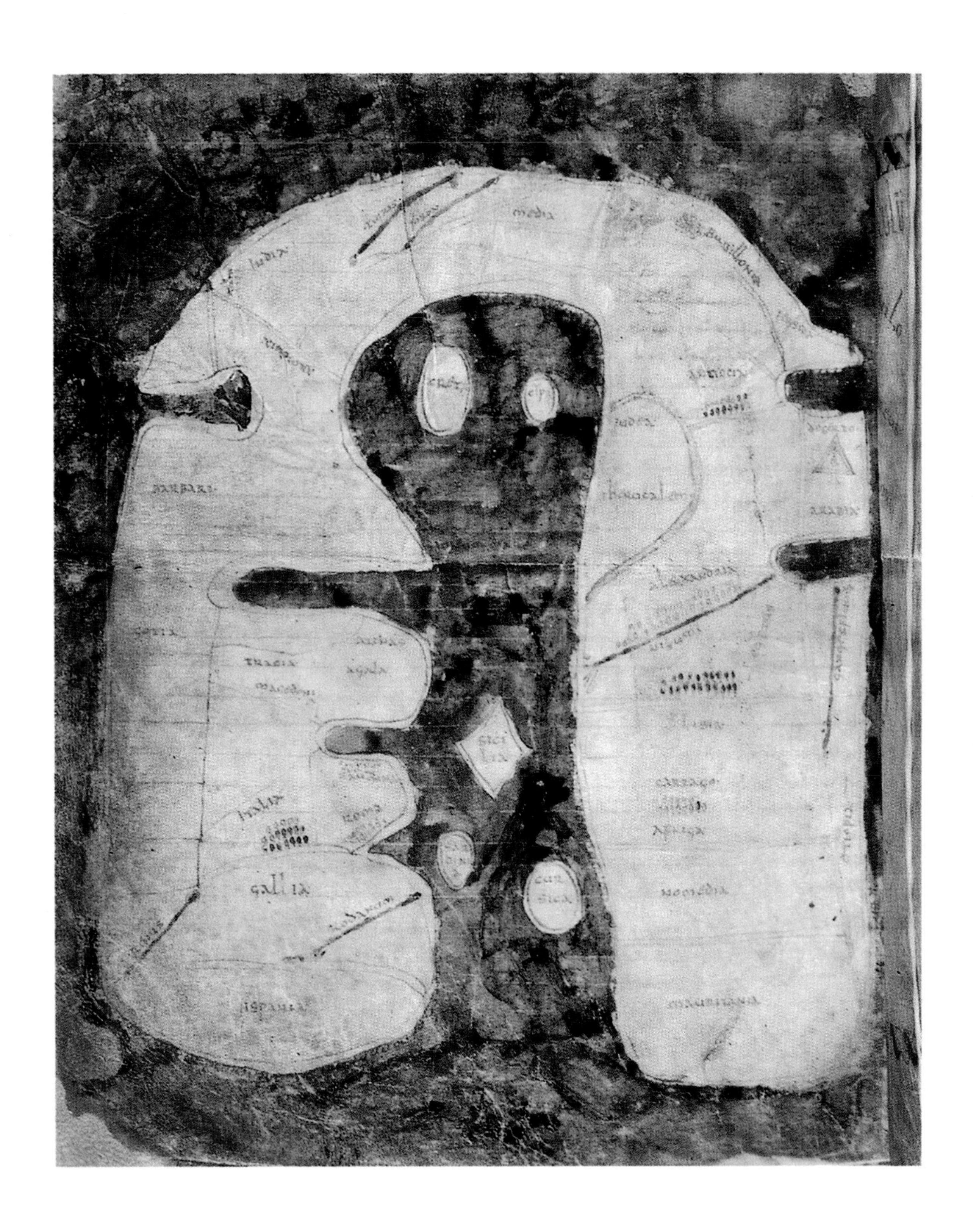
BARBARI
gallia
CARTAGO
MAURITANIA

A Mozarabic Spanish map reveals the inhabited world at the end of human time, predicted as AD 800.

Awaiting the Last Judgement

Right World map illustrating Beatus of Liebana's *Commentaries on the Apocalypse*, Silos, 1106.

THIS MAP ILLUSTRATES a section of a commentary on the Apocalypse as described in the Book of Revelation that had originally been compiled, from late classical sources, by a Spanish monk, Beatus of Liebana in 776. It was copied, in the Mozarabic style blending Arabic and Christian elements, in the northern Spanish monastery of San Domingo de Silos in 1106.

The commentary was put together at a time when the Arabs had driven the Christian rulers of Spain almost to the Pyrenees. It deals mainly with what was expected to happen at the end of human time in (so Beatus believed) 800, when vengeance would descend upon Muslims, sinners and other enemies of Christ. The map was intended to display the 'cornfield' of the inhabited world, or *ekumene*, that had been converted to Christianity by the Apostles. An essential prologue was the expulsion from Eden. That is why Adam and Eve stand at the top (east) of the map.

The Mediterranean in the middle of the map is reduced to a thin dark blue strip, indistinguishable from the rivers Nile on the right and the Don and Danube forming a triangle on the left. Mountains and mountain ranges resemble red or blue pine cones. Jerusalem is shown as a thin, spikey three-turreted shrine, at the top of the Mediterranean. The Italian city of Ravenna, the western capital of the Byzantine Empire before it was lost in the 750s, is mentioned. The map gives a lot of space to the Iberian peninsula, naming Beatus's home province of Galicia, and Tarragona which fell to the Moors in 711. In line with classical wisdom about the fringes of the earth, an inscription mentions the monsters to be found in Ethiopia, and the golden mountain is illustrated. The thick red strip of sea with fish at the far south (right) of the map is probably meant to be an inland sea which, like the Red Sea, could be crossed by humans. The inscription states that the land beyond was unknown because of the great heat of the sun.

The fifteen surviving Beatus maps were produced in Spain and southern France from 944 until well into the thirteenth century. Despite their common roots in classical literature, they developed completely separately from the world maps being produced in England and northern France.

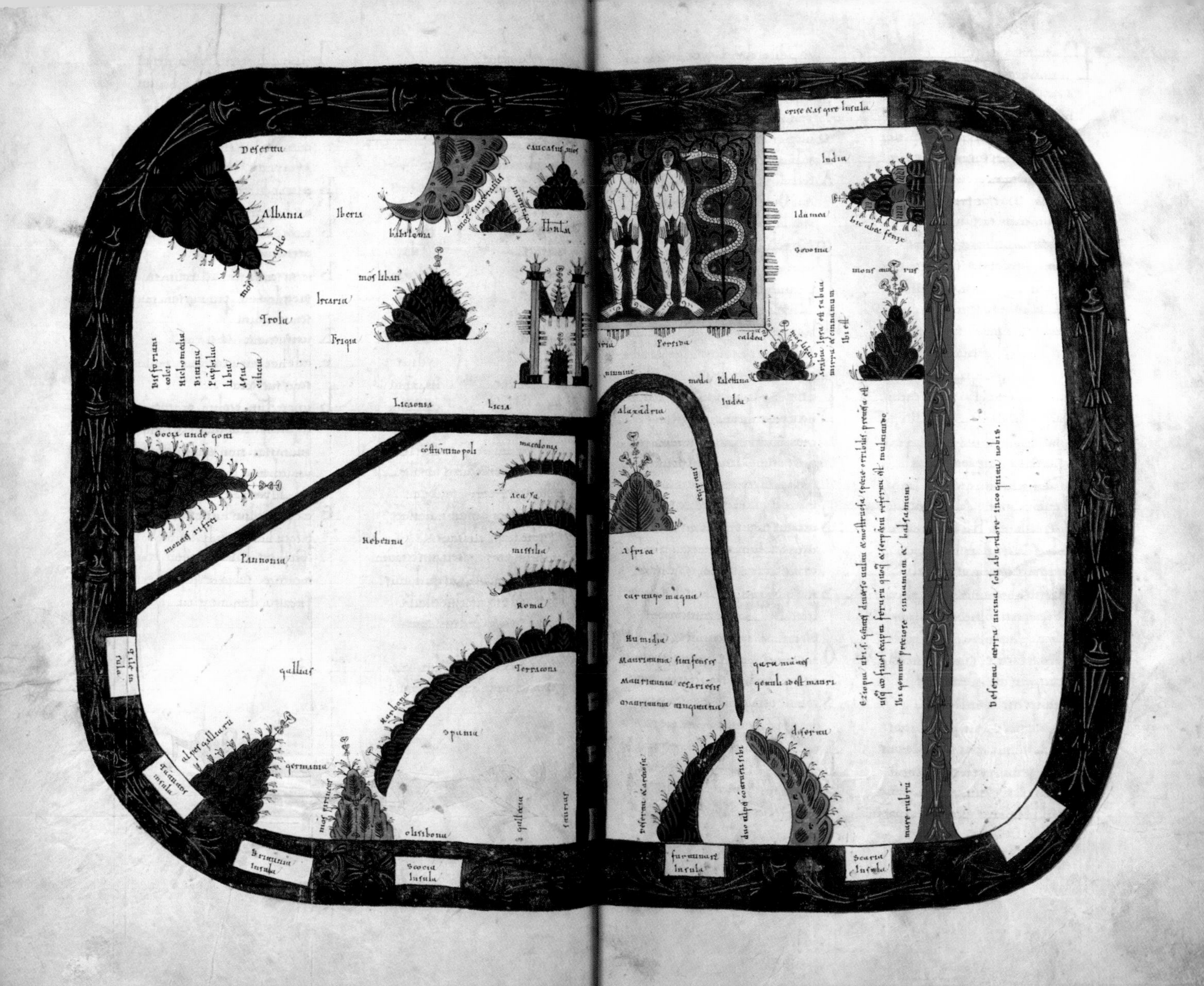

Albania
Iberia
babilonia
mos libani
Troia
Frigia
Licaonia
Licia
India
Idumea
Perside
Palestina
Iudea
Alaxadria
Africa
macedonia
Acaia
Rebenna
Roma
Pannonia
gallias
Terracona
Spania
germania
olissibona
Numidia
Mauritania sitifensis
Mauritania cesariensis
Mauritania tingitana
Scocia insula
desertu
mare rubru

A map, made in England, based on a much older Roman map of the world which was updated to show the current state of geographical knowledge.

'Here Lions Abound'

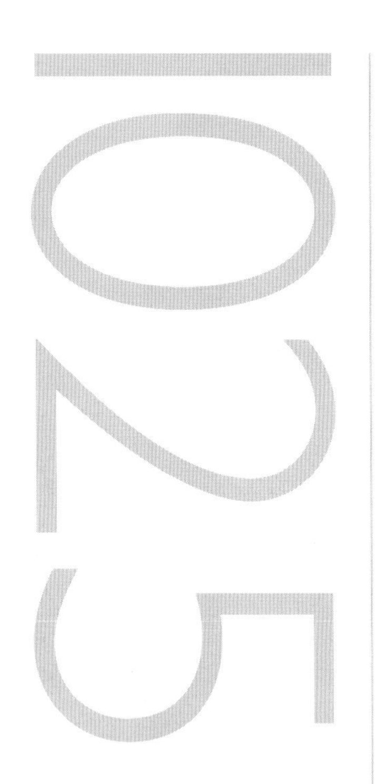

THE ANGLO-SAXON World Map gives a pale, but still the best surviving, impression of the possible appearance of the detailed Roman map of the world that the creators of all the great medieval *mappaemundi* believed formed the basis for their work. It has east at the top and, recalling ancient Greek tradition, is centred on the eastern Mediterranean – the first areas to see sunlight after the great flood of classical (as opposed to biblical) tradition. It attempts to portray the provincial boundaries of the Roman Empire and names the centres of the four successive great civilizations, Babylon, Medea, Macedonia and Rome. The coastlines of Britain, Denmark and the Peloponnese, France and Spain are truer to reality than on later medieval maps, though the French and Spanish outlines are compressed in order to fit into the available space.

The map lost precision in the process of being repeatedly copied. In the late eighth century and again two hundred years later, it must also have been substantially revised, with the names of then-significant places and regions being included. On these occasions biblical and encyclopedic information derived from classical writers was added. The map shows the travels of the Hebrews and mentions Bethlehem (but it does not show the Garden of Eden). It names the monsters Gog and Magog who were confined behind an impassable wall in the north-east of Asia by Alexander the Great. It also names animals, monsters and marvels, such as the dogheads (Cynocephali) and the griffons, in their appropriate places, though only a lion is depicted in the extreme north-east with the message '*hic abundant leones*' – 'here lions abound' – that may be a source for the oft-quoted, but unverifiable, map inscription 'here be dragons'.

England (bottom left), where the map was created, receives particularly detailed treatment. The border drawn to the west of 'Cantia' (Kent), the emphasis on the Cornish peninsula and the strange symbol which seems more like two figures fighting than a distorted town symbol, may perhaps refer to events of earlier centuries, immortalized in the Arthurian legends, when the ancient Britons briefly halted the Anglo-Saxon drive to the west.

The map forms part of a volume containing texts emphasizing time, geography, marvels and astronomy and accompanies a Latin translation of a classical Greek poetical description of the world.

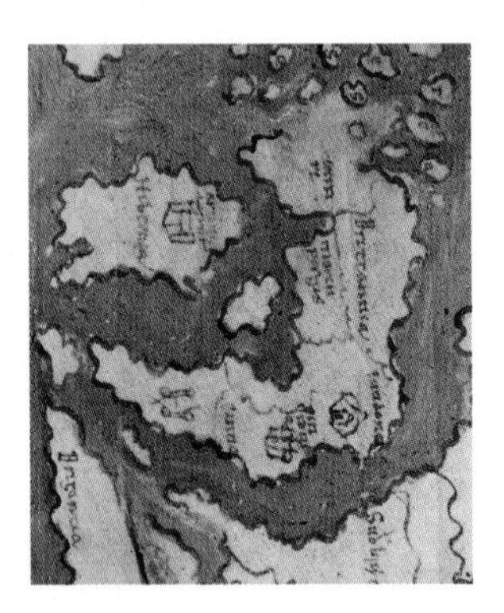

Right The Anglo-Saxon or Cottonian World Map. **Above** Detail showing the British Isles, rotated to a more recognizable orientation. **See also** 1770.

hic abundant leones
India in qua sunt gentes xliiii
mons taurus
mare caspium
media
arabia
mesopotamia
chaldea
siria
mons sina
arabia deserta
egiptus superior
hiberia
babilon
commagena
mons
Albanorum
colchorum
cappadocia
Antiochia
cesarea philippi
Isachar
trib. dan
tribus zabulon
galilea
iericho
asser
ebron
neptalim
amalech
Ierusalem
Alexandria
pentapolis
libia etiopum
cilicia
tharsos
mons olimpus
asia minor
ephesus
attica
macedonia
Tracia
Danubius fluvius
hunorum gens
pannonia
dalmatia
Histria
Roma
cartago magna
mauritania
Britannia
hibernia
Hispania citerior

A charmingly geometric conception of the Mediterranean – one of the earliest Islamic maps – provides much detail for contemporary traders of anchorage for their ships.

Assisting Arab-Byzantine Trade

THE DARK GREEN sea in this diagrammatic map of the Mediterranean is crammed with 118 islands. Those to the far left are each labelled merely 'island', whereas the remainder have names of islands in the eastern Mediterranean, with those belonging to the Cyclades in the middle and islands near Italy and Anatolia on the right. Sicily and Cyprus are represented as rectangles. Around the periphery, 121 anchorages on the mainland are described, with information on winds and landmarks.

The Egyptian mapmaker concentrated on the Islamic and Byzantine shores of the Mediterranean. Not only the Christian West, but also Islamic Spain, are all but ignored. The straits of Gibraltar are indicated by a thin red line at the far left of the oval. The next seven ports, proceeding clockwise, are anchorages past the straits on the Atlantic coast of Morocco. Thereafter the mapmaker only briefly alludes to the ports of al-Andalus (Spain), the Galicians, the Franks, the Slavs, and the Lombards. Beginning at the upper left corner and continuing to the far right of the oval there are details of Byzantine anchorages around the Aegean and eastern Mediterranean. Along the bottom of the oval sea are anchorages in the Islamic lands extending from Syria through Egypt to Tunisia.

No other medieval Islamic map provides such detail regarding the Byzantine shores of the Mediterranean. Trade with Byzantium and around the eastern Mediterranean, as well as between Egypt, Tunisia and Sicily, was of primary interest to the author of the treatise containing this charmingly geometric conception of the Mediterranean.

It is one of fourteen unique maps illustrating an anonymous Arabic cosmography titled *Kitab Ghāra'ib al-funūn wa-mulah al-'uyūn*, loosely translated as *The Book of Curiosities of the Sciences and Marvels for the Eyes*, composed in Egypt between 1020 and 1050. The original does not survive, but the Bodleian Library in Oxford has the only known copy made between 1200 and 1250. Amongst the other maps illustrating the author's knowledge of the earth are a map of Sicily, another of Cyprus, a map of the city of Tinnis in the Egyptian delta (then an important commercial centre of the cloth trade but destroyed in 1227 during the Crusades), five maps of the great river systems (Nile, Tigris, Euphrates, Oxus and Indus), a map of the Indian Ocean, and a rectangular map of the world having a graphic scale at the top.

Right The Mediterranean from *The Book of Curiosities.* **Above** The world map (with south at the top) from *The Book of Curiosities.* **See also** 650, 775, 1025, 1200, 1265, 1250, 1300, 1770.

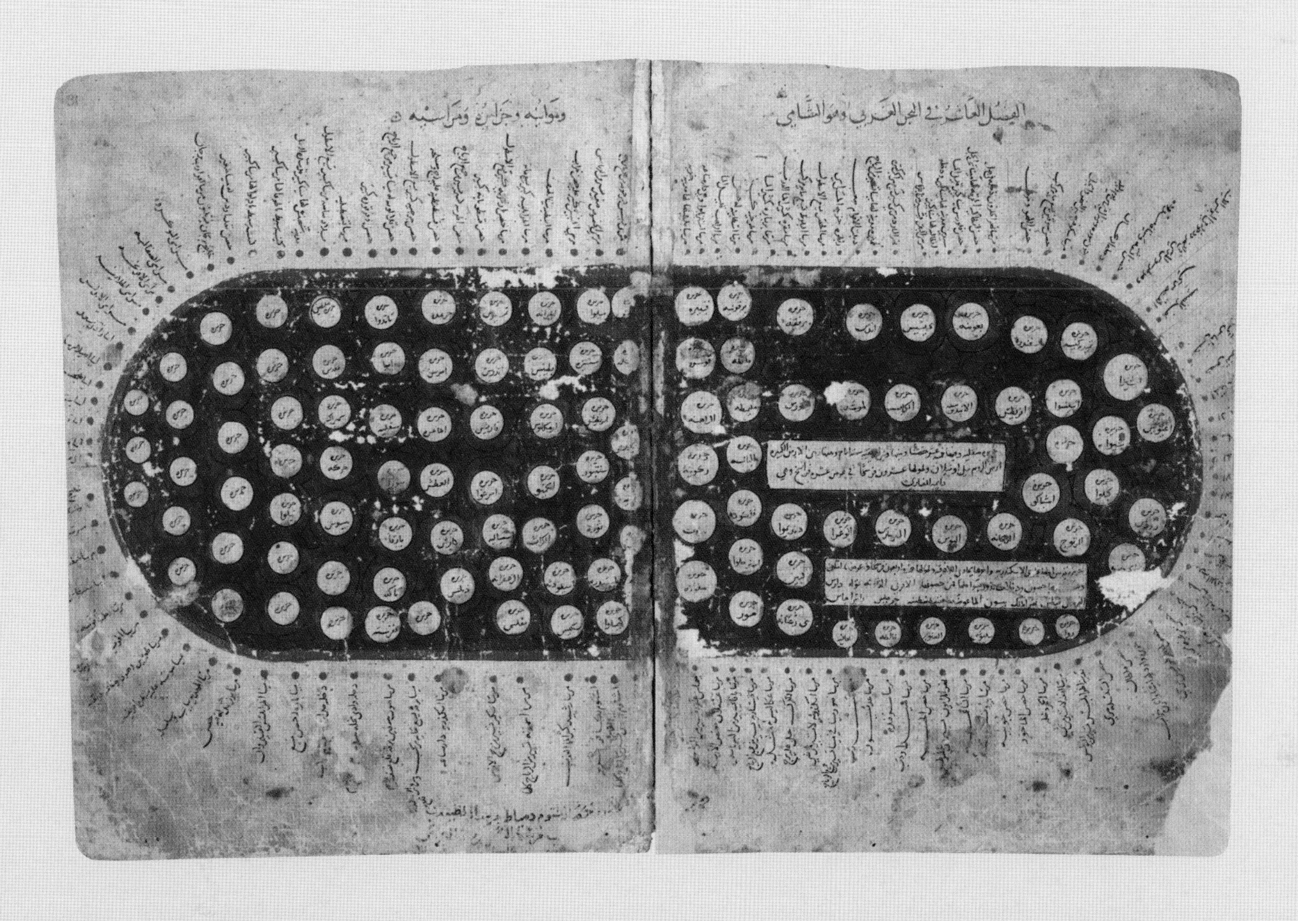

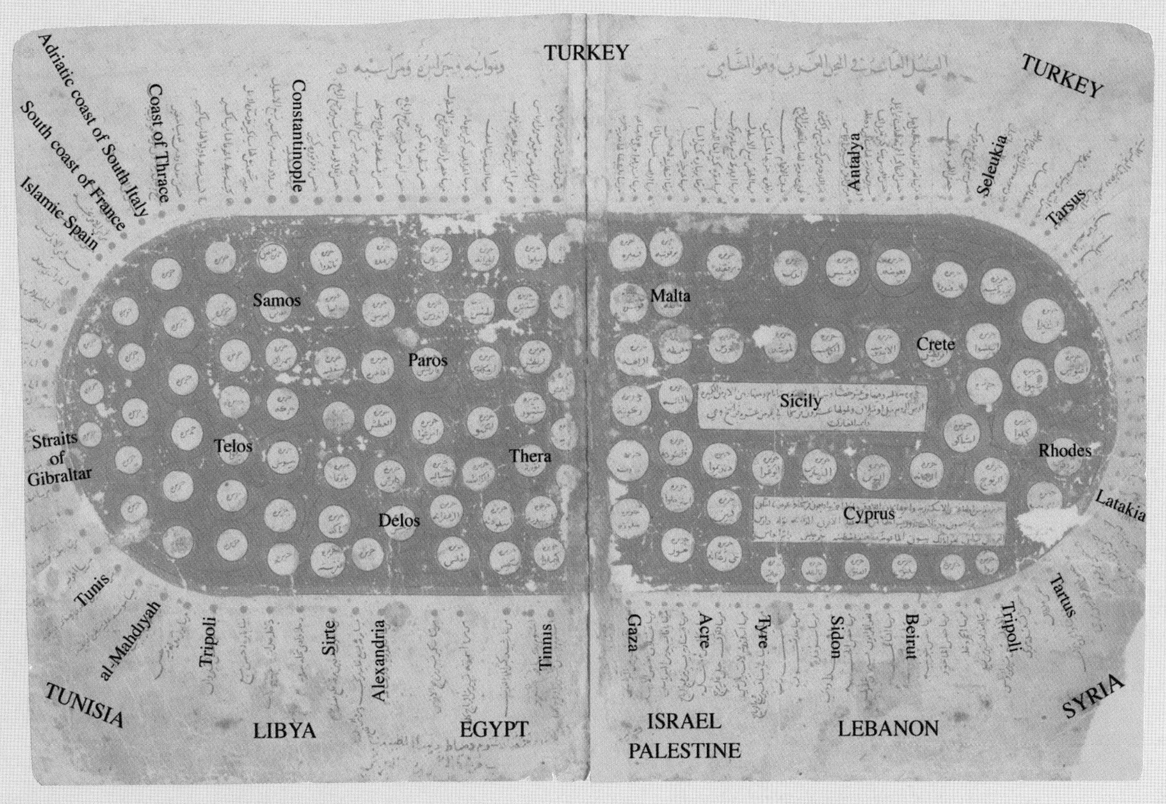

TURKEY
TURKEY
Adriatic coast of South Italy
South coast of France
Islamic Spain
Coast of Thrace
Constantinople
Antalya
Seleukia
Tarsus
Samos
Malta
Paros
Crete
Sicily
Straits of Gibraltar
Telos
Thera
Rhodes
Delos
Cyprus
Latakia
Tartus
Tunis
al-Mahdyah
Tripoli
Sirte
Alexandria
Tinnis
Gaza
Acre
Tyre
Sidon
Beirut
Tripoli
TUNISIA
LIBYA
EGYPT
ISRAEL
PALESTINE
LEBANON
SYRIA

A precursor of the famous Hereford Mappa Mundi and originally the frontispiece for a popular medieval historical text.

A Real Original?

Right The Sawley World Map. **See also** 1300.

THE SAWLEY WORLD MAP, traditionally though inaccurately known as the Henry of Mainz map, was created about a century before the Hereford World Map but it more closely resembles it in structure and content than any other surviving early map. It is smaller, though, and much less detailed. Its undulating line echoing the simplicity of Romanesque art, its restrained colouring, the absence of the elaborate scenes that surround the map on the Hereford Mappa Mundi and its excellent state of preservation tend to further disguise the similarities.

Yet its geographical outlines and the arrangement of its information, be it river systems, islands, animals and marvels are virtually identical to Hereford's. Even the Sawley's few illustrations, like the Basilisk and the barns of Joseph (or pyramids) in Africa, or the fabled rocks of Scilla and Charybdis near Sicily in the Mediterranean, are very similar. Its spiritual emphasis, like the Hereford map, is on the insignificance and end of human time, and it is flanked by the four angels holding back the destructive winds mentioned in the Book of Revelation.

The Sawley map is placed at the start of a popular medieval historical text, the *Imago Mundi* by Honorius Augustodunensis, and is strongly historical and biblical in character. East is at the top. In accordance with ancient Greek conventions, Delos, surrounded by the Cyclades, rather than Jerusalem (as on the Hereford map) is at the centre but it also depicts the boundaries between the eleven tribes of Israel. The map has relatively few references to contemporary Europe. The important German cities of Mainz and Cologne and the Italian commercial hub of Pisa are shown but France, and particularly the Angevin dominions there, come in for particular attention. It seems likely that the creator of the model for the Sawley map studied in Paris – which is shown on the island in the Seine that it then occupied – but that he came from provinces ruled by Henry II of England.

The Sawley map is likely to be a reduced copy of a large world map owned by Durham Cathedral during the period when Hugh Le Puiset (d. 1195) was bishop. In other words, Durham possessed a map that bore a very strong resemblance to the Hereford map more than a century before that map was created. It reminds us of how careful we must be about assuming that ancient maps are necessarily original maps.

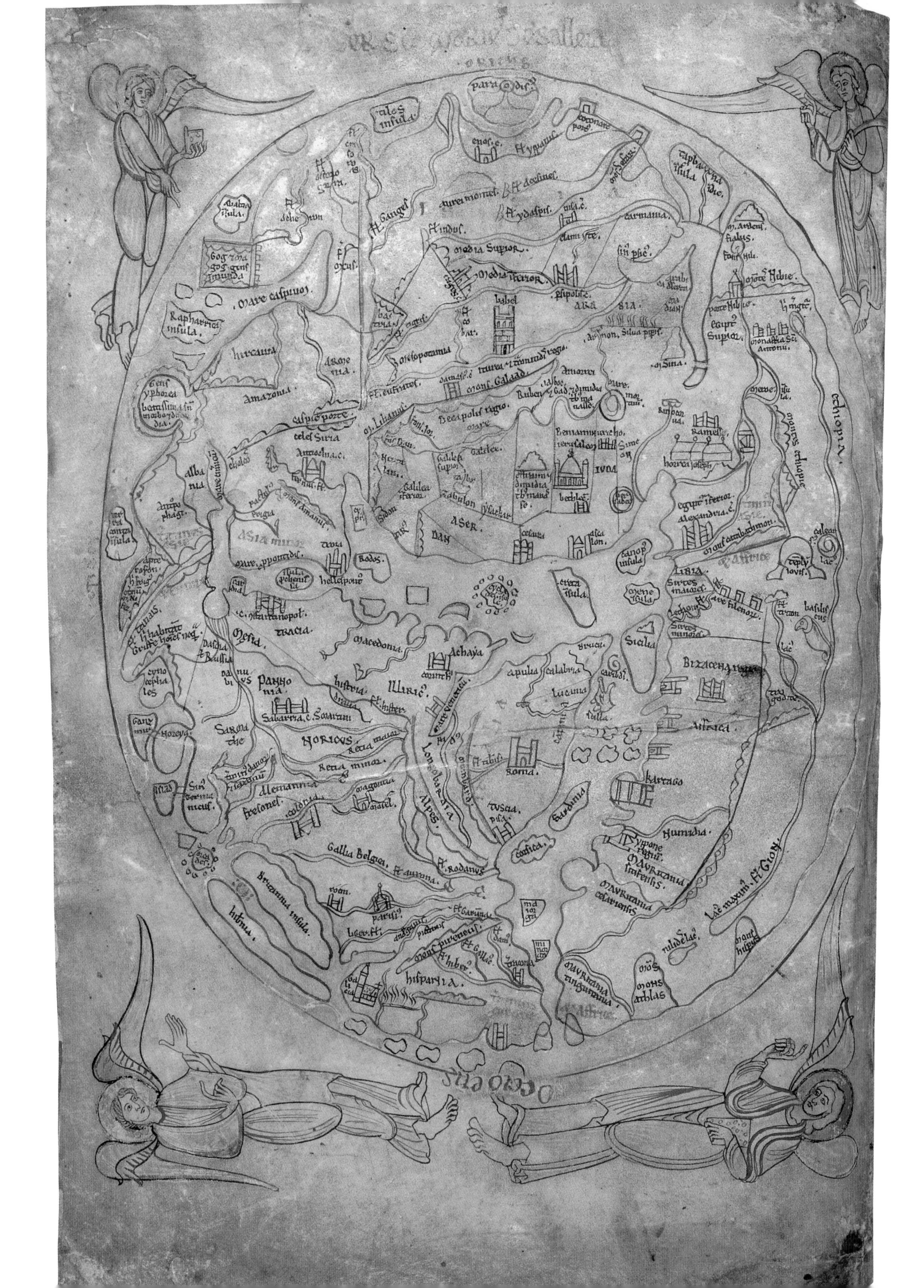

babel
Roma
Sicilia
Kartago
Numidia
Tracia
Macedonia
Achaia
Sardinia
Corsica
Rodos
Tuscia
Apulia
Calabria
Gallia Belgica
Britannia insula
hibernia
hispania
Alemannia
Noricus
Armenia
Amazonia
Mare caspium
Iudea
Aser
Dan
Asia minor
Libia
Affrica

A personalized map by Giraldus Cambrensis shows amongst other things the path to Rome from Britain, a journey he made four times in eight years.

Do All Roads Lead to Rome?

BORN IN PEMBROKESHIRE, Giraldus was both Welsh and Norman, and was fascinated from childhood with the Church. Educated in Paris, he returned (1175) to Wales hoping to become Bishop of St Davids. He entered King Henry II's service in 1184 and in connection with this wrote two books on Ireland – both are in this manuscript. In 1188 he preached the Crusade alongside the Archbishop of Canterbury in Wales and then wrote two books on Wales. Again in 1199 the canons of St Davids chose him for bishop, but his masters, the King and the Archbishop, rejected him. So began four years of bitter legal disputes between Giraldus and the Archbishop to obtain the see. Giraldus even travelled on three occasions (1199, 1201 and 1203) to Rome to plead his case, but eventually acknowledged defeat. The last twenty years of Giraldus's life are obscure. We know he continued his literary pursuits, and that he made a final trip to Rome in 1207 as a pilgrim, after which he retired to a living in the diocese of Lincoln.

This manuscript was kept for years by Giraldus as his working copy of his two Irish works; however, careful examination of the codex shows that the map was not intended to illustrate those works but was a distinct intellectual product. It is a map of Europe combining literary accounts of where regions are located, along with Giraldus's experience of travelling to Rome.

Right A map of Europe drawn under the direction of Giraldus Cambrensis. **See also** 300, 1300, 1675, 1790.

Divide the map into three equal vertical strips and you see the middle strip begins at the viewer's feet in the British Isles, and as you move directly up the map you follow a route to a final destination – the same convention used on motorway signs today. From Britain you cross the Channel; then Flanders; then Paris (note the island in the Seine); then Lausanne; then the Great St Bernard (note the unique V-shape amid the swirls representing the Alps); then Pavia; then over the Apennines into Tuscany and on to Rome.

Giraldus was a disappointed man – and it is visible. Canterbury never rewarded him; his heart's desire, St Davids, eluded him. The map identifies eight sees in the British Isles, but look what is omitted: both Canterbury (centre of the Norman Church world) and St Davids (centre of Giraldus's Church world). The map records Giraldus's experience in what it includes and in what it ignores!

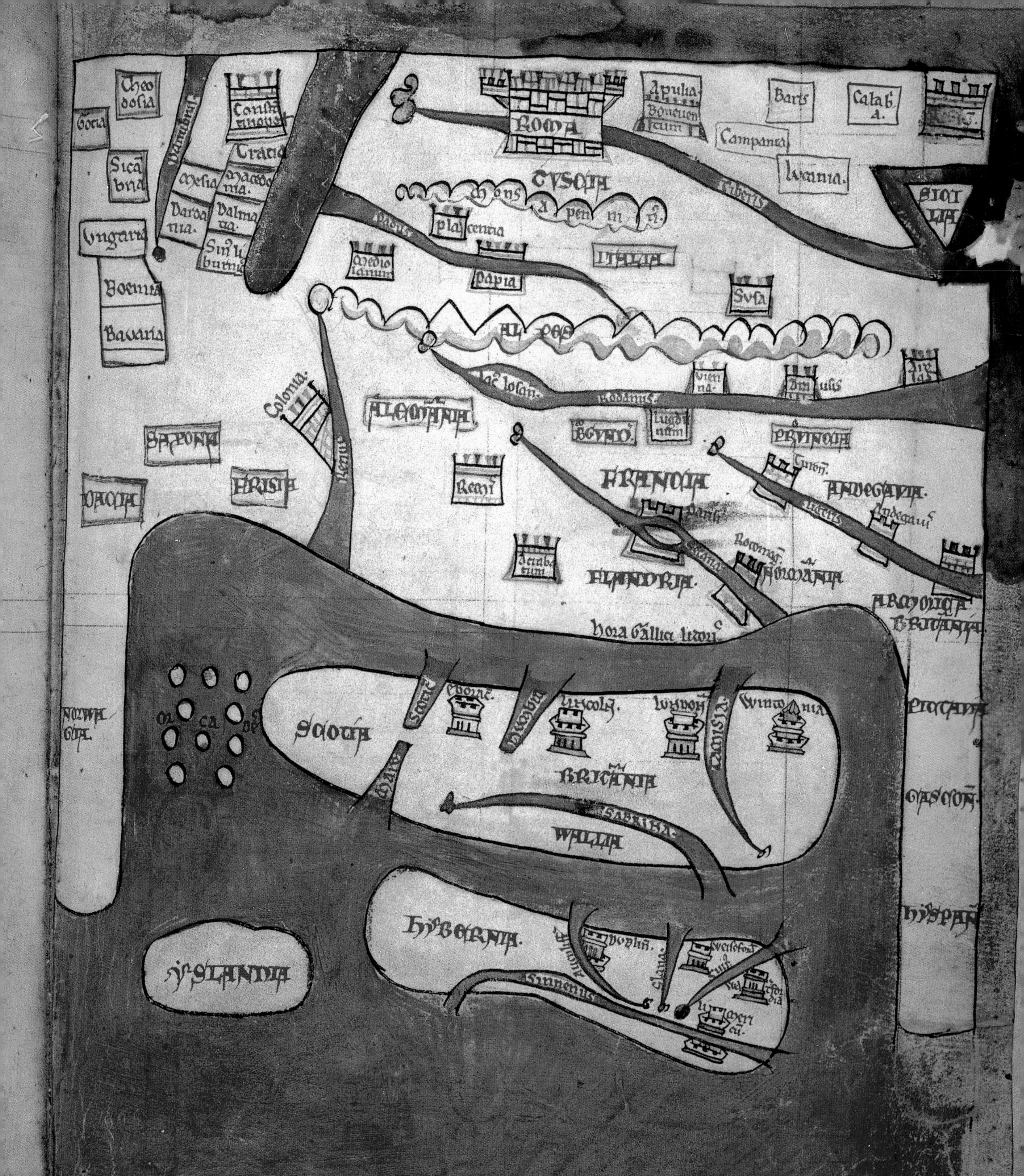
Theodosia
Goria
Ungaria
Boemia
Bavaria
Roma
Apulia
Baris
Campania
Lucania
Tuscia
Tiberis
Padus
placentia
Papia
Italia
Susa
Rodanus
Colonia
Remi
Francia
Andegavia
Flandria
Scotia
Britannia
Sabrina
Wallia
Hibernia
Yslandia
Hispania

This rare survival of a sketch map of Crusader Palestine was found bound into a thirteenth-century Bible.

The Holy Land of the Crusaders

MATTHEW PARIS, WHO DIED in 1259, is one of the few medieval mapmakers we know by name. A Benedictine monk at St Albans Abbey, England, he is the author of chronicles that are among our most important sources for the history of his own time. A skilled artist, he was interested in all kinds of graphic representation, and we have a number of maps in his hand.

What may be the earliest is this map of Palestine, the Holy Land. It is unlikely that he compiled it himself. The map shows every sign of having been copied in haste, for it gets less detailed, less careful, as one moves from top left to bottom right. It was drawn on a piece of parchment that must simply have come readily to hand. On the other side are two earlier pictures by an unknown artist, of the Deposition and of the three Marys at the Sepulchre, finely drawn but unfinished; their frames and outlines show through the parchment. One-quarter of the same side as the map had been used to copy documents that date from 1246; this explains the map's curious L-shape.

North is at the top, the Mediterranean coast on the left, with along it a series of towns from Seleucia (Portus Sancti Simeonis) in the north to Alexandria and Cairo in the south. On the right the River Jordan flows from twin sources in the Mountains of Lebanon, through the Sea of Galilee to the Dead Sea. All these and other major features probably derive from contemporary world maps. Much of the other information on the map comes from the accounts of the Holy Land that were circulating, in differing versions, in the mid-thirteenth century: 'The tomb of Abraham, Isaac and Jacob in a double cave', 'The land of the Old Man of the Mountain', and so on. A star shines above Bethlehem, at Cana are two jars of the water that Christ turned into wine, and near the Dead Sea we see Lot's wife turned to salt. Crusader castles include Krak des Chevaliers in the north, Kerak and Montreal in the south.

Possibly soon after the map was drawn, the parchment was bound with a mid-thirteenth-century Bible so that the pictures on the other side could serve as a frontispiece. That the map has thus also survived is simply a lucky accident.

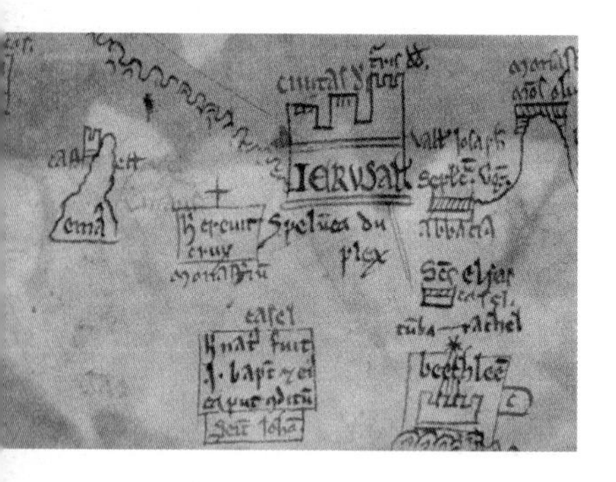

Right Map of Palestine in the hand of Matthew Paris. **Above** Detail of surroundings of Jerusalem with Bethlehem.
See also 565, 750, 1695.

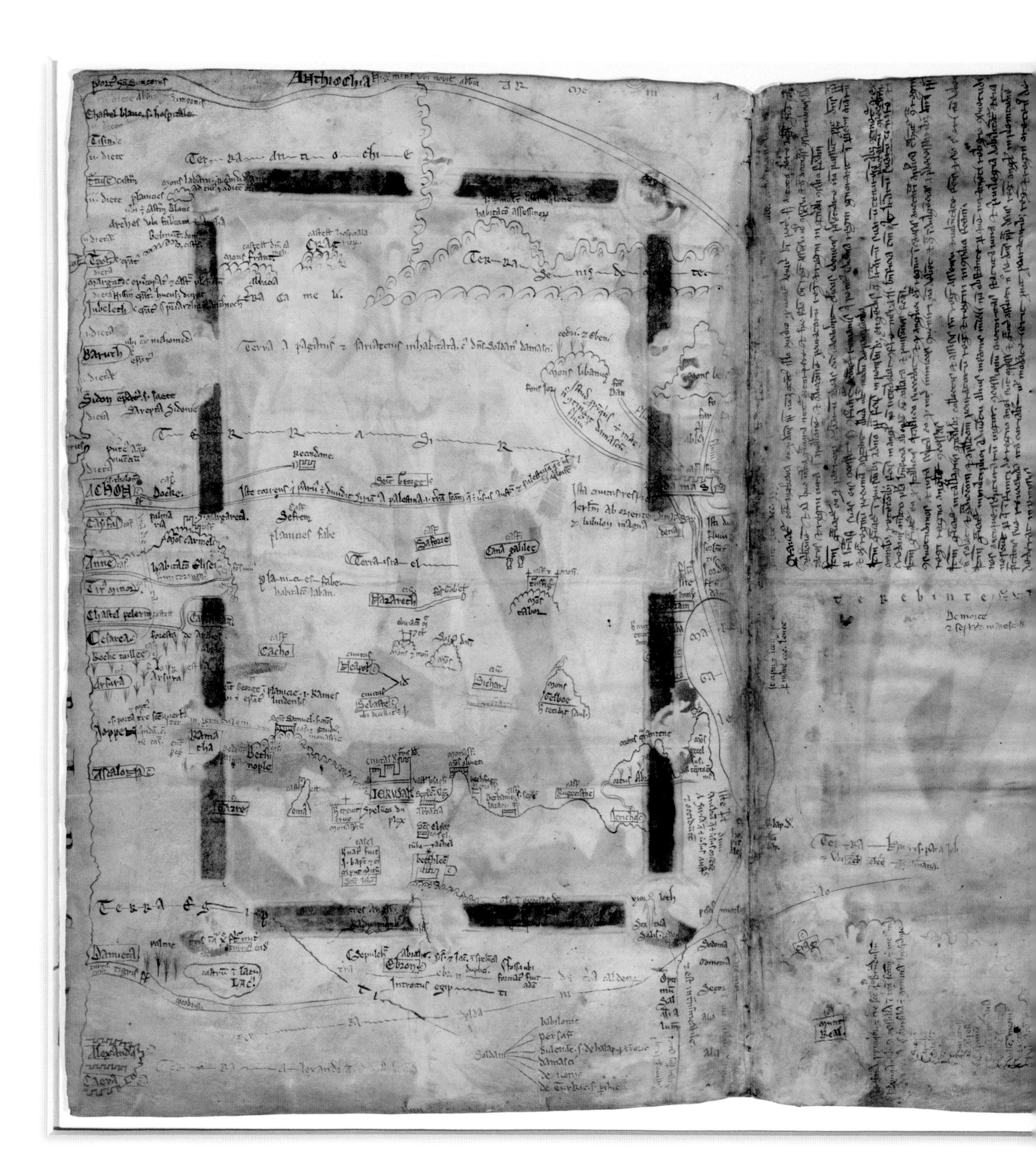

A world map in a prayer book from Westminster Abbey, packed with contemporary detail from the known world of the mid-thirteenth century.

A King's Map in Miniature?

DESPITE ITS DIMINUTIVE size, this map, occupying a leaf in a prayer book or 'psalter', has long been recognized as one of the 'great' medieval world maps. There is a lot to see, some of it barely visible such as the amazingly accurate depiction of the British Isles at the bottom left. Its vivid colour gives an impression of the original splendour of the great world maps which were as much status symbols intended to impress onlookers as expressions of piety.

The overall structure of all of them was guided by the theories of Hugh of St Victor, a German theologian and teacher in Paris in the 1130s. The imagery outside the map is, however, distinctive. Unlike the Ebstorf map, where the world is shown as though it were God's body, and the Hereford map with its emphasis on the Last Judgement, here God, holding a map-like orb in his left hand and flanked by two incense-swinging angels, blesses the world. In the more diagrammatic map on the next page (see the small illustration, left), God is seen crushing the little dragons, or wyverns, to be seen at the bottom of this map.

The content reflects the usual elements in the large, encyclopedic medieval world maps. East, and the Garden of Eden represented by the heads of Adam and Eve separated by a tree bearing a miniature apple, are at the top. Unlike earlier world maps, Jerusalem is placed, like a bull's-eye, at the centre. There are plentiful references to the successive world empires, classical learning and legend and to the Bible. The right edge of the map is filled with strange peoples derived ultimately from the writings of the Greek historian Herodotus and Roman writer Pliny. Yet the naming of the modern towns of Paris, Lyon, Barcelona, Cologne, London and perhaps Salzburg and references to the Crusades – particularly Damietta, captured by the crusaders in 1219 and 1249 – reflect the thirteenth-century world. The mentions of Normandy and Aquitaine, the lost ancestral French dominions of Henry III, who was then King of England, are particularly significant. The map is known, on stylistic grounds, to have been drawn in Westminster Abbey: it could well be a close copy of the well-known but now lost great world map which adorned the King's audience chamber in Westminster Palace, just a few hundred yards away.

Right The Psalter World Map. **Above** The diagrammatic world map at the back of the Psalter World Map.

The largest known European medieval world map that was destroyed by bombing in the Second World War.

Of Rulers, Nuns and Salvation

THE EBSTORF WORLD MAP, on thirty goatskins, was the largest of the great medieval world maps. The original was destroyed by bombing in 1943 and it is now known only from a series of copies ultimately based on detailed photographs taken in 1891. Place names, the emphasis on northern Germany and the style suggest that the original was a slightly modified copy – perhaps created in about 1290 by the nuns of the convent in Ebstorf in Lower Saxony, where the map was found in 1830 – of a map made for the Lüneburg palace of Otto the Child, Duke of Brunswick in about 1240 or possibly a few decades later. The features emphasized on the map seem to correspond to the political situation in northern Germany at that time, with a particular emphasis on Otto's relatives and allies.

Though the identity of the mapmaker is unknown it is generally accepted that the map belongs to the Anglo-French school of medieval *mappaemundi*. In overall appearance it is close to the Psalter World Map. Its relationship to the Hereford World Map is more distant, but both were meant for public display, are derived from the same classical and biblical sources, deal with the contrast between human and divine time, emphasize the centrality of Jerusalem and are built around ancient Roman and medieval commercial itineraries. An inscription on the Ebstorf map indeed mentions that the map could be used for the practical purpose of route-finding – a precociously pragmatic note to find on a medieval European map.

The biggest difference between the Hereford and Ebstorf maps, however, lies in the their cosmological frames. While the Hereford map emphasizes the Last Judgement, the Ebstorf map more optimistically focuses on salvation. To symbolize His experience on earth, the world is shown as the body of Christ whose head, hands and feet emerge from the map with the rivers and seas serving as His blood vessels. It has been suggested that this hopeful message is continued into the myriad playful depictions of the creatures of the world. While on the Hereford map they are static, rather resembling academic specimens, on the Ebstorf map they are shown engaging with other creatures – such as the bull-like Bonacon (possibly originally meant as a symbol for an oriental tribe) repelling its enemies with its wind or the antelope gambolling among the trees of the gardens of Babylon.

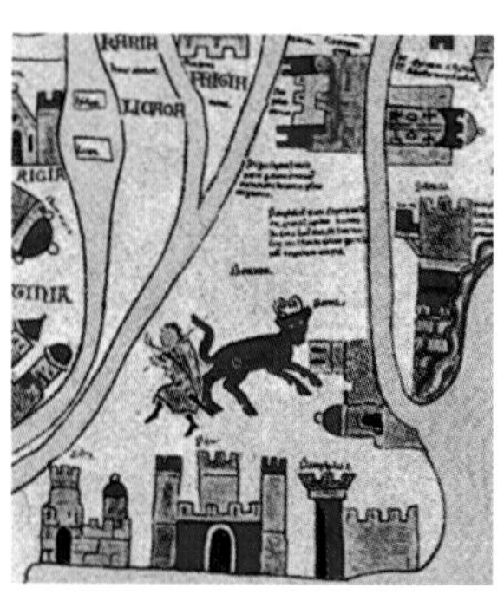

Right The Ebstorf World Map. **Above** The Bonacon. **See also** 776, 1025, 1200, 1265, 1300, 1415.

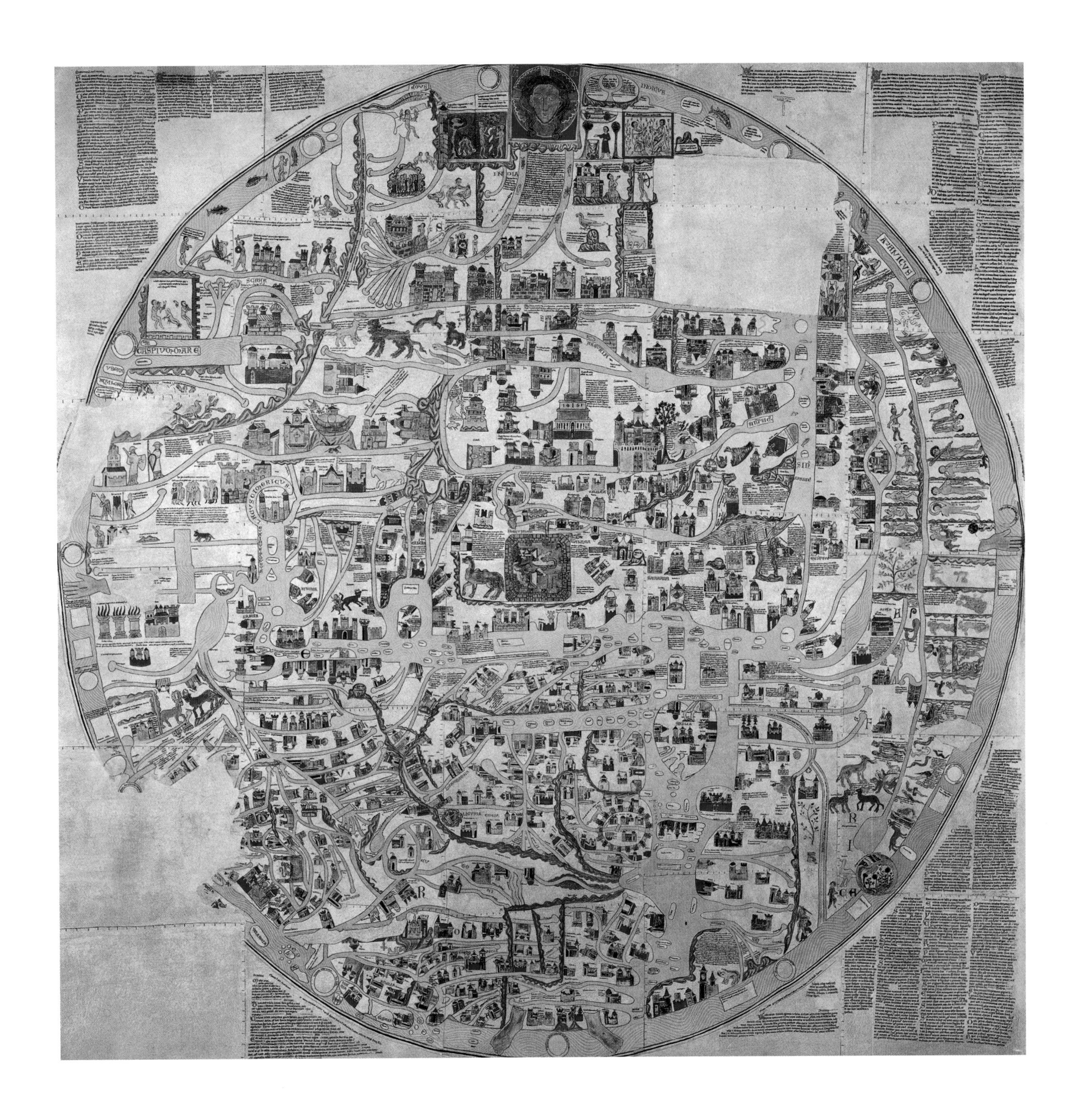

The largest surviving medieval world map, probably copied from a prototype in Lincoln Cathedral.

Divine and Human Time and Space Contrasted

THE HEREFORD MAPPA MUNDI, the largest surviving medieval world map, represents the culmination of a tradition of mapmaking stretching back to classical antiquity. The scene at the bottom left of 'Caesar Augustus', portrayed as a pope, ordering the surveying of the world, reflects a confused awareness of its beginnings. Closely related to the Sawley World Map, it may have been copied from a set of late twelfth-century written instructions for constructing world maps, the *Expositio Mappe Mundi.*

The dominant message of the map is the contrast between divine timelessness and infinity and the transience and limitations of human time and space. At the top one sees the Last Judgement, with the saved to the left and the damned to the right, and a bare-breasted Virgin pleading for mankind. At the bottom right a mounted huntsman looks wistfully back at the earthly world but his page urges him to move on. The map of the world, like a colossal wheel of fortune, is held down by four thongs containing letters which together spell out 'MORS', or death. The map itself has Jerusalem, surmounted by a depiction of the crucifixion, at the centre. Place names echo the four empires of human history, the travels of the Apostles and pilgrimage routes. The imagery depicting, for instance, the story of the golden fleece and the monstrous races on the fringes of the earth reflect the influence of the Alexander legend and the writings of Herodotus, Pliny, Orosius, Isidore of Seville and Physiologus.

Right The Hereford World Map.
Above Lincoln on its hill.

But the thirteenth-century world is also well represented. The image of England and Wales contains depictions of Lincoln Cathedral, where the map's prototype was probably to be found, and of the Welsh and English castles recently constructed for Edward I. Contemporary trade routes are shown. The depiction of the huntsman is perhaps also a reference to a contemporary hunting dispute. The scene with Caesar Augustus may similarly be a reference to a dispute over taxation. The map may have been commissioned in the hope of furthering the canonization of Thomas Cantilupe, Bishop of Hereford (d. 1283) and the career of his successor Bishop Swinfield. The map has numerous layers of meaning, from the divine to the emphatically mundane, which scholars have only begun to reveal over the past few decades.

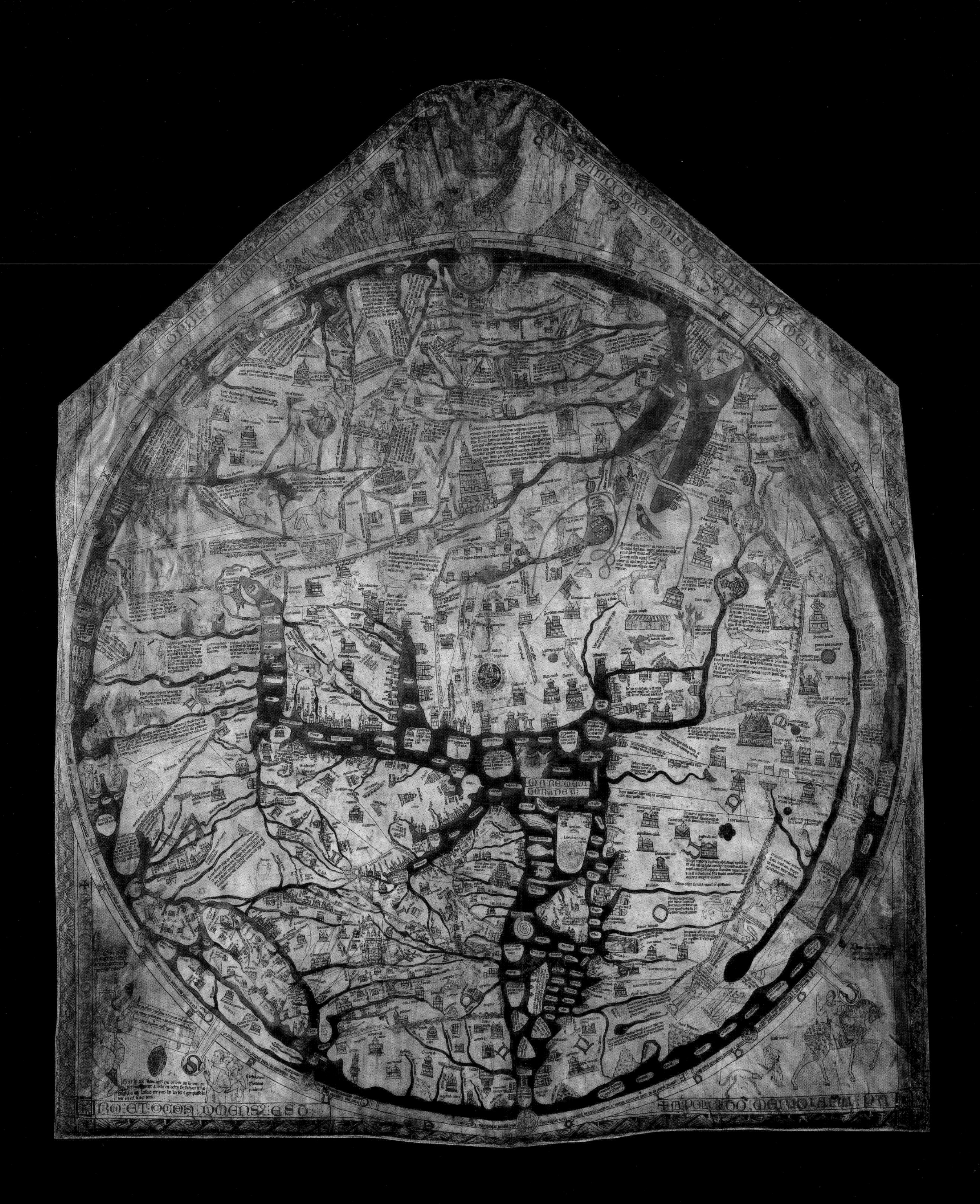

An early sea chart showing coastlines, ports, hazards and wind directions, originally used by medieval mariners.

For Practical Use

1325

THE EARLIEST SURVIVING European sea (or 'portolan') chart was created on the shores of the Mediterranean between Barcelona and Genoa in the late thirteenth century, at about the time that the great medieval world maps were being produced in England and France. The two could not have been more different. While *mappaemundi* were encyclopedic and deeply imbued with learning and spirituality, portolan charts began as practical instruments for trade and navigation. They earned their name from the written lists of ports used as navigational aids by Mediterranean sailors from the earliest times. It is not known when these *portolani* began to be illustrated by drawings of coastlines, but the full-fledged charts could not have been created before the invention of the magnet compass in the twelfth century. The earliest instructions for constructing such charts may date from then. Originally their coverage was restricted to the Mediterranean and the Black Seas.

The essence of the chart was the depiction of coastlines, with the shapes of important capes and harbours exaggerated. Place names were positioned at right-angles to the coastline with important ports written in red. Hazards such as rocks and sandbanks were marked by dots or crosses. Because the Mediterranean is so narrow, lying between only a few degrees of latitude, and comparatively short in breadth, coastal outlines could be drawn with relative accuracy without the need for a mathematically determined projection to take account of the earth's curvature, the calculation of which would have been beyond the capabilities of most medieval sailors. Navigation was possible using only compass, dividers and a ruler – known as 'dead reckoning'. The fretwork of intersecting, or 'rhumb', lines, linking points along the circumferences of invisible circles, served as copying aids for the chartmaker and also as navigational aids, with each line drawn a different colour representing a different wind.

All surviving examples, whether single charts or in atlases, were made for merchants' studies or collectors' libraries, the working charts having been used to destruction. They fall into two categories: the so-called 'Italian' charts with no interior detail but for rulers' banners, and the more decorative so-called 'Catalan' charts, though occasionally 'Catalan' charts were made in Italy and vice versa. This 'Italian' chart, created at about the time that the first Mediterranean fleets reached England, is probably by the earliest named chartmaker, who was born in Genoa but worked mainly in Venice.

Right A portolan chart of the Atlantic coasts of Europe and Africa attributed to Pietro Vesconte. **See also** 1465, 1558.

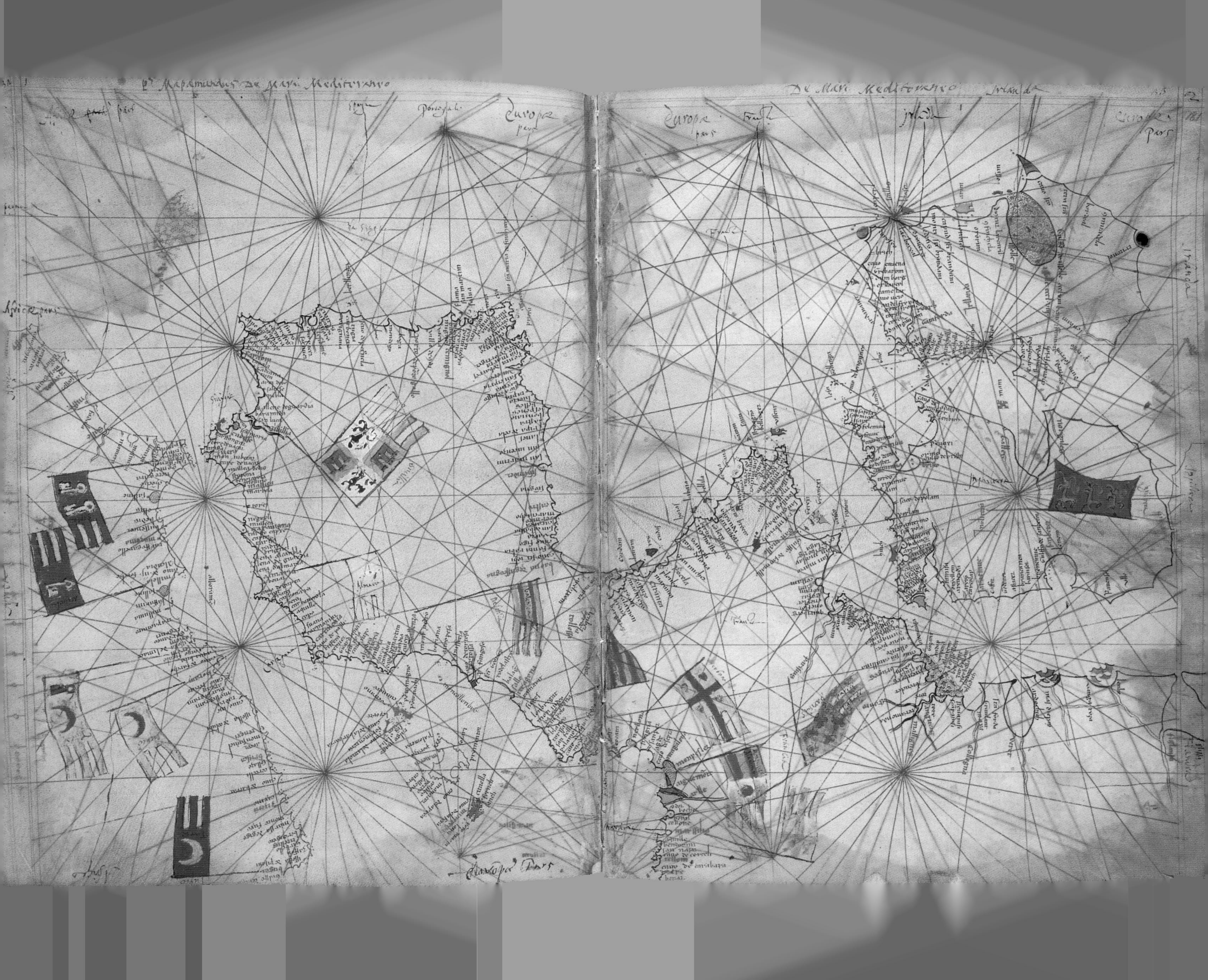

De Mari Mediterraneo
Europe pars
Portogali
Irlanda

The first detailed map of England and Wales, complete with Edward I's newly built castles, which served as an accurate representation of the realm for two hundred years.

A Monarchy of the Whole Island

MAPS OF REALMS or countries – pre-eminent among cartographic genres nowadays – did not become common until the sixteenth century. But incipient secular states like England began to articulate their territorial coherence much earlier. From the mid-thirteenth century, the concept of the State as a political and territorial 'body' came increasingly to be set against the ecclesiastical concept of the Church as a corporate body, or universal community of believers.

The 'Gough' map of Britain departs from the conventions of medieval Church cartography. Although oriented east and containing some legendary and pictorial material, Gough's geography is recognizable. It looks like a modern map: island-outline is a major concern, and sites within Britain are located with extraordinary precision. Most of the map is devoted to towns, coasts, roads and rivers; less important are topographical barriers like hills, lakes and forests. Written legends are few, place names ubiquitous.

Different areas possess these features in different proportions. South-east England is dense with towns and routes, while Scotland has few settlements and no roads but room for highland icons, local fauna ('here dwell wolves') and natural marvels. The Gough map's London-centred highway system – whose mileages remained viable for centuries – may contribute to its accuracy in outline and town-location. Basic to the Gough map's geographical argument is that it represents not just England, but the entire island of Britain. The map was probably compiled or used initially by English royal clerks, whose purview included diplomacy, taxation and commodity purveyance. The map's route-network suggests ties to mercantile interests, especially the woollen industry upon which the national economy rested. But above all the Gough map speaks to the imperial ambitions of English kings and their clerks. Edward I subjugated Wales in a series of campaigns 1277–95; Gough registers this conquest with a road linking castles newly established along the north-west coast. In the 1290s, Edward turned his attention – first legal, then military – to Scotland. As had King Arthur and Britain's founder Brutus, Edward I asserted his ancestral right to a 'monarchy of the whole island'. One of the map's few written legends appears off Devon: 'here landed Brutus with the Trojans'. Gough's mysterious North Sea illustrations – a wrecked ship (of state?), a mariner taking on cargo and battling sea-monsters oriented like England, Wales and Scotland – may relate allegorically. Gough dates *c.*1280–1360, but this template served, via copies and derivatives, as the standard map of the realm until the mid-sixteenth century.

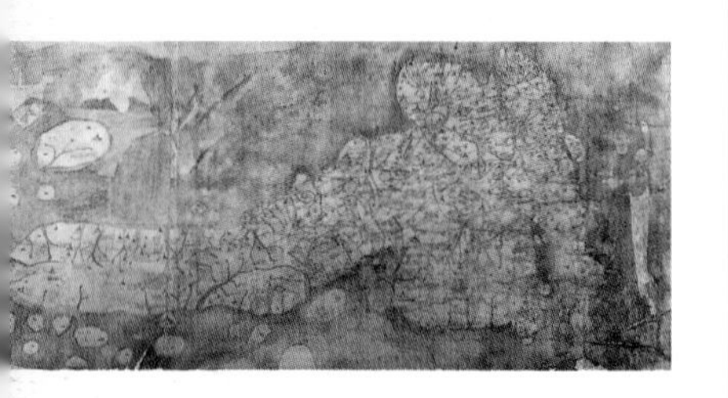

Right Southern England and Wales from the Gough map. **Above** The Gough map of Britain.

A medieval English, general-purpose world map which celebrates in size and apparent importance England and its commercial interests.

God for Harry, England and St George

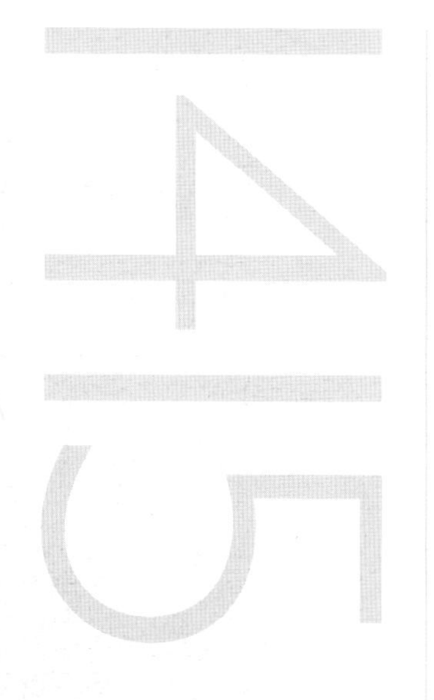

Right The Evesham World Map.
Above Taddiport, Devon.
See also 650.

THIS MAP ILLUSTRATES the durability of a certain vision of the world. An English map of the late middle ages, it bears a considerable resemblance in general form and particularly in the shapes of the Mediterranean and the Red Sea, in the relative sizes of the continents, and in the form of what, here, are the British Isles, to one of the earliest detailed world maps, probably created in southern France in the late eighth century. This conservatism is probably due to the fact that the Evesham map was not supposed to be anything special. It was a standard map, intended to accompany sermons and lectures and to illustrate anything and nothing in particular. Such maps must once have been very common but this seems to be the only surviving complete example.

Yet, typically of medieval maps, a considerable amount of modernity is hidden within the antique forms. If the upper, eastern part of the map, with Paradise at the top, the Tower of Babel beneath and a splendid many-towered Jerusalem at the centre, speaks of tradition, the lower western part proclaims itself to be a product of the England of the time of the Hundred Years' War. England extends from Scandinavia to the Mediterranean; the tower representing Calais, a glorious English conquest, is larger than Rome and challenged only by the tower representing St Denis, the burial place of the kings of France (which the kings of England claimed to be). Paris is marked by an insignificant tower to the left of St Denis. England's commercial partners, Bruges and Cologne, are honoured with towers. Place names mirroring trading routes are scattered throughout England, with particular attention paid to the villages near the wool-rich abbey of Evesham. We do not know the draughtsman but it is a reasonable guess that he came from the tiny hamlet of Taddiport in Devon which is proudly named.

India
Babilonia
Tigris fluvius
Mare rubrum
Galaad
Cedar
Rodus
Roma
Campania
Sicilia
Sardinia
Cartago
Numidia
Alpes
Hispania
Gades
Insule

A medieval map of Pinchbeck Fen in Lincolnshire records the rights of various villages to graze sheep – a vital staple of the economy.

Of God, Man and Sheep: A Medieval Landscape

THIS LARGE MAP on parchment shows a central area of Lincolnshire fenland painted vivid green. It is surrounded by twenty-eight churches, which represent the villages, monasteries and wayside chapels that lay on higher ground around the fen, visible for miles across the flat marshy countryside. These churches were carefully drawn in perspective, many apparently from life, and outlined in red. The lack of secular buildings suggests that the mapmaker was not a layman, but probably belonged to one of the monastic houses shown on the map. Most likely, he came from Spalding Priory (near top right, above the river), since this area is shown in greatest detail, including Spalding parish church (below the river).

Despite the profusion of churches, the map was probably not drawn primarily to record them. Down the left-hand side are written, in Latin, two legends which tell us that five named villages had common in the marsh of Pinchbeck, but that another village was excluded from the fen. So the main reason for making the map seems to have been the wish to record commoning rights in Pinchbeck Fen. These rights to graze animals, especially valuable wool-bearing sheep, were an important part of the medieval economy.

The map was drawn about 1430: we know this because it shows a chapel which burnt down not long after that. At that time, it probably seemed to the monastic mapmaker that the Church to which he belonged was set to dominate the scene in perpetuity. This small area of about 10 square miles (16 square kilometres) contained not only Spalding Priory but also an important abbey at Crowland to the south, prominently shown on the map standing by a three-cornered bridge across two rivers. To the north-west lay Sempringham Priory, chief house of the Gilbertines, the only English monastic order; this map bears probably the only extant contemporary picture of it. In the fifteenth century it would have been bustling with white-robed monks and nuns. Just over a century later, all of these houses were to be dismantled as part of Henry VIII's dissolution of the monasteries. The map apparently passed to the Duchy of Lancaster at the dissolution.

There is now barely a trace of Spalding Priory, Sempringham Priory lies levelled under long grass, and at Crowland the ruined abbey buildings lie around the abbey church, now the parish church.

Right Map of Pinchbeck Fen.
Above Detail showing Crowland Abbey.

Southbrune
Weylaunn

The earliest 'modern' European world map, probably commissioned by the mercantile state of Venice to update its knowledge of the known world.

Challenging the Ancients

THIS ENORMOUS AND well-preserved map may be the original commissioned in 1448 by the Venetian senate from Fra Mauro – a mapmaker and monk in the Camaldolese monastery of St Michael on the Venetian island of Murano – and the chartmaker Andrea Bianco, completed by 1453.

At first glance the circular form, the plentiful illustrations and the two thousand-odd inscriptions all around recall medieval world maps. But where they relied on tradition, in his inscriptions Fra Mauro questioned it, on the basis of his own observations, reports received from abroad and from his contacts with contemporary merchants, mariners and visitors to Venice. The map is oriented with south and not east at the top (though Jerusalem does receive an especially large vignette). The texts are in the Venetian dialect and not in Latin. There is no spiritual or cosmological framework around the map's circumference. The Earthly Paradise, beautifully depicted by, perhaps, Leonardo Bellini, is exiled beyond the map to the bottom left where an inscription tries to establish its actual physical location. No monstrous races are depicted in Africa since, Fra Mauro tells us, none have ever been seen, with the exception of the Dog Heads (Cynocephali). Indeed figures of any sort are missing though the map does depict cities, buildings, ships and fish.

Right The Fra Mauro World Map. **Above** Detail showing the rivers Senegal and Gambia. **See also** 1500, 1507, 1525, 1544 (p. 102), 1597 (p. 134), 1599.

The European coastlines are copied from portolan charts. The expanded depiction of Africa, showing Timbuctu and the rivers Senegal and Gambia, reveals knowledge of the ongoing Portuguese voyages which, with the help of Italian navigators (who had no scruples about passing on information to their compatriots in Venice), were rapidly revealing the true course of the African coast. Fra Mauro was the first mapmaker to make extensive use of the writings of Marco Polo and Niccolo de Conti for his portrayal of the Asian interior and the islands of the Indian Ocean. He also tried as far as he could – for instance in his depiction of India which was not to be reached by the Portuguese until 1498 – to include information from Claudius Ptolemy's *Geographia*. However, as Fra Mauro noted, modern discoveries had shown some of Ptolemy's information to be incorrect and had uncovered important places completely unknown to him.

Though often acclaimed as representing the culmination of medieval mapmaking through its synthesis of the traditional, portolan and Ptolemaic, the Fra Mauro map in truth represents the first great map of a new tradition.

SEPTENTRIO
POLVS ARTICVS
POLVS ANTARTICVS
ETHYOPIA
AFRICA
CIRENAICA
ASIA
PERSIA
TARTARIA
ROSSIA
EVROPA
NORVEGIA
CHATAIO

A late medieval diagrammatic world map, consciously unrealistic in its form and content but potent with symbolism.

Mapping a Civilization

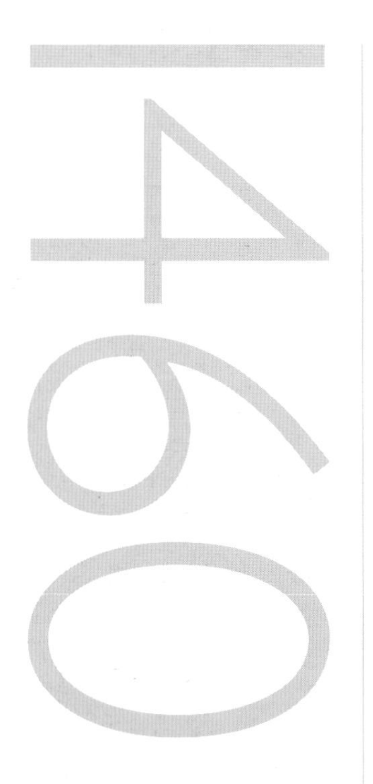

THIS IS ONE of the most colourful depictions of the medieval symbol for the world, the T-O map, which can be traced back to ancient Egypt. The ocean-sea surrounding the inhabited world, or *ecumene*, forms the O. The cross-stroke of the T symbolizes the Don (Tanais) and the Nile, and the downstroke the Mediterranean. Asia, by far the largest continent, is at the top (east), Europe at the left (north) and Africa at the right (south).

This example, in a popular universal history in French created for the Duke of Burgundy, accompanies a description of the world based on the Etymologies of Isidore, Bishop of Seville in the early 700s. It demonstrates why the T-O symbol became so popular in the middle ages. The three sons of Noah, whose descendants peopled the world, could be associated, as here, with the individual continents, Shem in Asia, Japheth in Europe and Ham in Africa. The number three had connotations of the Trinity and the shape of the waters probably reminded viewers of the crucifixion.

This depiction adds further reminders of the Bible and the classical and medieval worlds. The imaginary landscape and towns, more reminiscent of Flanders than the Mediterranean, symbolize places associated with human and biblical history. Noah's Ark is stranded on top of Mount Ararat. Historic civilizations are recalled by the naming of Nineveh (Babylon), Athens (placed in Africa!) (Greece) and Mainz which is presumably a reference to the (Holy) Roman Empire. The wondrous creatures of the east are represented by a small phoenix, placed next to Noah's Ark near a large lake that may represent the Black Sea.

By the time that this map was drawn, its viewers would have known that it was not a realistic picture of the world. They would have gone elsewhere for that. They would have appreciated the map as a symbol for their world and a civilization that was as old as time and was laden with layer upon layer of meanings beyond the purely geographical. Christopher Columbus's son Ferdinand, a pious Catholic as well as one of the most skilled makers of sea charts of his time, is known to have had several printed maps resembling this one. He doubtless appreciated that good maps do not necessarily have to reflect the prosaic reality.

Right World map by Simon Marmion. **Above** World map with surrounding text. **See also** 500.

Orient
Asie

A 'Catalan'-style chart, probably commissioned by a wealthy Spanish merchant, combines realistic coastlines with elements taken from classical authors and more recent travellers.

A Map from Majorca

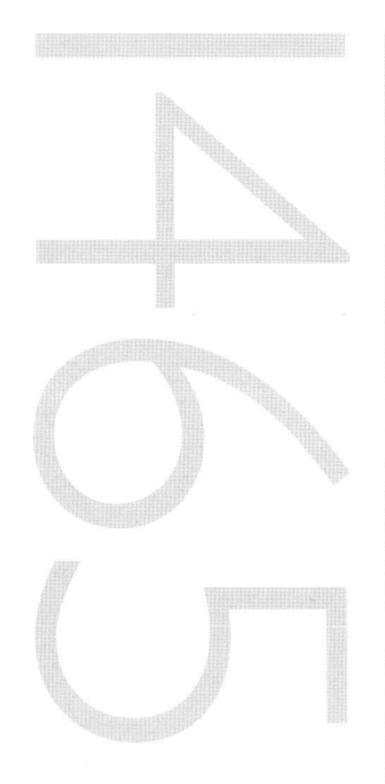

THE 'CATALAN'-STYLE sea or 'portolan' chart was much more decorative than its 'Italian' counterpart and bore a passing resemblance to the *mappaemundi* of northern Europe. While 'Catalan' like 'Italian' charts emphasized coastlines and the names of ports, 'Catalan' charts also depicted inland towns, rivers, mountain ranges, rulers' arms and non-European potentates. Unlike 'Italian' charts, they showed the Baltic even though it had long been closed to Mediterranean and all other non-German seamen by the powerful Hanseatic League of German merchants. Conversely in the fifteenth century they showed no hint of the Portuguese penetration from 1415 of the African coast beyond Cape Bojador, though Italian chartmakers, supplied with information from the navigators who had been employed on Portuguese ships, often did.

From the mid-fourteenth century wealthy Italian and Spanish merchants presented the kings of England and France with ornate 'Catalan'-style world maps and atlases in the hope of receiving commercial favours in return. These combined realistic coastal outlines in Europe, Africa and western Asia with *mappamundi*-influenced depictions derived from the classical authors and the accounts of Marco Polo and other travellers, in central, southern and eastern Asia. They were produced in workshops in Barcelona and in Palma on Majorca, where the maps were created by a (usually Jewish) chartmaker and the decoration by a professional artist. The only surviving example of such a world map is the magnificent *Catalan Atlas* of about 1375 which is now in the Bibliothèque Nationale de France.

Roselli, who was possibly Jewish-born, belonged to this tradition, having been apprenticed to a leading earlier chartmaker, Battista Beccari, whose family had been associated with the Cresques family, the probable creators of the *Catalan Atlas*. He produced this rather damaged and cut-down chart and ten other surviving examples between 1447 and 1469. Europe is thick with mountains, rivers and walled cities while Africa is dominated by the Atlas mountains, towns, the odd elephant and camel and local sultans, seated under splendid tents. The Red Sea is painted red, with a white strip representing the passage of the Israelites and there is some emphasis on the religious sites of the Holy Land. The grid-like pattern superimposed over the rhumb lines may be a polite, but insincere, nod in the direction of the mathematical mapmaking popularized following the translation into Latin of Ptolemy's *Geographia* in 1407.

Right Portolan chart by Petrus Roselli. **Above** Egyptian sultan.
See also 1325, 1558.

Map Signs

Every map starts as a blank surface on which is marked the distribution of selected features, real or imagined, on or above the surface of the earth or below its waters. Indeed, the marks, or signs, are the map. Yet for more than four thousand years, whoever drew a map was free to select whatever signs was thought appropriate. Only in the case of highly specialist maps, such as charts for navigation, was there some sort of authority to insist on particular signs for specific features, such crosses for rocks or points for shoals. The modern notion of a prescribed or 'conventional' sign was defined only at the start of the nineteenth century in connection with the first national survey of France by the French Army. International mapping conventions are now regulated through commissions for the different kinds of map (geological maps, marine maps, nationally surveyed maps, etc.).

From the start, two kinds of map signs have been used either exclusively or, more commonly, together: pictorial and non-pictorial signs. Pictorial signs resemble the feature they represent. When presented in profile (as if viewed from ground level or a little above) the subject of a pictorial sign is quickly recognized, like the huts on some prehistoric maps (1500 BC). On the surviving papyrus fragment of the earliest known Roman map (early first century BC, probably of Iberia), towns and cities are represented by sketches of angular buildings or groups of buildings, and what seem to be the smaller settlements are marked by non-pictorial signs (rectangles). Later in Roman times, similar pictorial signs illustrated the surveying manuals of the *Corpus Agrimensorum* (AD 350), and were used on the Madaba map (565) and the Peutinger map, a medieval copy of Roman origin (300). Echoes of the characteristic Roman place-signs are found on medieval maps, such as the maps made, or copied, in Byzantium from the thirteenth century onwards for Claudius Ptolemy's *Geographia* (150) and Matthew Paris's maps of Palestine (1250). Pictorial signs continued to be used on maps after the sixteenth century, even when major cities were sometimes singled out by being shown in plan (as if viewed from high above; see 1590), as were pictorial signs for other landscape features. The ancient way of indicating mountains pictorially, as inverted Vs or 'fish-scales' (as on a Babylonian clay tablet of *c.*2300 BC from Yorgan Tepe, Kirkuk, Iraq), or with realistically craggy outlines, or as smoothed 'mole-hills', remained more popular than any kind of plan view (as in fifteenth-century printed editions of Ptolemy's *Geographia*) until the mathematically defined contour line came into widespread use in the eighteenth century. Many vegetation signs are still pictorial, even on official national survey maps such as those of the Ordnance Survey (1914, p. 304).

Signs representing the landscape feature in plan are generally avoided because the unfamiliar perspective is less easily recognized. The problem of interpretation is compounded when the object is also unusual. E-shapes marked along rivers on a map of Moravia (1626, Jonas Scultetus) are baffling until the key reveals that they represent logging weirs. Non-pictorial signs always need explanation, either oral or written. The significance of squares, rectangles, circles, dots, points, etc., would otherwise be incomprehensible, especially as each mapmaker tended to impart his own meaning to the different shapes. A rectangle, a floret and a black dot on one map may all indicate settlement of one category or another on one map, but stand for bridges, towns and mineral deposits respectively on another map. When not incorporated into a pictorial sign, the locational dot (the point or circle used since the earliest Ptolemaic maps to indicate the exact, measurable, location of a place) can be confused with a non-pictorial sign.

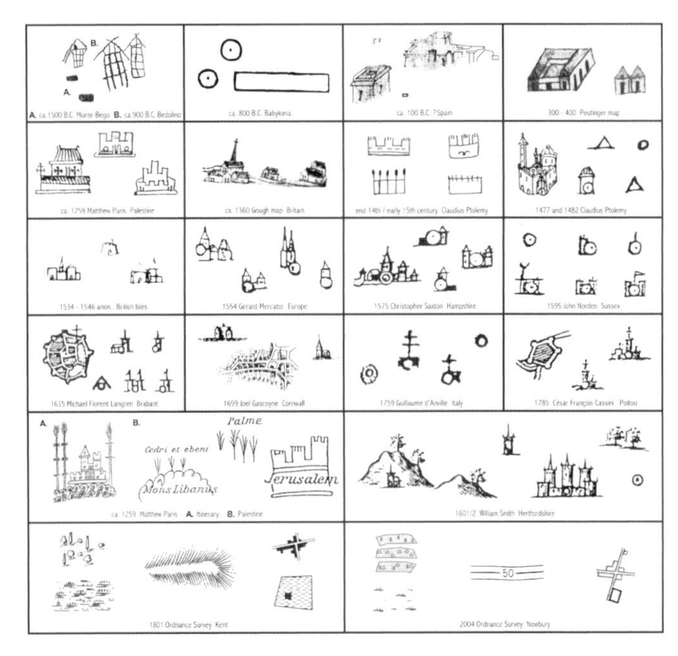

Signs on maps from the earliest times.

Stylistic differences affect pictorial map signs in particular. Most pictorial signs bear only a general resemblance to the feature they represent however 'realistic' in appearance. Many mapmakers, moreover, at all ages simplified their pictorial signs, making them schematic. Compare, for example, the town signs on the Peutinger map and Byzantine versions of Ptolemy's map. By the early seventeenth century, a small black square or a circle, surmounted by a cross, indicated a church. Highly stylized pictorial signs are still used. On modern Ordnance Survey maps the meaning of the church signs has been refined, so that the square sign stands for a church with a tower, the circular sign a church with a spire, minaret or dome, and a cross alone a church without such additions.

The mixing of pictorial and non-pictorial signs was an easy way of extending the range of signs available to the mapmaker to allow more information to be included on the map. Another way was to add a cipher to the sign. Coded pictorial signs became particularly common in the seventeenth and eighteenth centuries as the quantity of information that individual mapmakers wanted to included on maps was vastly increased. Almost any mark could be used as a cipher. Asterisks, pennants, crowns, alchemists' signs were all pressed into service by one mapmaker or another. Commonly, for example, a crozier was added to indicate that the town in question was the seat of a bishopric, a mitre identified an archbishopric, a crown a royal town, and a coat-of-arms a baronial town. Again, however, explanation was essential, for there was no consistency. A crescent added to a town sign on a map of Hungary was used to denote a place under Turkish (Islamic) rule (Lazarus Secretarius, 1528); on a map of France it signified the seat of a regional parliament (Jean Jolivet, 1560); and on a map of the English county of Hampshire (John Norden, 1596), it indicated the presence of a regular market.

The first map to show the Americas depicts the discoveries of the great explorers, including John Cabot and Christopher Columbus, of the late fifteenth century.

A New World?

1500

THIS PORTOLAN WORLD map is the very first to show the new lands discovered by the Europeans. It brings together two distinct maps, each with a different scale as can be proved by comparing the outlines of Cuba and the Iberian peninsula. It is signed and dated '*Juan de la Cosa la fizo en el Puerto de Santa Maria en anno de 1500*' – 'Juan de la Cosa made it in the port of Santa Maria in the year 1500'.

The information about Europe derives from portolan charts and that of Asia from the accounts of Marco Polo and other Italian travellers, but the African coasts are based on Portuguese sources. Nevertheless the importance of the map lies in the depiction of the newly discovered lands about which Juan de la Cosa had confidential information. In the Antilles, the insular nature of Cuba, of which Columbus had been informed by the indigenous peoples during his first voyage and which de la Cosa himself proved in 1499, stands out. The five banners with the English royal arms and the inscription '*tierra descubierta por ingleses*' – 'Land discovered by the English' – to the north indicate the discoveries made by John Cabot in the course of his first voyage in 1498. The map also shows the discoveries made during Columbus's voyage of 1498, that of Ojeda, Amerigo Vespucci and Juan de la Cosa himself in 1499 and the discoveries of Yañez Pinzón ('*este cabo se descubrió en el año de 1499 por Castilla, siendo descubridor Vicens ians*' – 'this cape was discovered in the year 1499 for the crown of Castile, Vicens Ians being the discoverer') and the land discovered by the Portuguese Cabral to the south (now Brazil) which is depicted as a blue island because he considered it to be one and named it Vera Cruz or Santa Cruz.

Juan de la Cosa accompanied Columbus on his first two voyages and it is believed that he made a total of seven voyages to the Indies, dying at the hands of the indigenous Americans at Cartagena (now in Mexico) in 1510. The map was probably made to show Ferdinand and Isabella of Castile and Aragon the new discoveries in relation to the lands that were already known.

Right The world map of Juan de la Cosa. **See also** 1450, 1507, 1525, 1544 (p. 102), 1597 (p. 134), 1599.

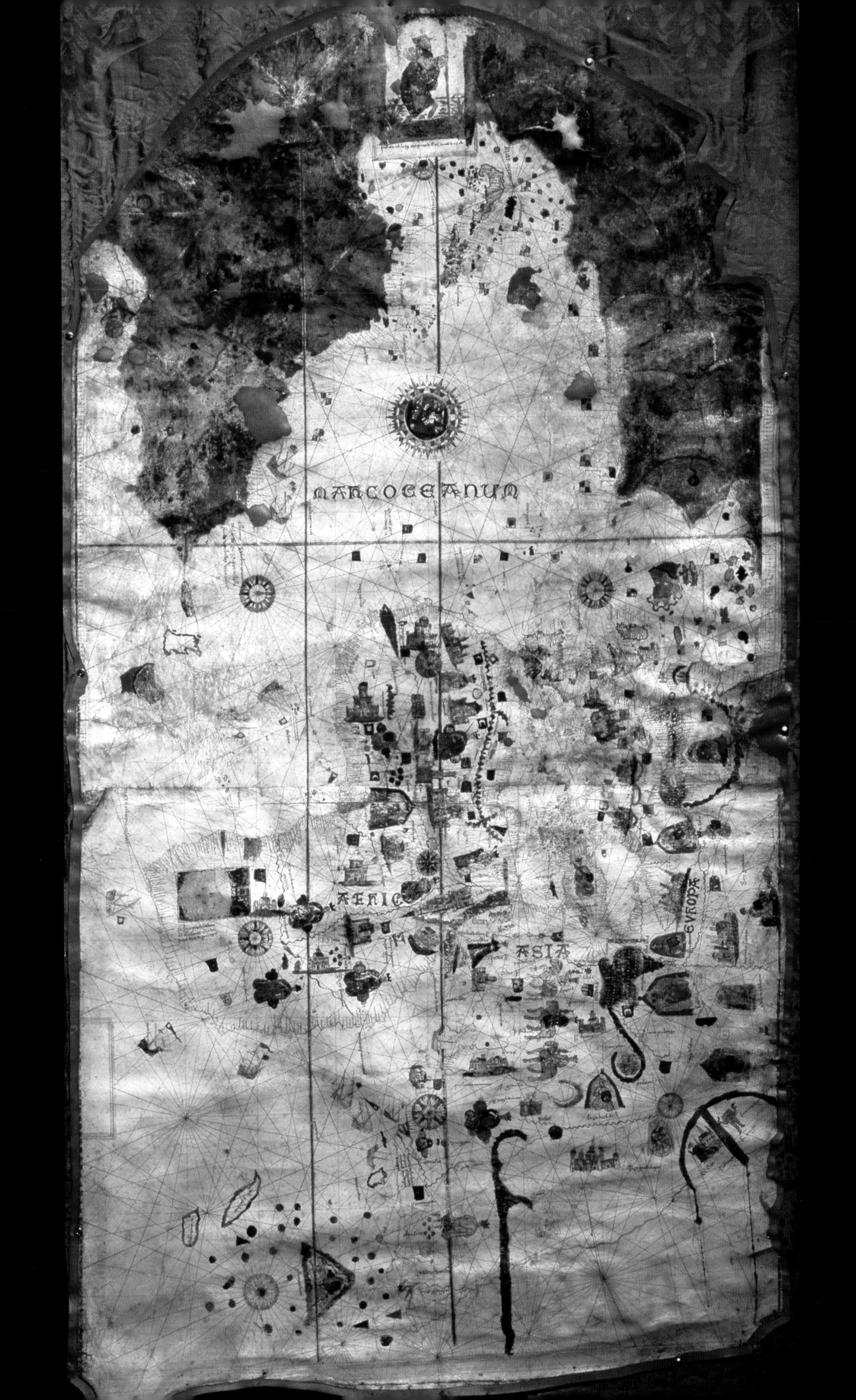
ASIA
EVROPA

The first map to name America was an educated guess and created far ahead of the discovery that the continent stood alone and independent of Asia.

A Lucky Guess

AMONG THE COUNTLESS Europeans passionately interested in the great discoveries was a small group of mainly youthful German scholars based in a town in the Vosges Mountains, now in France. The 'Gymnasium of St Dié' included the humanist Matthias Ringmann and the cartographer Martin Waldseemüller. Thanks to the patronage of Duke Reé II of Lorraine, they were free from commercial pressures to produce conventional, if pretty, maps.

Ringmann was particularly fascinated by one of the bestsellers of the time: an account by Amerigo Vespucci, a Florentine-born navigator and associate of Columbus, of his voyage of 1501–2 along the coasts of what had hitherto been considered the Indies from 5 to 52 degrees south of the Equator. Instead, Vespucci declared, these lands constituted a separate, fourth continent.

In 1507, as part of an ambitious project to produce maps with 'modern' outlines to replace the outdated images of Ptolemy, Ringmann got Waldseemüller to create two maps, one for a small globe and the other a very large wall map on twelve sheets. Pushing Vespucci's words to their logical conclusion, these showed the new lands as a continent, separated from Asia by an ocean. This was purely speculative and totally at odds with contemporary knowledge, reflected in maps, which, ignorant of the existence of the Pacific, kept open the option that the new lands were somehow connected to Asia. In the accompanying booklet *An Introduction to Cosmography* (*Cosmographiae Introductio*), which also included Vespucci's travel accounts, Ringmann suggested that since all known continents were named after women, this new continent discovered by Amerigo Vespucci, should be feminized and named in his honour. 'America', the 'fourth part of the world', was duly named on the two maps. Although only four examples of the globe sheets, and only one of the large map are now known, they were very influential at the time.

Right Detail of the Waldseemüller and Ringmann World Map showing South America. **Above** Amerigo Vespucci. **Following pages** The map in full. **See also** 1450, 1500, 1525, 1544 (p. 102), 1597 (p. 134), 1599.

Waldseemüller himself, however, was not convinced. What may be his own trial map of 1507 showed the conventional outlines, though it named 'America'. His later world map, created in 1516, five years after Ringmann's death, did so again – and omitted the word 'America'. Ringmann's and Vespucci's speculations ultimately proved correct and it is perhaps just that the continent be named after the man who grasped its true nature rather than Columbus, who died convinced that he had simply reached the East Indies.

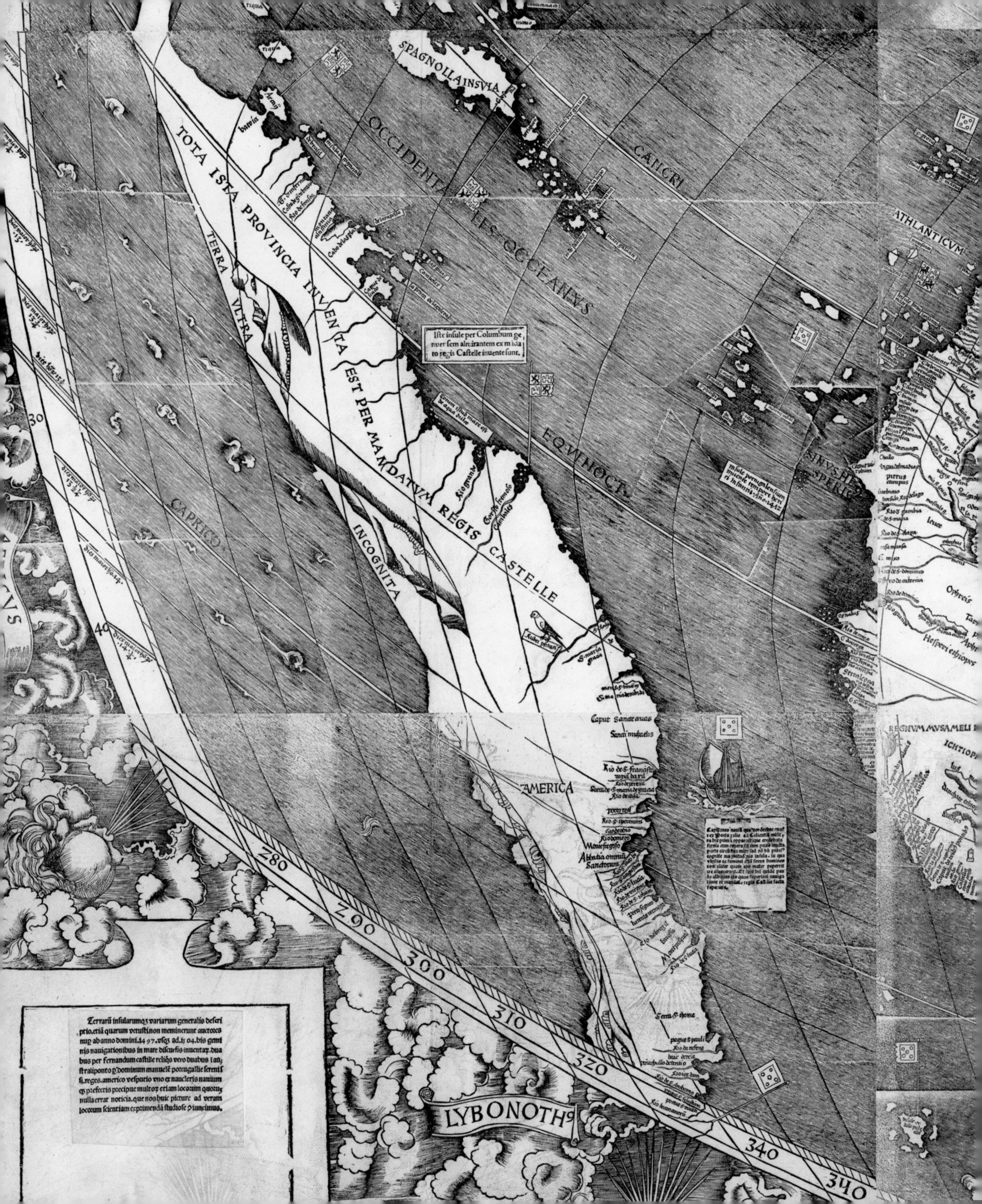
SPAGNOLLA INSULA
OCCIDENTALES OCCEANUS
CANCRI
ATHLANTICUM
TOTA ISTA PROVINCIA INVENTA EST PER MANDATUM REGIS CASTELLE
TERRA ULTRA INCOGNITA
Iste insule per Columbum genuensem almirantem ex mandato regis Castelle inuente sunt.
EQUINOCT
SINUS HESPERIC
CAPRICO
AMERICA
REGNUM MUSAMELI
280
290
300
310
320
340
LYBONOTH
Terrarũ insularumq3 variarum generalis descriptio. etiã quarum vetusti non meminerunt auctores nup ab anno domini .14 97. usq3 ad .15 04. bis geminis navigationibus in mare discursis inventa. duabus per fernandum castille reliqs vero duabus iam strali ponto p dominum manuelẽ portugallie serenissi. reges. americo vespucio uno ex naucleris navium q3 prefectis precipue multo etiam locorum quorum nulla errat noticia. que nos huic picture ad veram locorum scientiam exprimendã studiose coniunximus.

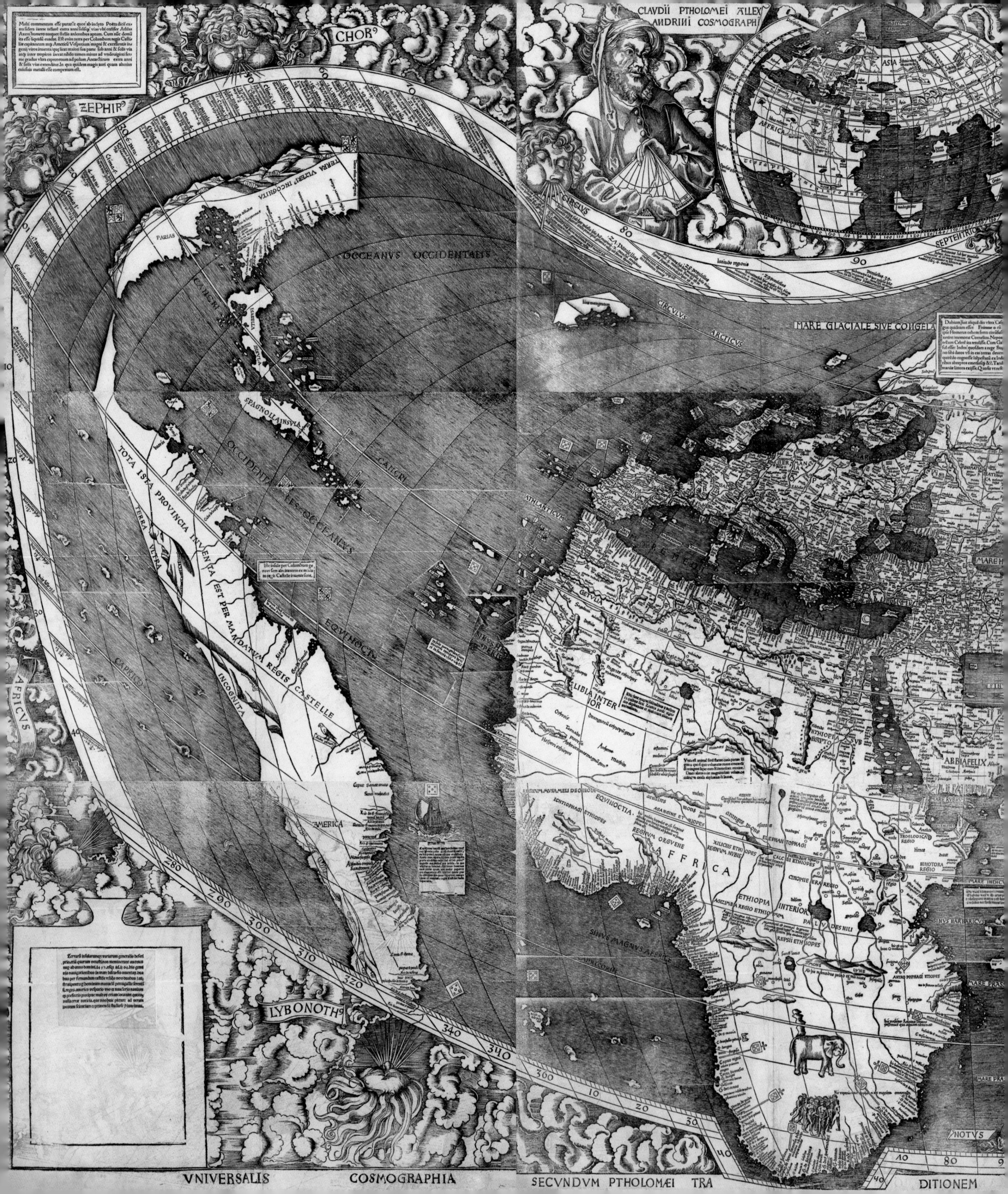

CLAVDII PTHOLOMEI ALLEX
ANDRINI COSMOGRAPHI
CHOR9
ZEPHIR9
AFRICVS
LYBONOTH9
NOTVS
SEPTENTRIO
ASIA
AFRICA
OCCEANVS OCCIDENTALIS
TERRA VLTERI' INCOGNITA
PARIAS
ISABELLA
SPAGNOLLA INSVLA
CANCRI
CIRCVLVS ARCTICVS
MARE GLACIALE SIVE CONGELA
OCCIDENTALIS OCCEANVS
TOTA ISTA PROVINCIA INVENTA EST PER MANDATVM REGIS CASTELLE
TERRA VLTRA INCOGNITA
EQVINOCT
CAPRICORNI
AMERICA
ATHLANTICVS
LIBIA INTERIOR
AFFRICA
ETHIOPIA INTERIOR
PALVDES NILI
SINVS MAGNVS AFFRICE
SINVS BARBARICVS
ARABIA FELIX
VNIVERSALIS COSMOGRAPHIA SECVNDVM PTHOLOMÆI TRA DITIONEM

AMERICI VESPVCII
CECIAS
AQVILO
TIPVM ORBIS GENERALEM DESCRIBENDO: VE
TERVM INVENTA PONERE/ ET EA QVE A NEO
TERICIS INTERIM REPERTA SVNT (SICVT EST
CATAIA REGIO) CONIVNGERE PLACVIT . VT
TALIVM RERVM STVDIOSI DVM VARIA COGNO
SCERE CVPIVNT VOTORVM COMPOTES LABO
RI NOSTRO SINT CRATI FLAIRAQVE OMNIA
TAM PASSIM COGNITA QVAM NOVITER LVS
TRATA DILIGENTER AC DISTINCTE SVB VNO
ASPECTV COLLOCATA PROSPICIENTES.
SVBSOLANVS
latitudo regionis
Mare glaciale
ARCTICI
MARE GLACIALE
ARCTICI
CIRCVLVS
TAGV PROVIN
CIAMAGNA
CHA TAY
BALOR REGIO
THOLOMA PROVIN
CHAIRAM
THEBET PROVIN
INDIA SVPERIOR
SERICA REGIO
SCITHIA EXTRA IMAVM
SCITHIA INTRA IMAVM
TARTARIA
CIAMBA PROVINCIA MAGNA
ASIA
INDIA MERIDIONALIS
HYRCANIA
PARTHIA
ARIA
ARACHOSIA
CARMANIA
INDIA INTRA GANGEM
INDIA EXTRA GANGEM
REGNVM AVRVSVLI
LOACH PROVIN
SINVS GANGETICVS
TAPROBANA INSVLA
OCCEANVS ORIENTALIS INDICVS
IAVA MAIOR
VVLTVRNVS EVRVS
EQVINOCTIALIS
INDIA
REGNVM VARR
CAPRICORNI
PETAM
NECVRA
ANGAMA
OCCEANVS INDICVS MERIDIONALIS
SEYLAM
IAVA MINOR
TROPICVS CAPRICORNI
MADAGASCAR
EVRONOTVS
AVSTER
LICET PLAFRIQVE VETERVM DESCRIBENDI TERRA
RVM ORBIS STVDIOSISSIMI FVERINT : NON TAMEN
PARVM IPSIS EISDEM INCOGNITA MANSERVNT, SI
CVT EST IN OCCASV AMERICAE: AB EIVS NOMINIS
INVENTORE DICTA, QVE ORBIS QVARTA PARS PV
TANDA EST. SICVT ET VERSVS MERIDIEM APHRICE
PARS: QVAE SEPTEM PENE GRADIBVS CITRA CAPRI
CORNVM INCIPIENS/VLTRA TORRIDAM ZONAM ET
EGOCERI TROPICVM AD AVSTRVM LATISSIME PRO
TENDITVR. SICVT QVOQVE IN TRACTV ORIENTA
LI REGIO CATAIAE/ ET QVICQVID INDIAE MERIDIO
NALIS VLTRA CENTESIMVM ET OCTOGESIMVM LON
GITVDINIS GRADVM EST SITVM. QVAE NOS PRIO
RIBVS OMNIA ADIVNXIMVS: VT ISTIVSCEMODI RE
RVM AMATORES/ QVAECVNQVE SVB HANC DIEM
NOBIS PATENT OCVLIS INTVENTES / DILIGENTIAM
NOSTRAM POBENT. ID AVTEM VNVM ROGAMVS
VT RVDES ET COSMOGRAPHIAE IGNARI HAEC NON
STATIM DAMNENT ANTEAQVAM DIDICERINT CHA
RIORA IPSIS HAVD DVBIE POST CVM INTELLEXE
RINT FVTVRA.
80
90
60
40
30
40
100
110
120
130
140
150
160
170
180
190
200
210
220
230
240
250
260
270
ET AMERICI VESPVCII ALIORVQVE LVSTRATIONES

A fragment of an Ottoman world map of remarkable accuracy revealing the degree of cultural exchange between the Christian West and the Muslim East in the early sixteenth century.

Piri Reis

Right Fragment of world map by Piri Reis.

THE OTTOMAN EMPIRE was one of the greatest global powers of the early modern world. With the fall of Constantinople in 1453, the Ottoman sultans inherited many of the classical texts of the Graeco-Roman world, including manuscripts of Ptolemy's *Geographia*. In the Mediterranean, the Ottomans participated in the evolution of portolan charts and their mixed cultural traditions of Arabic, Spanish and Italian knowledge of the area's islands and coastlines. Drawing on these classical and practical traditions of mapmaking, a small but vigorous school of Ottoman cartography developed. This culminated in the work of the finest of all Turkish mapmakers, Piri Reis, and his greatest surviving achievement, this fragment of a world map, one of the most important and enigmatic maps of the Age of Discovery.

The fragment that survives is part of a much larger vellum manuscript map of the world, the rest of which is lost. What remains is a remarkably accurate depiction of the Iberian peninsula, north-west Africa and the coastline of the recently discovered Americas. The map's twenty-four legends, written in Turkish, provide invaluable information on its creation and significance. Completed in Gallipoli in April 1513, the map was later ceremonially presented to Sultan Selim I, which suggests it had a diplomatic rather than navigational function. Piri Reis claims he used Ptolemy's *Geographia*, Christian *mappaemundi*, Indian maps, Portuguese charts and maps drawn by Christopher Columbus. The map incorporates medieval *mappaemundi* in its depictions of exotic rulers and animals (including anthropophagi in the Americas), as well as more practical portolan-style features as compass-roses and rhumb lines.

The map's style and detailed knowledge of the Americas has close affinities to the most up-to-date Portuguese and Spanish discoveries represented in maps like the *Cantino Planisphere* (1502). It accurately relates Columbus's discovery of America (1492), the Treaty of Tordesillas (1494) and Cabral's discovery of Brazil. Even more remarkable is Piri Reis's claim to have seen maps of the New World drawn by Columbus, none of which now survive.

The map is one of the earliest and most detailed representations of the Americas, and its remarkable accuracy continues to astonish modern historians of cartography. Although the Ottomans never reached the New World, this map shows their political interest in the geographical discoveries of their Christian neighbours, and it reveals the widespread scientific and cultural exchange between the Christian West and the Muslim East in the Age of Discovery.

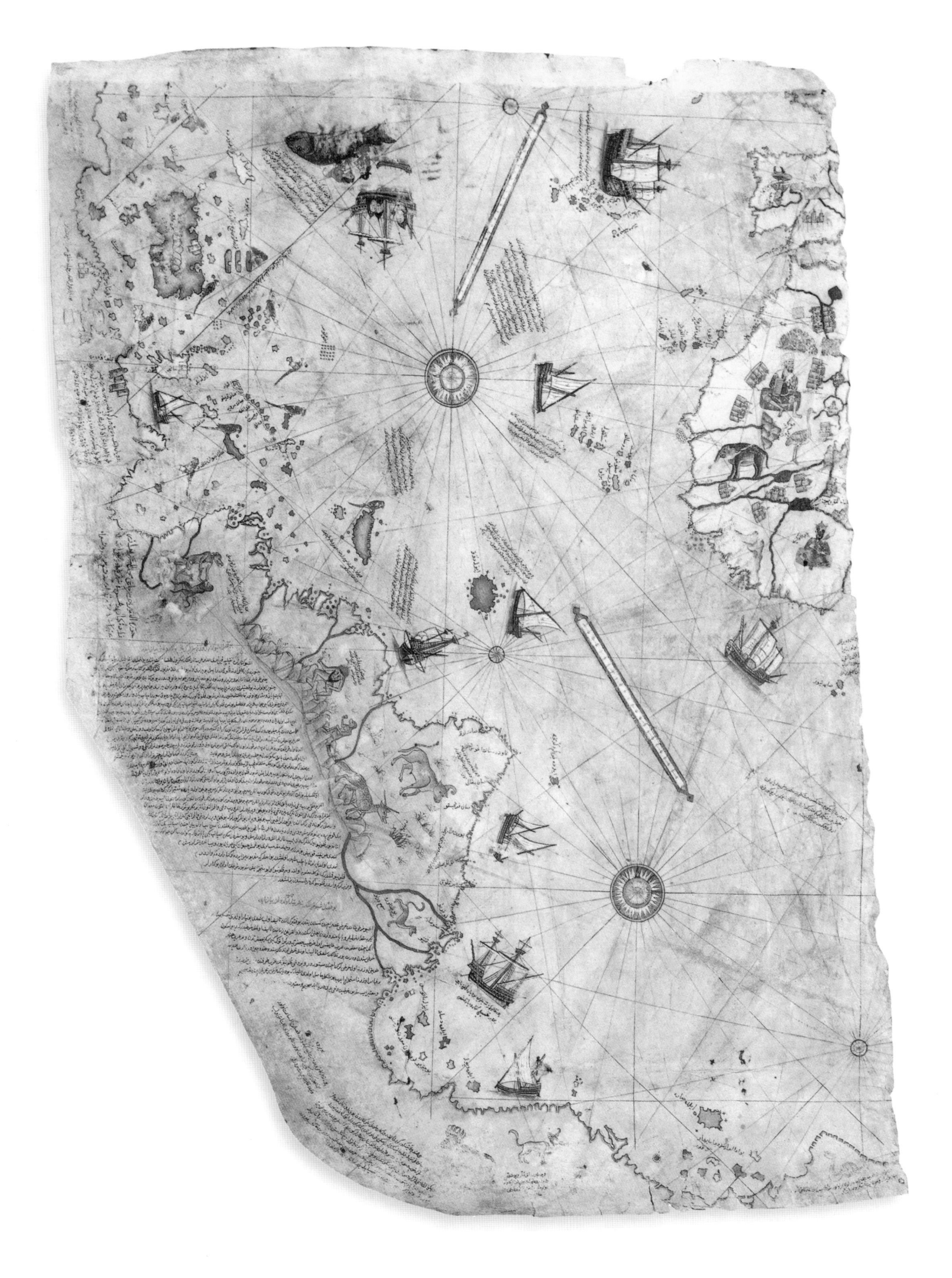

The earliest printed north European measured town plan shows the City of Augsburg surveyed with mathematical accuracy and depicted with great artistry.

A Merchant City

FROM ABOUT 1475, town views became all the rage with noble Italian families. Popes, doges and the cultured wives of municipal despots fought to outdo each other in the quality of the urban prints and paintings adorning their audience chambers. Most prized of all were views that combined elevations of the principal buildings with a mathematically determined ground plan surveyed in accordance with theories enunciated in the 1440s by the leading Renaissance architect, Leon Battista Alberti. By the 1520s the fashion had spread to Germany, and from the audience chambers of princes to the no less opulently furnished great chambers of merchants.

This apparently purely pictorial view of Augsburg is the first known mathematically measured town plan from north of the Alps. Surveyed between 1514–16 by Jörg Seld, an Augsburg goldsmith and military engineer, it was cut into wood in eight blocks by the important Strasbourg artist Hans Weiditz and printed in numerous copies (of which only two survive) by Grimm and Wirsung. The sharp-eyed viewer will see that not only the buildings but the citizens themselves are portrayed, walking or riding through the streets and lanes, shopping, and even attending a funeral. Augsburg looks what it was: one of the wealthiest cities in Germany. Information panels emphasize the mathematical accuracy underlying this feast for the eyes, but above all they eulogize the city and the Holy Roman Empire to which Augsburg belonged and to whose Emperor its burghers looked for protection against envious neighbours like the dukes of Bavaria. The splendid depiction of the double-headed imperial eagle at the top left with the arms of Emperor Charles V beneath emphasizes this. At the centre front of the view, however, are the arms of the city itself.

We do not know who paid for this plan-view, but it is quite likely that it was directly or indirectly subsidized by the immensely wealthy Fugger family. The emperors themselves were dependent on their financial support and they dominated the city. It is possible that the plan may have been created to commemorate the foundation, in 1519, of the Fuggerei, the regimented settlement shown to the north of the city's inner town walls. A mini-economic zone with almshouses, it survives little altered to this day.

Right A detail of the city of Augsburg by Jörg Seld and Hans Weiditz. **Above** The double-headed imperial eagle of the Holy Roman Empire with the arms of the Emperor Charles V beneath. **Following pages** The map in full. **See also** 1574, 1587.

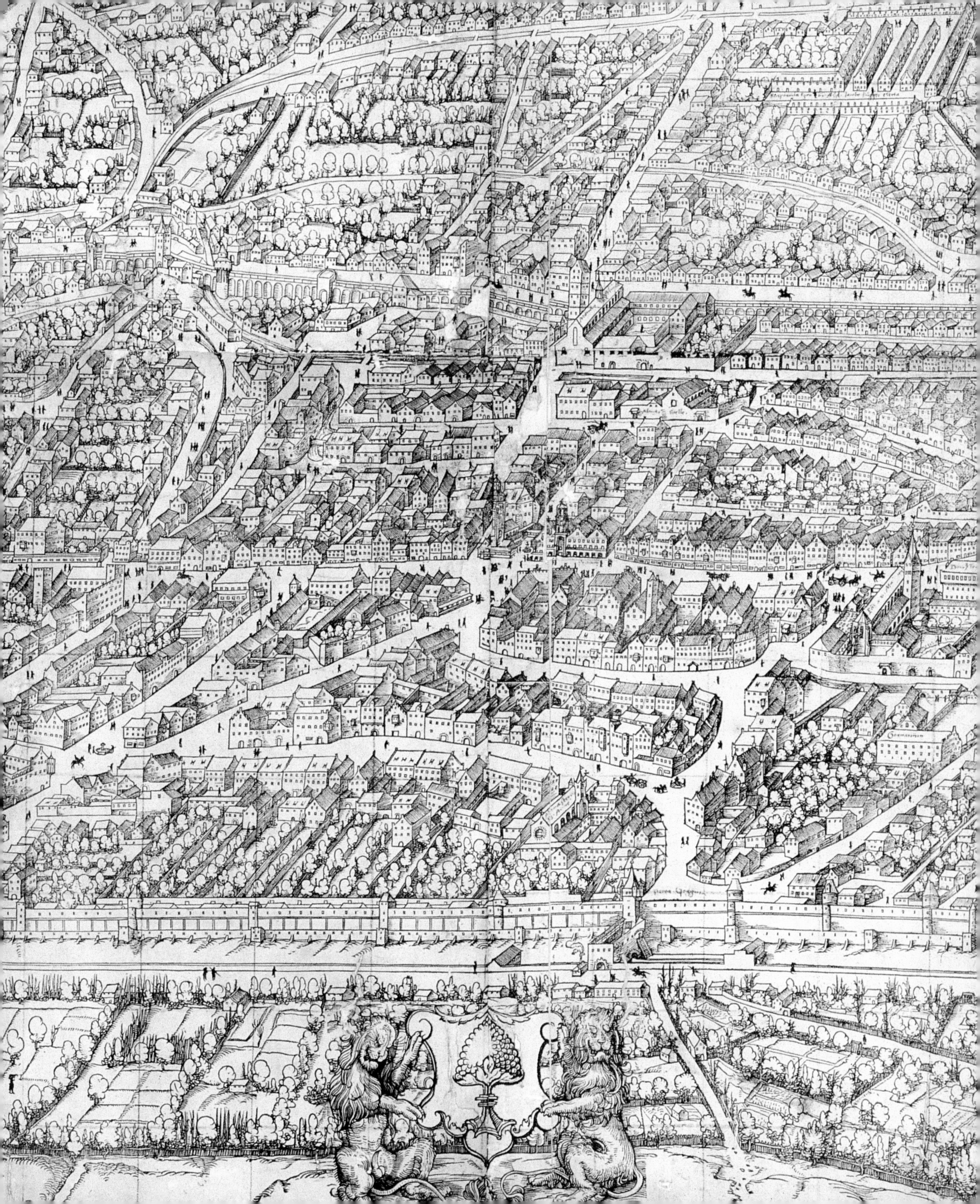

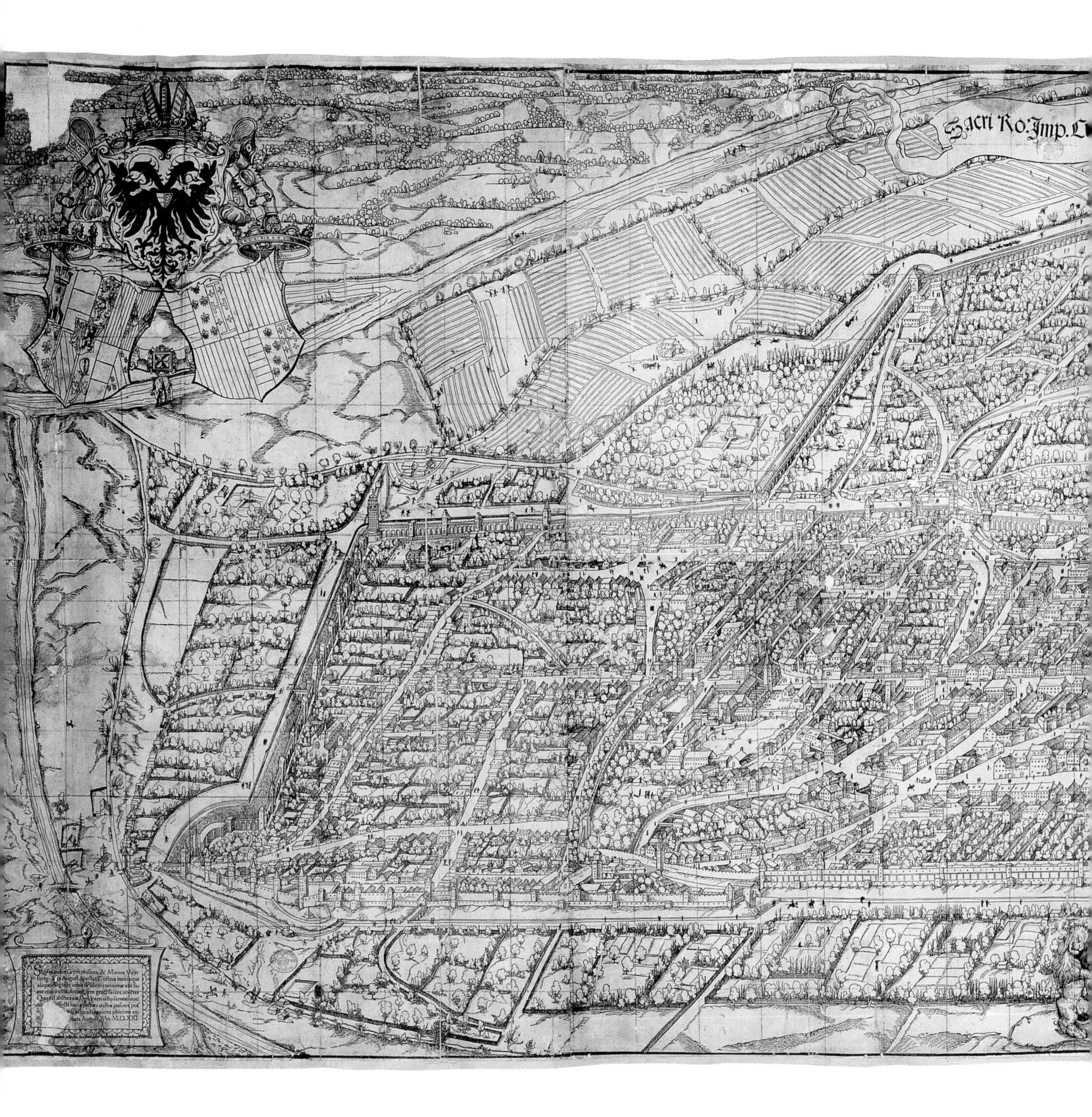
Sacri Ro: Imp.
Sigismundus Grymm phisicus, & Marcus Vuirsung
An. M.D.XXI.

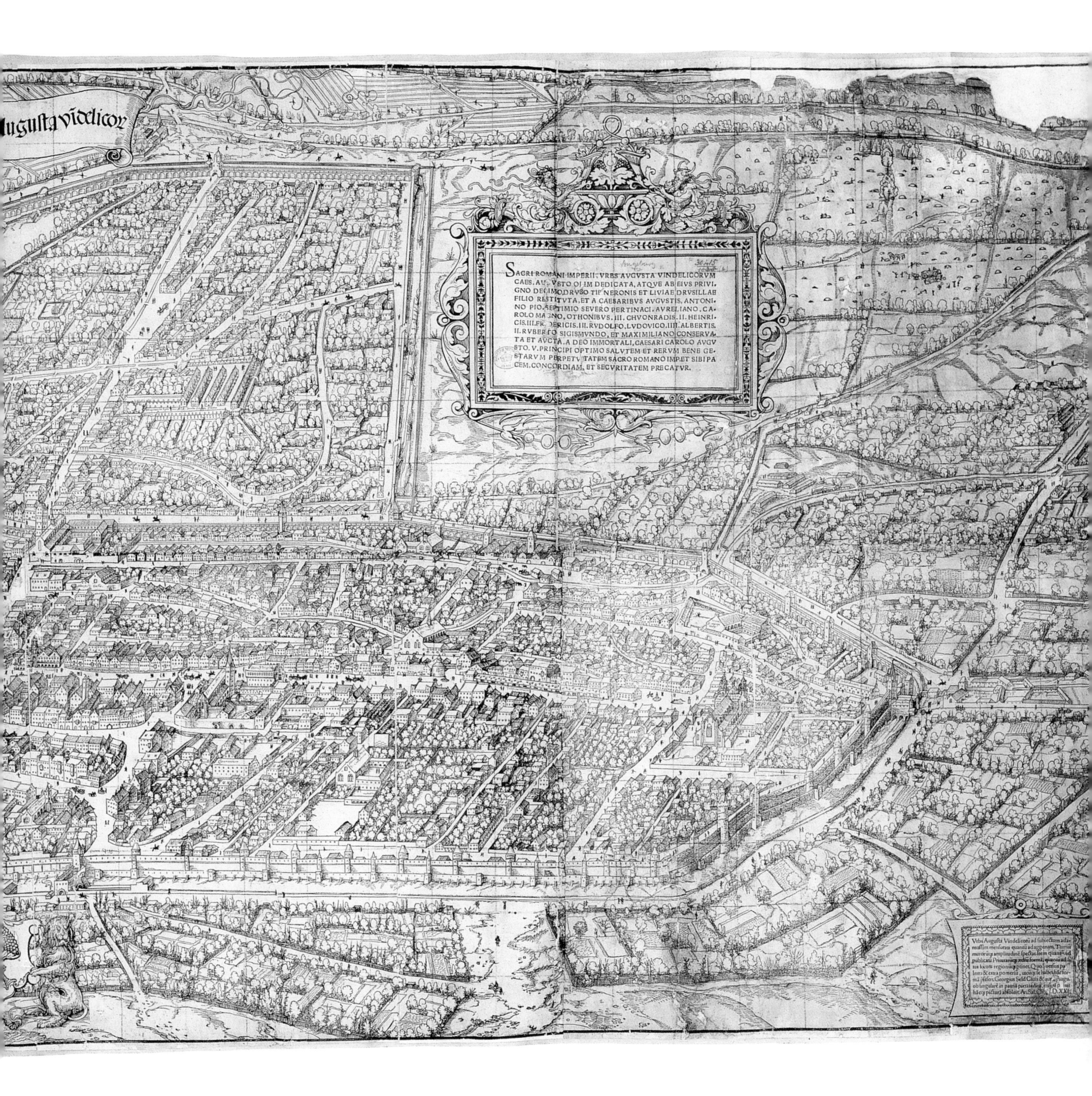
SACRI ROMANI IMPERII, VRBS AVGVSTA VINDELICORVM
CAES. AV VSTO OI IM DEDICATA, ATQVE AB EIVS PRIVI.
GNO DECIMO, DRVSO TIF NERONIS ET LIVIAE DRVSILLAE
FILIO RESTITVTA, ET A CAESARIBVS AVGVSTIS, ANTONI.
NO PIO, SEPTIMIO SEVERO PERTINACI, AVRELIANO, CA.
ROLO MAGNO, OTHONIBVS. III. CHVONRADIS. II. HEINRI.
CIS. III. FR DERICIS. III. RVDOLFO, LVDOVICO. IIII. ALBERTIS.
II. RVBERTO SIGISMVNDO, ET MAXIMILIANO CONSERVA.
TA ET AVCTA. A DEO IMMORTALI, CAESARI CAROLO AVGV
STO. V. PRINCIPI OPTIMO SALVTEM ET RERVM BENE GE.
STARVM PERPETV TATEM SACRO ROMANO IMP. ET SIBI PA
CEM, CONCORDIAM, ET SECVRITATEM PRECATVR.

The earliest printed plan of an American city made by a European artist and based on Hernán Cortés's eyewitness account.

Map of Aztec Tenochtitlan

THIS IS THE EARLIEST map of the great Aztec capital of Tenochtitlan to be published in Europe. Made by a European artist to accompany Hernán Cortés's eyewitness account of the city, it captures what would fascinate foreigners about the Aztecs: their ordered urban fabric, as well as their sacrificial rites. Tenochtitlan, an island city crosscut by canals, was much like Venice. It appears here ringed by neighboring cities in the Valley of Mexico. At its height in the sixteenth century, this region housed as many as 200,000 people, making it one of the most populous urban centres of its time.

The sacrifice of captives was meant to appease the Aztec gods and ceremonies took place in public. Here, the city is dominated by its enclosed temple precinct, which appears at a much larger scale than the rest of the map. Visible are twin temples that were dedicated to the deities Tlaloc and Huitzilopochtli (temples that Mexican archaeologists have discovered beneath the main square, or Zócalo, of Mexico City). At their foot is a headless victim.

The map is so accurate that it was probably based on one that the Aztec Emperor Moteuczoma had his mapmakers make for Cortés. The central, circular perspective, though, was probably a European element, which is also to be seen on a plan of Vienna published in Nuremberg in 1529. By the time this woodcut map was published in 1524, Moteuczoma had been assassinated and Cortés and his allies had reduced the orderly city pictured here to rubble. This map, though, endured, and was the first of a long line of images that Europeans produced to satisfy their curiousity about the New World, being copied and reworked by at least six other publishers during the sixteenth and seventeenth centuries.

Right Cortés's Tenochtitlan.
Above Detail showing a decapitated sacrificial victim.

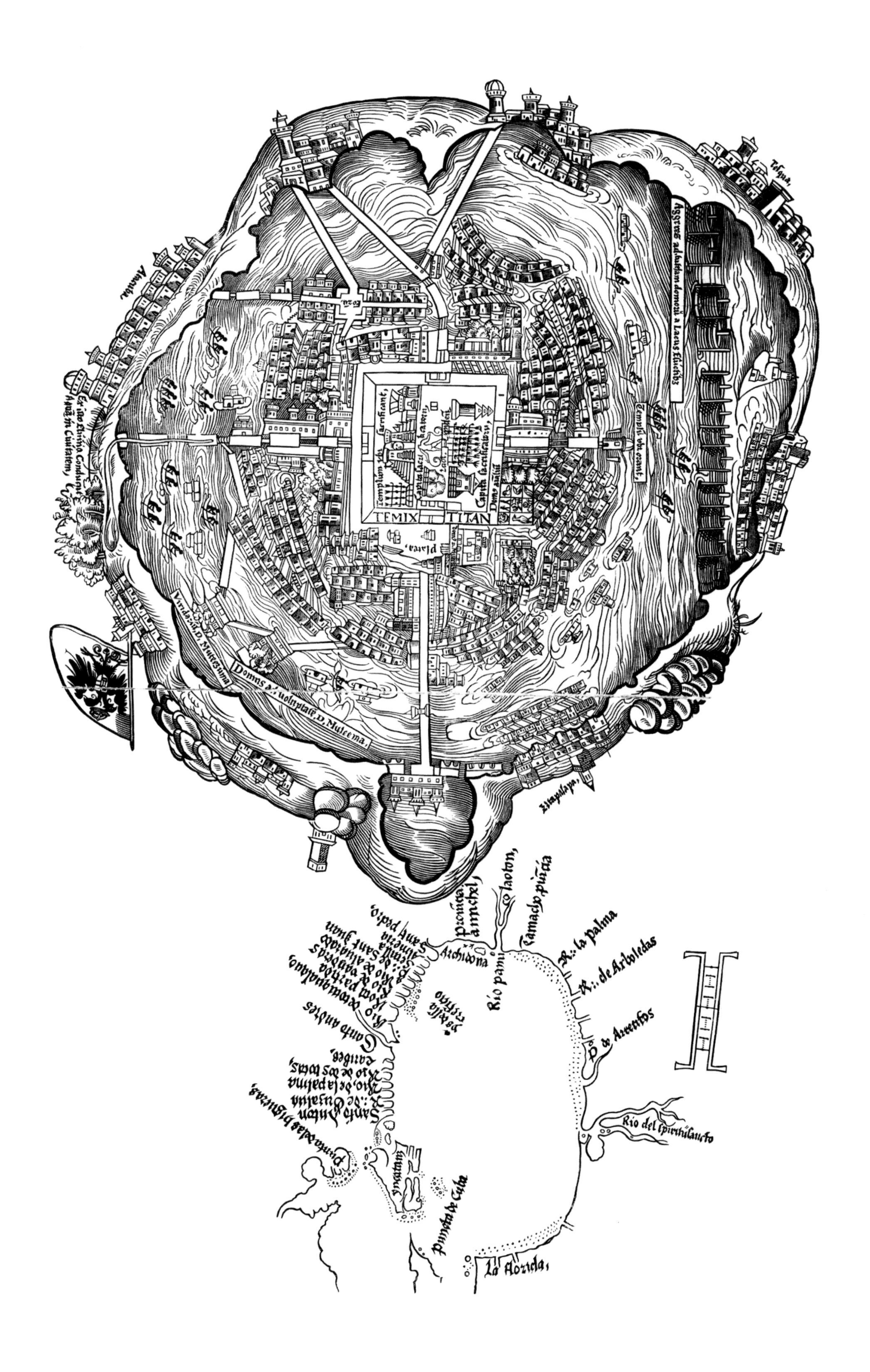

TEMIX
TITAN
Platea
Rio pami
R.. la palma
R.. de Arboledas
Rio del spiritusancto
Punta de Cuba
La florida,

A planisphere from Seville shows the latest geographical information available – and Spain's view of its rights outside Europe.

Maps and Diplomacy

Right A detail of the map. **Following pages** Nuño Garcia de Torreno, the complete world map. **See also** 1450, 1500, 1507, 1544 (p. 102), 1597 (p. 134), 1599.

THIS NAUTICAL CHART on three sheets contains the best and fullest geographical information of its time. The vessel *Victoria* appears twice – in the Atlantic and again in the Indian Ocean – adorned with the arms of Emperor Charles V and a Latin legend that proclaims it to be the only ship to have circumnavigated the world. The network of rhumb lines is organized around two circles of wind roses centred on points placed along the Equator in the Pacific (with a large wind rose) and the other in equatorial Africa. The Tropics and the Equator are also shown and along the latter there is a longitude scale of 5 degrees in 5. The Line of Demarcation, dividing the extra-European world, established by the Papacy under the terms of the Treaty of Tordesillas in the 1490s, contains a vertical inscription reading '*linea de repartimeinto entre Castilla y Portugal*' – 'line of division between Castile [i.e. Spain] and Portugal'. It is shown twice: once in the Atlantic and once in the Pacific.

This world map was a gift from Charles V to Cardinal Giovanni Salviati, the papal nuncio in Spain from 1525 to 1530. It was possibly intended to explain to the Pope Spain's territorial claims against Portugal and to support the Spanish negotiating position in the Spanish-Portuguese conferences about sovereignty over the all-important Spice Islands or Moluccas ('*Malucos*' – now part of Indonesia) at Elvas-Badajoz. For these purposes the lands discovered by Ferdinand Magellan during his circumnavigation of the globe (1519–1521) are strategically placed west of the Line of Tordesillas to show they were owned by Spain.

The sumptuous decoration of this map can be seen in the beauty of the wind roses and the combination of colours. It is very similar to the only map signed by Nuño Garcia de Torreno in the Royal Library in Turin. Until his death in 1526 Nuño Garcia de Torreno was, with Diogo Ribeiro, responsible for constructing navigation charts at the organization responsible for Spain's overseas empire, the Casa de la Contratacíon in Seville. In 1519 he made the charts for Magellan. For these reasons we can confidently attribute this map to him and date it to about 1525.

EVROPA
MARE MAGNVM
MARE HIRCANVM
ASIA
SINVS PERSICVS
AFRICA
MARE RVBRVM
CIRCVLVS EQVINOCIALIS
TROPICVS CAPRICORNII
HIC RATIS EQVINQVEST TOTVM QVI CIRCVIT ORBEM

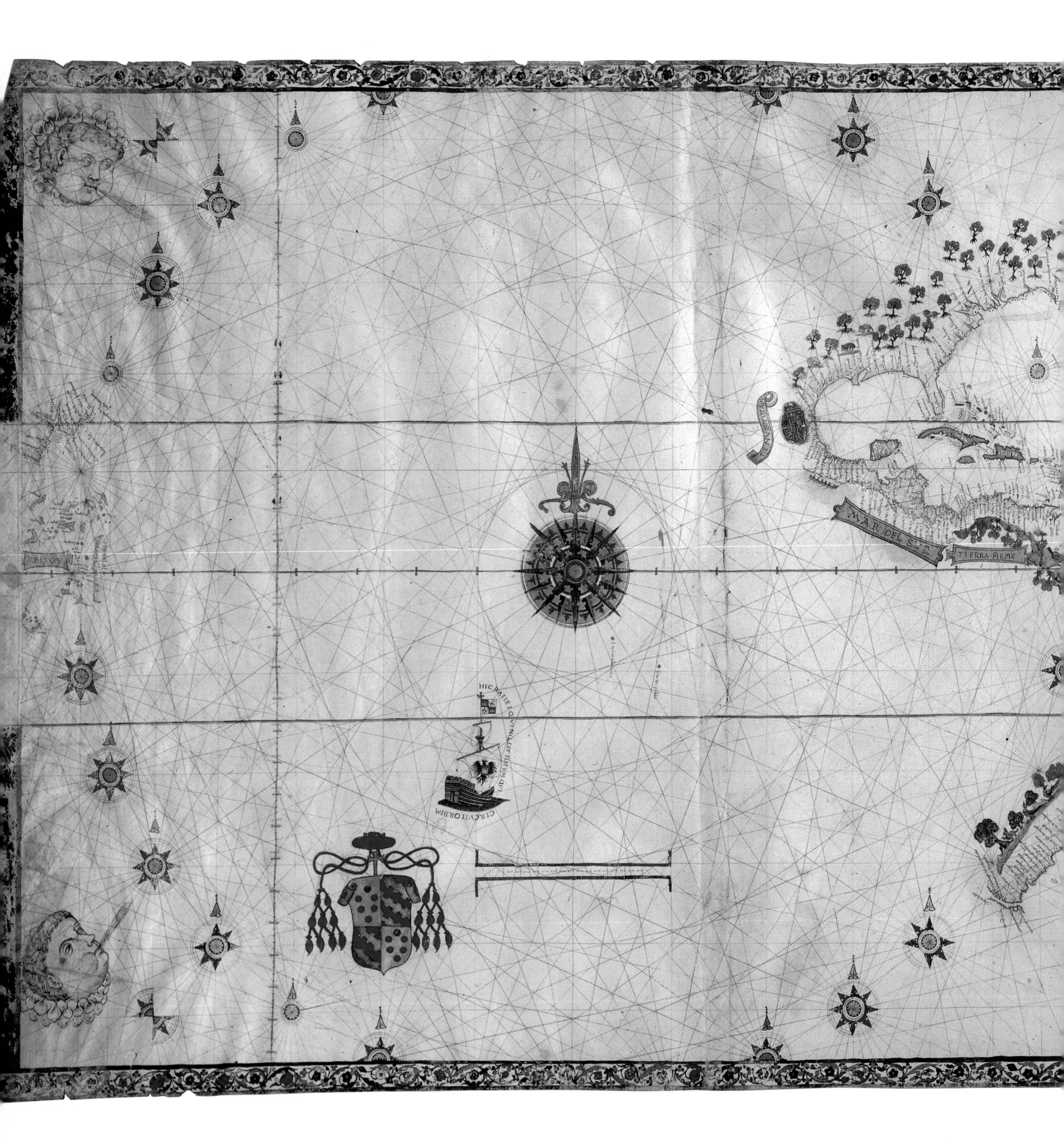
MAR DEL SVR
TIERRA FIRME
HIC RATIS EQVINOCTIALIS LAVDEM
CIRCVIT ORBEM

EVROPA
MARE MAGNVM
ASIA
AFRICA
TROPICVS CANCRI
CIRCVLVS EQVINOCIALIS
TROPICVS CAPRICORNII

The first realistic printed map of Scandinavia created by a Catholic priest and intended to influence the Papacy in the fight against the spread of Lutheranism.

Scandinavian Life and Fishy Tales

IN THE EARLY SIXTEENTH century a priest named Olaus Magnus was, for a number of years, employed collecting funds for the Catholic Church in the far north of Sweden and Norway. During his travels he made copious notes and sketches concerned with the way of life of the native inhabitants and the flora and fauna.

In 1528, when the King, Gustavus Vasa, embraced Lutheranism, Olaus was out of the country and was not allowed to return. He settled in Italy where he opposed the Reformation in every way he could. He became determined to demonstrate to the Pope, the Catholic Church and the peoples of southern Europe that Scandinavia was a great and glorious part of the known world which should be saved for Catholicism.

He set up a printing press and engaged draughtsmen to compile a massive map of the northern states and surrounding seas. It was to be extensively adorned with woodcut illustrations derived from sketches he had made during his northern travels. The *Carta Marina et Descriptio Septentrionalium Terrarum* was published in 1539, whereupon Olaus began to work on a monumental *Historia* of Scandinavia.

The joy of the *Carta Marina* is that it is not just a mere chart, for the whole of northern life is there – the natives in their domestic surroundings, travelling by sledge, skis or on horseback. There are agricultural and hunting scenes, churches, cities and fortifications and very much more.

Right Olaus Magnus's *Carta Marina.* **Above** The Spouter, a mythical beast which terrified sixteenth-century mariners.

The north-west portion of the chart shows the sea between Norway and Iceland merging with the '*Mare Glacial*' beset with ice floes on one of which a polar bear is marooned. The sea itself is occupied by a number of monsters which could not have been sketched by Olaus, but derived perhaps from one of his acknowledged sources, Pliny's *Historia Naturalis* (AD 77). No doubt the Norwegian fishermen, whom Olaus claims to have interviewed, were only too happy to embellish the monster stories.

One of the most fearsome monsters was the Spouter, which, having taken on a great draught of water through tubes above its head, could rise above the upper yards of a ship and pour down a great cataract onto the decks below sufficient to drown the sailors and sink the vessel. A method of deterring an attack by monsters, illustrated on the chart, was to throw overboard empty casks for them to play with while sounding a trumpet from the poop.

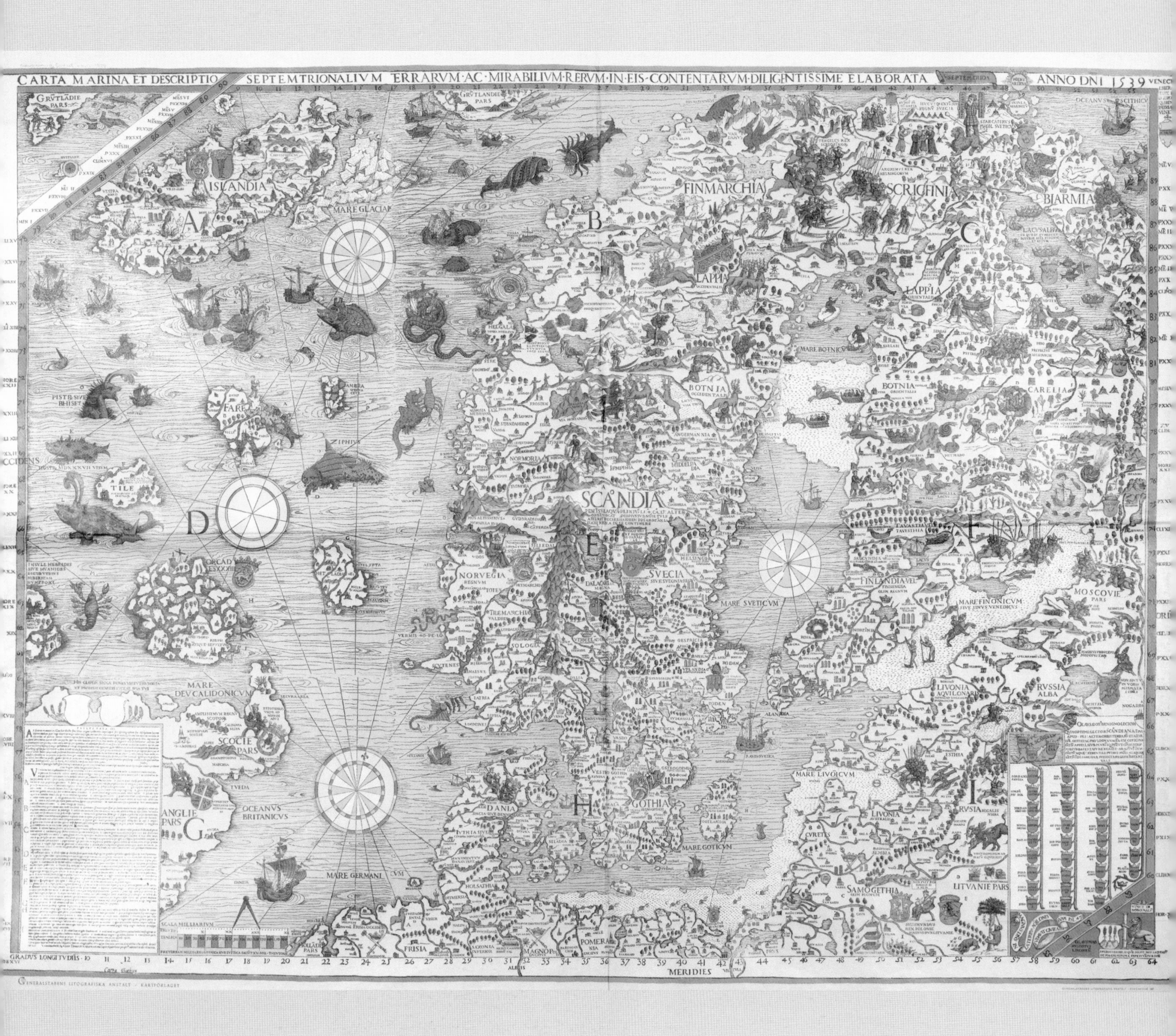

Generalstabens Litografiska Anstalt – Kartförlaget

One of the first maps to depict a dramatic event for a new audience interested in current affairs.

A New Vesuvius

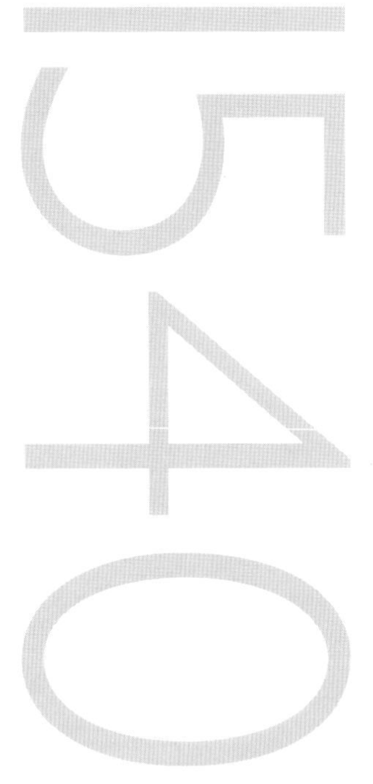

IF SOME CONTEMPORARY made a map of the damage to Pompeii in AD 79, it has not survived. However, there is contemporary cartographic evidence of a volcanic eruption that occurred a few miles north-west of Mount Vesuvius some 1,450 years later. At eight o'clock at night on 29 September 1538, a crack opened up in the earth and flames were visible. 'In a short time,' an observer recorded, 'the fire ... burst open the earth in this place and threw up so great a quantity of ashes and pumice-stones mixed with water, as covered the whole country, and in Naples a shower of these ashes and water fell a great part of the night.' Within a few days, a conical hill 123 metres (403 feet) high had been formed and christened Monte Nuovo. On 3 October, barely four days after the initial event, another gentleman 'went up with many people to the top of this mountain [and] saw down into its mouth, which was a round concavity about a quarter of a mile in circumference, in the middle of which the stones that had fallen were boiling up, just as in a great cauldron of water that boils on a fire'. The mountain remained relatively quiet for a few days and the inactivity drew many visitors to the new attraction. Then, on Sunday evening 6 October, about ten o'clock at night, the southern flanks of the cone suddenly exploded, releasing tons of bubbly, glassy lava. Twenty-four visitors were killed.

One of those who hurried to the site after the initial news arrived was Simone Porzio, a Neapolitan scholar and physician. Within a few weeks he had published a first-hand account of the new mountain, which he illustrated with a woodcut view. This is the earliest dated print of a volcanic eruption. Porzio's little book is exceedingly rare, our illustration being made from the copy in the New York Public Library.

The eruption of Monte Nuovo was also commemorated on several more conventional maps. This one, showing the whole Bay of Naples, from Capri to Ischia, is undated but was probably published shortly after the event. The map shows the bodies of two victims at the foot of the mountain. The appearance of this map is consistent with a broader interest among mid-sixteenth-century Italian printers and consumers in maps recording current events.

Right *'Il vero disegno... del monte nuovo'*, Rome (?), c.1540 (?).
Above View of the Monte Nuovo by Simone Porzio, 1539.

TRAMONTANA
BAIA
MONTE BAR BARO
NAPOLI
CAPRI
CASTELLO A MARE
Pozzuoli

An English local map made for Henry VIII as part of his review of the nation's defences following his break with Rome.

A Revolution in Mapmaking

THE QUARTER CENTURY after 1525 witnessed what could be considered a revolution in English mapmaking as government in the increasingly bulky form of Henry VIII woke up to the practical value of maps. Through maps he felt he could get a direct impression of the state of the country's defences. Foreign ideas contributed but the main impetus was provided by the threat of invasion that followed the King's divorce from Catherine of Aragon and the break with Rome. The revenues from the dissolved monasteries enabled Henry to employ talented Englishmen and foreigners to experiment with mapmaking techniques until they were able to provide him with precisely the information that he needed. In the years between 1538 and 1544 the shores of England and Wales were mapped with a thoroughness that was not to be seen again until the nineteenth century.

This presentation map, with east at the top, of the sandbanks and coastal forts at the mouth of the Thames and Medway was made by Richard Cavendish, a Suffolk squire and master gunner with a talent for mathematics. It incorporates the results of some of the written surveys he had undertaken in the early 1520s. A careful look reveals Cavendish experimenting with mapping techniques. The faint fretwork of lines spreading from circles are navigational and constructional aids copied from Mediterranean sea charts and of little use on land maps. The scale bar surmounted by dividers (bottom left) gives the impression that the map was to a uniform scale – then at the cutting edge of map technology. In fact the sandbanks are shown at a far larger scale than their surroundings. But perhaps Cavendish hoped Henry would be fooled.

The most significant decorative features relate to the mainland towns and villages. While most are shown simply as formalized rows of cottages around a church – or, simply, a church – the map includes what in most cases are the earliest known realistic depictions of the more important towns, like Maldon in Essex, and Sandwich and Canterbury in Kent. Cavendish probably copied them from neat versions of sketch maps and views of coastal towns, and those subject to sea-borne invasion, that Henry and his first minister Thomas Cromwell commissioned in the autumn of 1538 when the invasion crisis was at its height, and which had been sent to London from the provinces. The map summarizes the early stage of the mapmaking revolution.

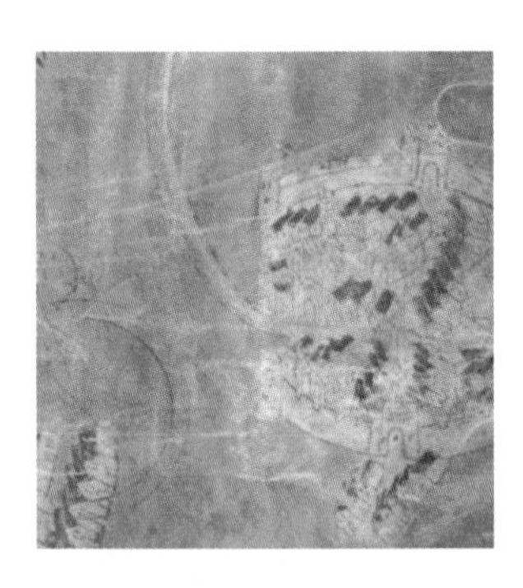

Right Richard Cavendish, map of the mouths of the Thames and Medway, c.1544. **Above** Detail showing Canterbury.

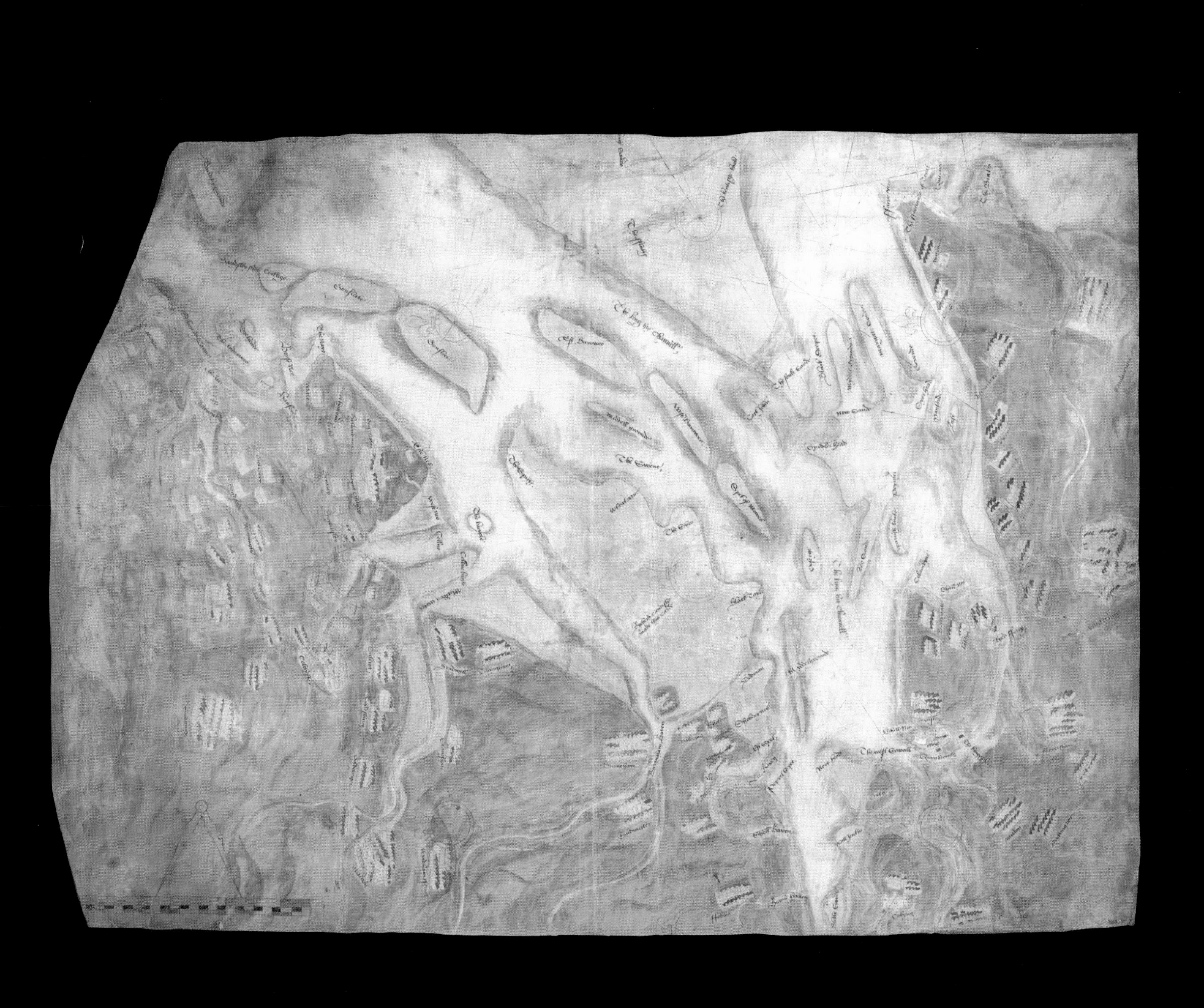

Est Barrowe
Middell grounde
West Barrowe

A world map by Sebastian Cabot, son of the celebrated explorer John Cabot, created when in employment as official chartmaker to the Spanish court.

The Encyclopedic Map

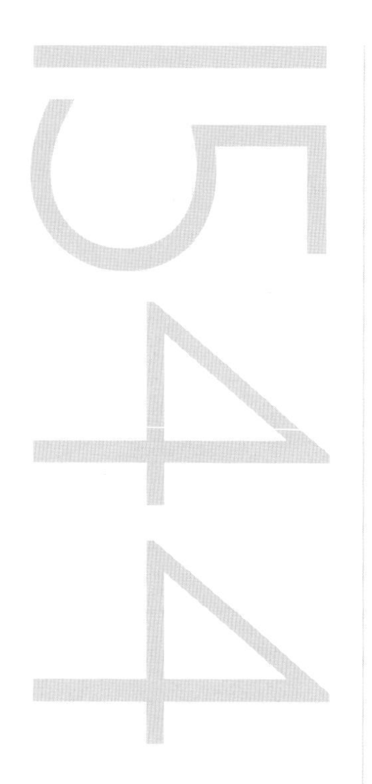

Right Sebastian Cabot, world map, Antwerp, 1544. **See also** 1450, 1500, 1507, 1525, 1597 (p. 134), 1599.

THE MAKER OF THIS eight-sheet map informs us that he constructed it by locating the landmasses according to magnetic declination and the graticule of meridians and parallels and then drawing in the rhumb lines as in a nautical chart. The geographical information about the new continent of America comes from the Casa de la Contratacíon in Seville, especially that about the Rio de la Plata and the discoveries of Ulloa in 1530 and of Vasquez Coronado in 1542 in the south-east of the present USA. The discoveries made by the mapmaker's father, John Cabot, at the latitude of about 50 degrees north are to be seen in the north-east as are also Jacques Cartier's discoveries in Canada of 1534–6.

This type of great map was made for public display and, in the tradition of medieval *mappaemundi* it contains a variety of historical, ethnographical and zoological information in addition to geographical and nautical. Twenty-two pieces of such information, linked to illustrations in the map, are to be seen in the left and right margins.

Sebastian Cabot, a controversial Venetian-born navigator and Third Piloto Mayor of the Casa de la Contratacíon at Seville (and as such responsible for the creation of its charts), sailed from Bristol in 1497 with his father, reaching the mouth of the St Lawrence. In 1512 he was named cosmographer to Henry VIII of England, before entering Spanish service where he was named Piloto Mayor of the Casa de la Contratacíon in 1518. Court intrigues did not leave him much time to exercise his official duties. He confidently asserted that he had discovered a new route to the Spice Islands (now part of Indonesia) and as a result in 1524 he was sent from Spain with a fleet to discover the lands of Ofir and Cathay, an adventure that he cut short on arrival at the Rio de la Plata. Having returned to Spain in 1530 he had to submit to legal punishment for his behaviour during the expedition and to the criticism of his colleagues for his conduct as Piloto Mayor. In 1548 he returned to England as an honoured guest. He became closely involved with Arctic navigation and trade with Muscovy (Russia) before dying in England in 1548.

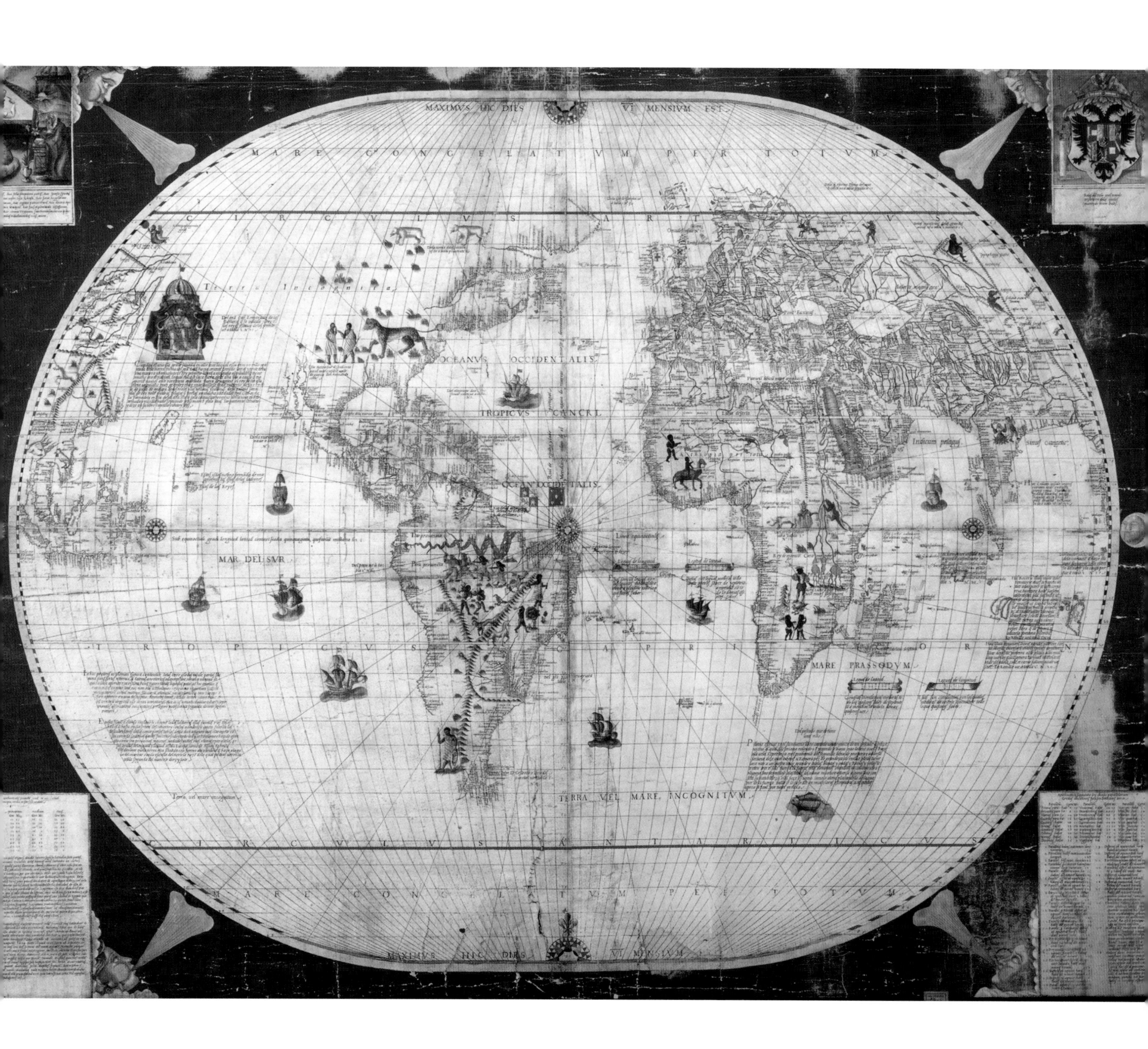

MAXIMVS HIC DIES VI MENSIVM EST
MARE CONGELATVM PER TOTVM
CIRCVLVS ARTICVS
OCEANVS OCCIDENTALIS
TROPICVS CANCRI
MAR DEL SVR
MARE PRASSODVM
TERRA VEL MARE INCOGNITVM
CIRCVLVS ANTARTICVS
Indicum pelagus

A colourful map from the Dieppe school of mapmakers by an equally colourful cartographer and explorer, Guillaume le Testu, eventually accused of piracy and beheaded by the Spanish.

Maps of Exploration – Australia

1555

UNDER THE NAME OF 'Jave le Grand', the north-western coast of Australia may be represented on French maps from as early as 1530, often with the description 'unknown land'. The earliest distinct reference to Australia is a passage in a book printed at Louvain in 1598: 'The Australis Terra is the most southern of all lands, and is separated from New Guinea by a narrow strait…' It is likely that a wealthy Dieppe shipowner, Jean Ango, was the prime mover behind what some argue were the first contacts with Australia and the creation of the first maps. It was at his house that leading cartographers and travellers met and shared their knowledge of what they knew of the mysteries of the southern oceans. Some two hundred maps from 1537–87 exist, many of them from the famous Dieppe school which flourished in the middle of the sixteenth century and at which Guillaume le Testu was trained in cartography and navigation.

Le Testu met an untimely end, beheaded by the Spanish after a joint raid with Sir Francis Drake on a Spanish mule train taking gold to Nombre de Dios in Panama. His head was prominently displayed in the marketplace as a deterrent to other 'pirates'. Was he a pirate? Certainly Sir Francis Drake was not averse to a little trading on his own account. Whatever he was doing in Panama, as captain of an 80-ton warship, Le Testu had travelled widely. He was born around 1509 at Le Havre in Normandy and after training became a pilot in Dieppe and later Le Havre before embarking on voyages of exploration, notably to Brazil in 1551 when he was given a commission to compile a map of southern Brazil, and with an expedition which founded a colony near Rio de Janeiro in 1555. His great world atlas, *Cosmographie Universelle selon le navigateurs, tant Anciens que Modernes,* consisted of fifty-six maps and was made for French Admiral Gaspard de Coligny, later a leader of the French Huguenots, and was based on his collection of charts from French, Spanish and Portuguese sources. The maps vary widely in accuracy and contain inhabitants, animals and plants, many of them quite imaginary.

Right One of the sheets from Le Testu's great atlas, the *Cosmographie Universelle* of 1555.

R: mai
ILLES DES GRIFONS
MER DE LINDE ORIANTALLE
Cap de non
TERRE AUSTRALLE

A decorative navigational chart by Portugal's leading chartmaker shows clearly the reach and influence of the Portuguese Empire.

The Portuguese Golden Age

1558

FROM 1500, AS A RESULT of their voyages along the African and Asian coasts and their use of native south-east Asian charts, the Portuguese became the leading European chart-makers. Their charts differed from those of their predecessors in one essential way. Unlike the Catalans and Italians, the Portuguese used scientific instruments and mathematics to measure the angle of the sun above the horizon at different altitudes and thus calculate latitude. This made possible the relatively accurate mapping of the world beyond the Mediterranean (though the continuing inability to determine longitude at sea caused distortions in the east–west depiction of coastlines). The horizontal lines of the Tropics and the vertical bars of latitude can be seen superimposed over the fretwork of rhumb lines, familiar from the older portolan charts, on this beautiful chart. In other respects, with its invisible circles, partly indicated by small wind roses, the emphasis on coastlines, the positioning of the place names and the repeated scale bars, the portolan chart legacy is unmistakable.

Extending from 61 degrees north to 40 degrees south, this chart, completed on 15 November 1558, is the only work to be signed by Sebastian ('Bastiam') Lopes (d. *c.*1595) who was to become the principal chart and instrument maker at the official Portuguese mapping establishment, the Armazém da Guiné e India, and one of the leading chart-makers of the golden age of Portuguese mapmaking. Its fine workmanship suggests that it was made for presentation. It was clearly intended to remind the viewer of Portugal's newly acquired power and empire. The presence of its rival, Spain, is alluded to only by the banners with Spanish arms along the western stretches of the Amazon. By contrast there is a realistic depiction of Lisbon in the middle of the Iberian peninsula. In northern Africa, two mounted Portuguese are shown routing a couple of Moorish horsemen. In Brazil, identified by the Portuguese arms, a native is shown chopping down a redwood which had already become an important export commodity, while a bird of paradise flies nearby. Sierra Leone in West Africa is indicated by a lion looking towards the principal Portuguese strongpoint, St George's Fort in Minah, which played an important role in the slave trade. Further to the south is a cross before a church façade, serving as a reminder of the importance of the missionary activity that accompanied Portugal's commercial and colonial expansion.

Right Bastiam Lopes, chart of the Atlantic. **See also** 1325, 1465.

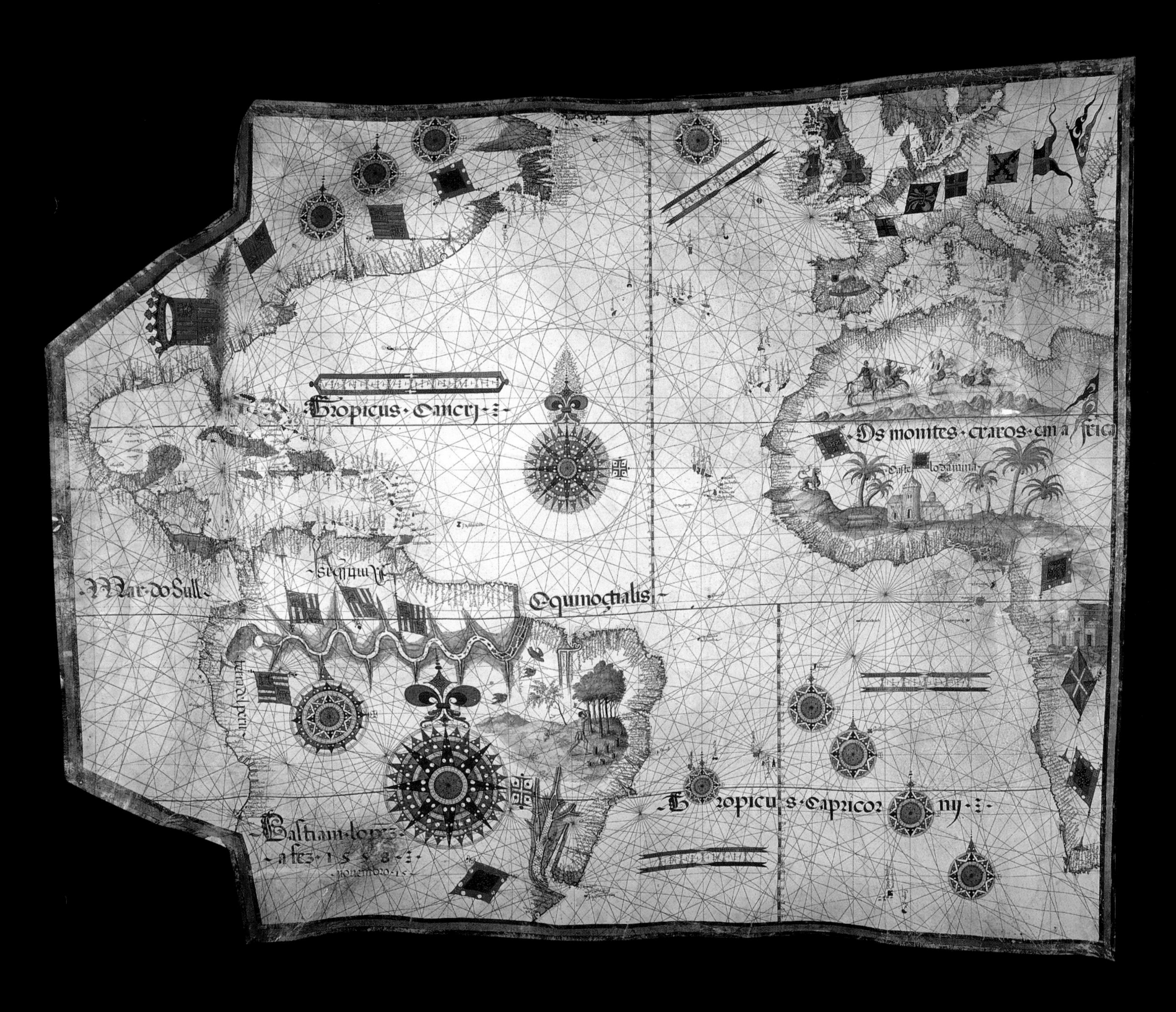
Tropicus · Cancrj
Equinoctialis
Mar do Sull
Os montes craros em africa
Tropicus Capricornij
Bastiam lopez
a fez 1558

A Venice-based cartographer's interpretation of the geography of the world at the time of the Italian Renaissance.

Guesses About the Furthest Frontiers of the World

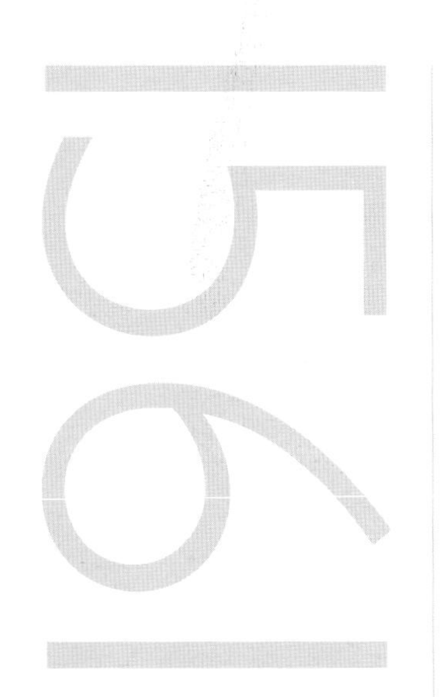

ALMOST ALL OF THE conjectures generated by the Europeans' geographical discoveries in the sixteenth century – conjectures that would continue to be accepted for the next one hundred and fifty years – make their first appearance in this 'universal' map. The world is entirely knowable, and open to the conquest of man – for which exclusive use, all believed at the time, it had been created. The sea is inhabited by mermaids and sea monsters, but European sailing ships are already traversing the globe.

The 'Strait of Anian' separates the unexplored extremities of Asia from the New World to the north of the Pacific, resolving the fifty-year-old debate between those who believed that the two continents were separate and those who considered them a single mass: the strait is slim enough to have allowed animals and humans to cross, and propagate, after the universal flood. Navigation between different parts of the world is therefore possible and not too distant: the English have already started to search the northern seas for north-western and north-eastern passages towards the Pacific Ocean. A southern continent, still to be discovered but surely enormous, rich in exotic animals and monsters, is inhabited also by men, albeit savage, who could become Christians and subjects of the European kings, as had happened in America. In the unknown interior of central and southern Africa, the cartographer speculates on the internal structure of the vast river network, of which only the mouths are known.

Right Detail showing the Atlantic with enthroned Philip II. **Above** The Gulf of Anian. **Following pages** Giacomo Gastaldi and others, *Cosmographia universalis*.

The map is a synthesis of the opinions of the principal cartographers and geographers of the time, and based primarily on the works of that most important and prolific of the cartographers of the Italian Renaissance: Giacomo Gastaldi, active in Venice between 1546 and 1560. This map is technically very refined, and the conjectures it presents are based on the most recent geographical discoveries: its recognized sources, other than the maps drawn by Gastaldi himself in the preceding years, are the studies of non-European lands contained in *Navigazione e viaggi* (1550–59), published in Venice by Giovanni Battista Ramusio.

In the four corners are small maps of the celestial hemispheres – including constellations – and of the terrestrial hemispheres: at the bottom right, a Mediterranean wind rose; at bottom left, an explanation of why Magellan's ships, circumnavigating the earth in a westerly direction, lost a day. The seven empty squares were probably destined for geographical symbols and stamps, and dedications.

GRVTLANDIA INCOGNITA
MARE OCEANO
TERRA INCOGNITA
Qui uenero otto Bretoni per discoprir le gente uno scanpo a ochelai e ghaltri furno mangiati da gli paesani
In questo Regno uestono di corio cotto e di pelle di orsi sono bella gente
PICNEMAAY REG
TERRA DEL LABORADOR
C. Fredo
C del laborador
LA NOVA FRAN
CANADA PROVIZ
TERRA DE NORVNBEGA
SOGVENAI REGNO
TERRA DE BACALAOS
Golfo calore
Hodielaga
ochelai
A Lago
Starnatana
GOLFO DI CAST
Y. Rotonde
C Real
Golfo Triagolare
Brisa
C Bioto
C. Loiena
C. Darena
Y. darema
Ysola de demoni
Yorbelande
C de Marzo
C del gado
C Rosso
MARE DEL LA NOVA FRA
Passo di Britoni
La Trinita
Ardon
Y de Brasil
C de molte Isole
C. Rosso
C des Iacomo
C de Arenas
Li bermuda
ARE DEL NORT
CIRCOLO DI CANCRO
RE PHILIPO
Gratiosa
S. Giorgio
Y di Fiori
Terzera
S Michael
Lo Faral
Lo Pico
AZORES
S. Maria
Formige
P. Santo
Y de Medera
Deserte
ISOLLE CANARIE
Y de palma
Comera
Y del Ferro
Tenarifes
Saluage
Gran canaria
Forte ventura
S. Vicenzo
S. Lucia
Y del sal
Y de S. Antonio
Y de S. Nicolo
Bonauista
ISOLE DI CAPOVER
Virgine
Anegada
Sombrero
Anguila
Lasbarbudas
Bortolamio
Saba
S. Martin
Guadalupe
Deseada
Marigalante
Barbudas

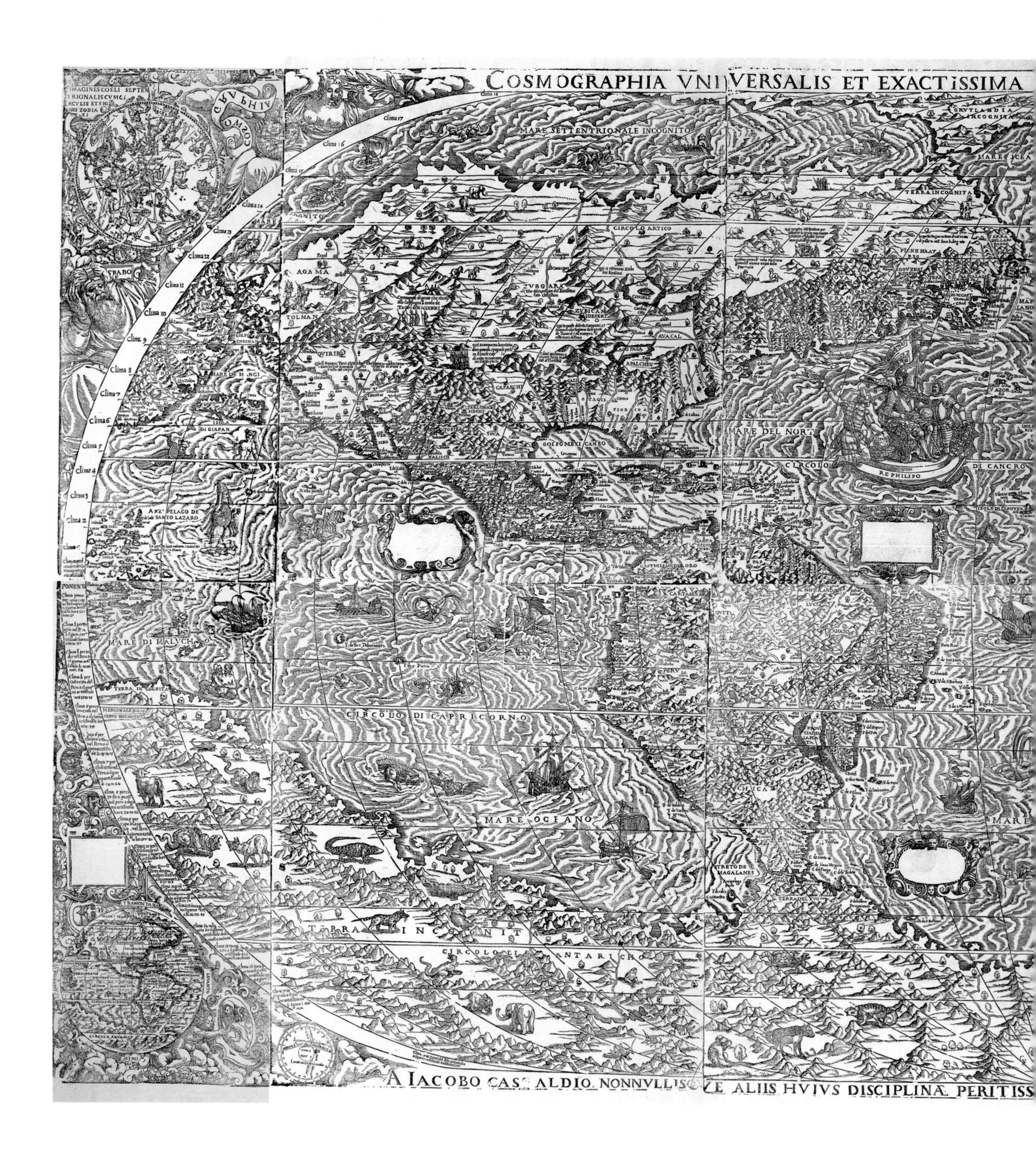
COSMOGRAPHIA VNIVERSALIS ET EXACTISSIMA
MARE SETTENTRIONALE INCOGNITO
TERRA INCOGNITA
CIRCOLO ARTICO
MARE DEL NORT
CIRCOLO DI CANCRO
CIRCOLO DI CAPRICORNO
MARE OCEANO
TERRA INCOGNITA
A IACOBO CASTALDIO NONNULLIS QUE ALIIS HUIUS DISCIPLINAE PERITISS

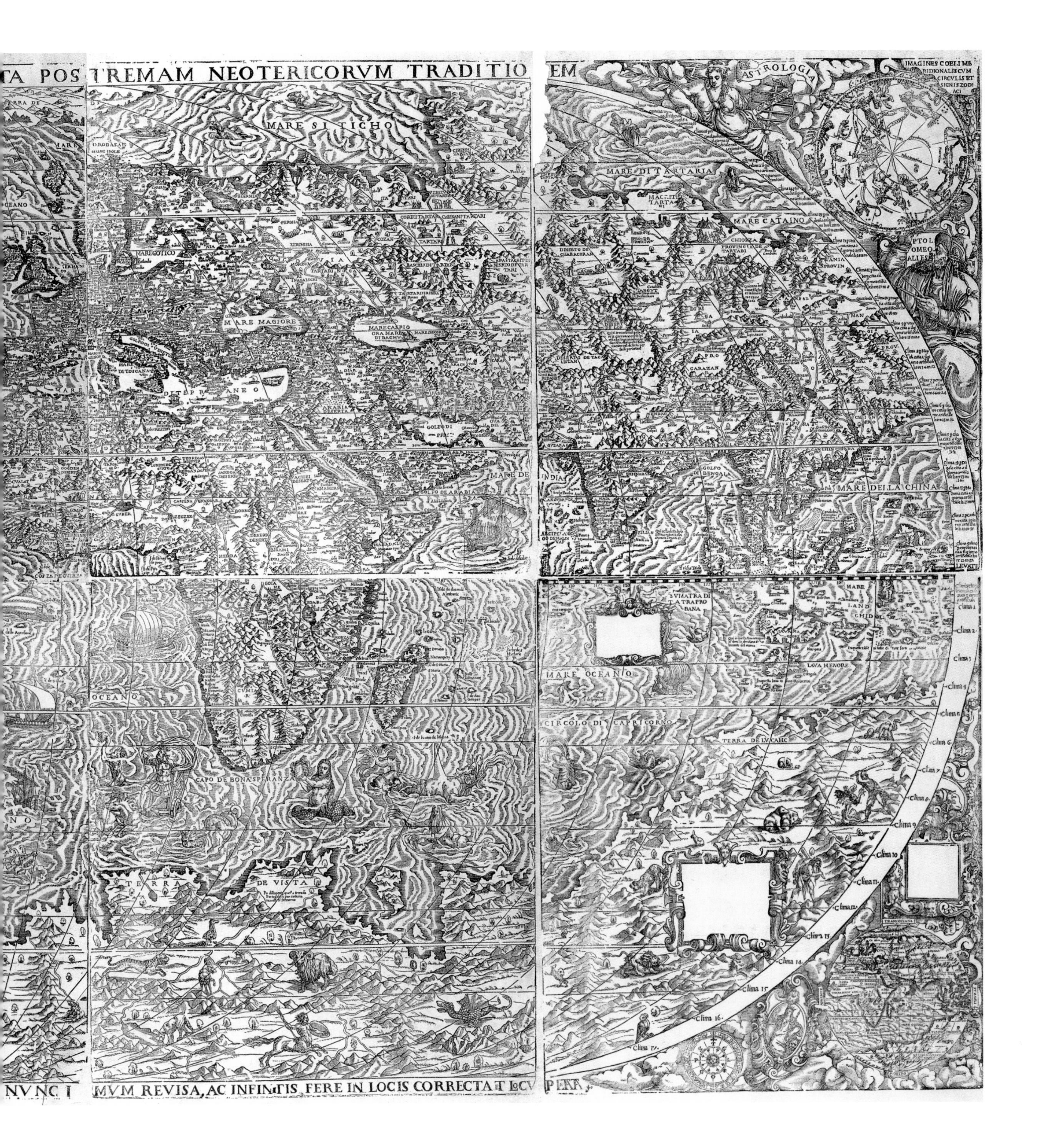
TA POS TREMAM NEOTERICORVM TRADITIO EM
NVNC I MVM REVISA, AC INFINITIS FERE IN LOCIS CORRECTA ET LOCVPLETA
MARE SI TICHO
MARE DI TARTARIA
MARE CATAINO
ASTROLOGIA
PTOLOMEO ALLES
MARE GOTICO
MARE MAGIORE
MARE CASPIO
GOLFO DI PERSIA
GOLFO DE ARABIA
MARE DE INDIA
GOLFO DI BENGALA
MARE DE LA CHINA
LEVANTE
MARE OCEANO
CIRCOLO DI CAPRICORNO
TERRA DE LVCAHC
CAPO DE BONA SPERANZA
OCEANO
TERRA
DE VISTA
Clima 2
Clima 3
Clima 4
Clima 6
Clima 7
Clima 8
Clima 9
Clima 10
Clima 11
Clima 12
Clima 13
Clima 14
Clima 15
Clima 16
Clima 17

One of the earliest 'modern' maps of England, Wales and Ireland with significant advances in accuracy over earlier maps.

An Elizabethan Renaissance Map

FOR MANY YEARS Laurence Nowell was identified as the Dean of Lichfield but it is now determined that he was a cousin of the same name, tutor to one of the wards of Sir William Cecil, later Lord Burghley. Nowell created this map expressly for Cecil, culling pieces of information from the libraries of Cecil and his friends, and may well have had access to state papers and official maps. The topographical works consulted are likely to have been those of Matthew Paris, Giraldus Cambrensis and the writings of Nowell's contemporaries such as John Leland. The map's accuracy for its time suggests that Nowell probably incorporated the results of local surveys and perhaps drafts of maps (now lost) by others such as John Elder or John Rudd. The two figures at the foot of the map are those of Nowell himself (left), wearily reclining on a pedestal bearing his initials and threatened by the bane of all surveyors, a barking hound. To the right an imperious William Cecil is resting on an hourglass, with the implication that the mapmaker's time is running out.

There are several striking features of Nowell's map. For Wales, the Caernarvon peninsula and Cardigan Bay are well pronounced for the first time. Further north the part of the Scottish coastline shown is drawn in advance of all previous maps. The margins show degrees of latitude and longitude, and the north–south orientation of England is much improved. For instance, Edinburgh is more correctly placed north of Cardiff compared to its alignment north of Oxford on earlier maps by Münster and Lily. The description of Ireland is advanced for its time, showing knowledge of tribal areas.

At one time Nowell was considered as the unidentified English source who despatched a map of the country to the geographer Gerard Mercator at Duisberg. From this a large map of the British Isles on eight sheets was printed in 1564. However, the Scotsman John Elder is now considered the most likely candidate although any map or draft he might have prepared is lost.

Right This manuscript mid-Tudor map is a unique work of national significance by Laurence Nowell, and originally owned by Lord Burghley, chief minister to Queen Elizabeth I. **Above** Laurence Nowell rests after his labours.

A general description
of England & Irelade
with ye costes adioy
ning:
Septentrio
Meridies
Occidens
Oriens
OC. HIBERNICVS
OC. GERMANICVS
Mare Hibernicum
Mare Gallicum
CONNACIA
LAGINA
MOMONIA
DVBLIN
GALOWAY
LOVDIAN
CVNINGAM
GALOWAY
CVMBERIA
ESSEX
KENT
Silley
Jersey
Garnsey
Alderney
Flanders
Pycardie
part of Brytain
part of Normandie

A plan-view of London, from the most famous atlas of city maps and views, which has selective distortions in favour of its supposed patron.

London Merchants Triumphant

IN 1572 A COLOGNE clergyman, Georg Braun, teamed up with a north German artist, Frans Hogenberg, to produce a book of town views and plans entitled *Civitates Orbis Terrarum* in emulation of Abraham Ortelius's bestselling atlas, the *Theatrum Orbis Terrarum*, that had been published in 1570. Such was Braun and Hogenberg's ultimate success that five more volumes followed before 1620.

This plan-view of London, which appeared in the first volume, was copied from an example of a superb fifteen-sheet wall map, completed in 1559, that Braun probably saw hanging in the home of his patron, Heinrich Suderman, the head of the Hanseatic League, a powerful grouping of north German and Baltic mercantile cities. Hanseatic merchants at the Steelyard in London (also known as German House) may have commissioned the original map, of which only three copperplates and no original impressions have survived, in the hope of flattering the government of Mary Tudor into supporting them against their rivals, the powerful merchants of the City of London, led by Thomas Gresham. Gresham was determined to obtain the revocation of the commercial privileges still enjoyed by the League in England. Partly to please the Queen and partly perhaps to encourage her to take on the merchants of the City, the original map emphasized Mary's power and underplayed that of the merchants by, for instance, showing the royal barge prominently in the middle of the map, and ignoring the existence of the splendid halls of the all-powerful City livery companies. They also inserted a text panel emphasizing the importance of the Hanseatic League to English commerce.

By 1570 the League had been humbled. Braun could not risk sales by ignoring this. On the other hand, he could not wilfully ignore the sensitivities of his patron. The result was a messy compromise: a close copy of the original map-view with what are evidently supposed to be English merchants in the foreground, a corrupted and abbreviated text panel (bottom right), still giving prominence to the League, and with the royal barge at the centre. It seems that this image met with criticism. Two years later, Braun produced a revised version, including a view of Gresham's newly opened Royal Exchange, a symbol of English mercantile power. St Paul's, in contrast, is still shown with the steeple that had been destroyed in 1561. It must have been unpleasant for Suderman but in the map world too, business was business.

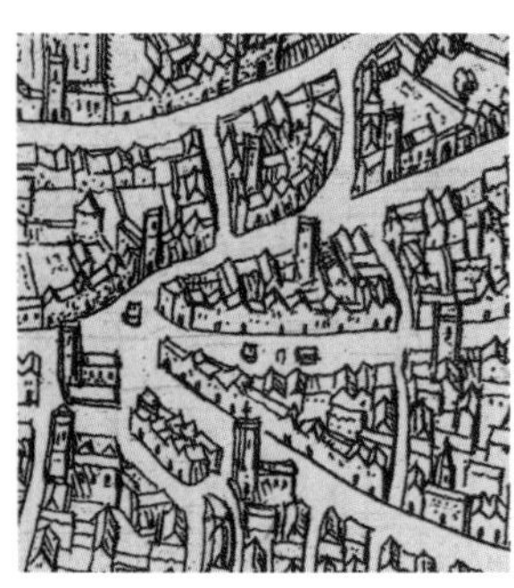

Right Georg Braun and Frans Hogenberg, *Londinium feracissimi Angliae regni metropolis*. **Above** Detail from the first edition of 1572 showing Cornhill without the Royal Exchange.
See also 1587.

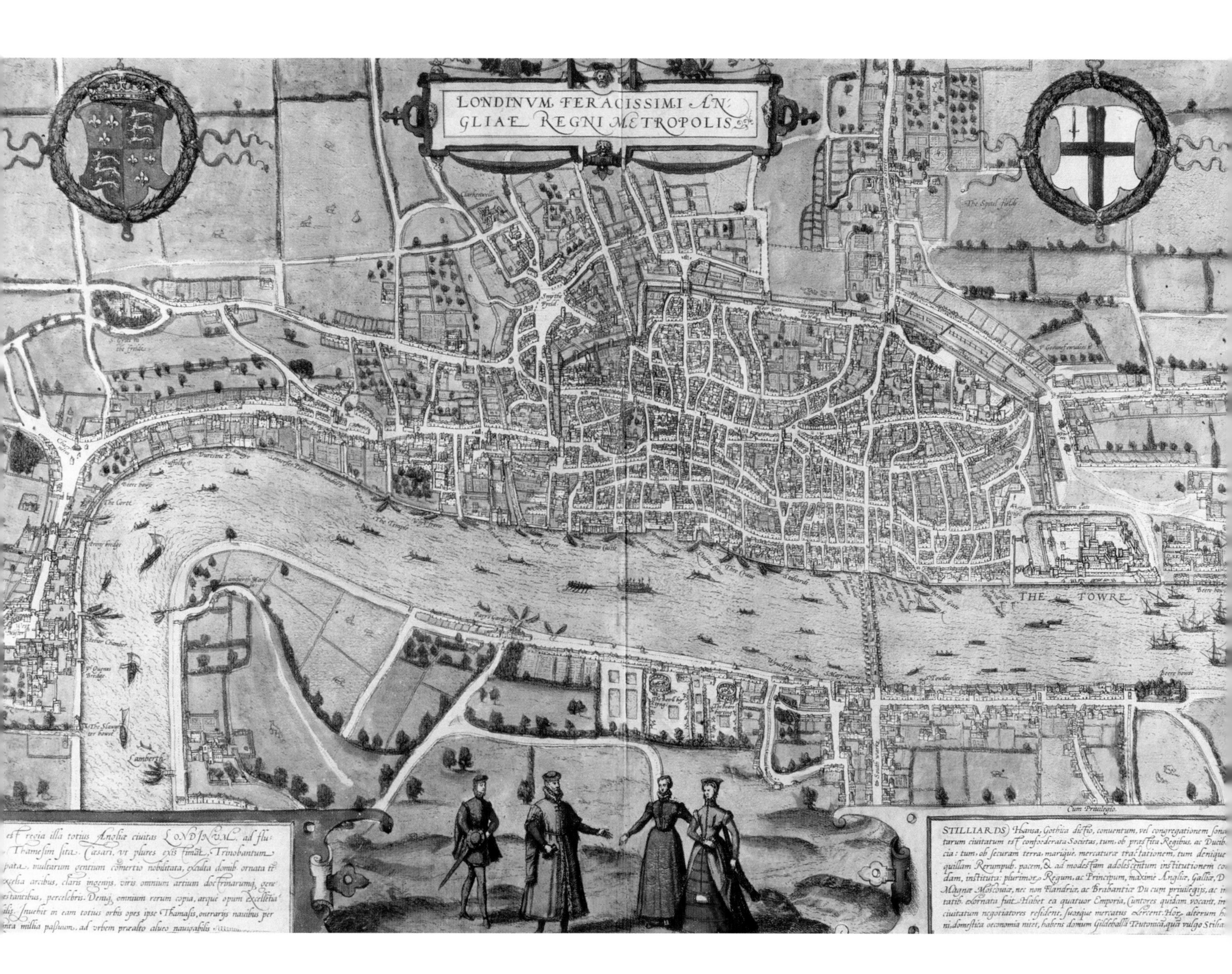
LONDINVM FERACISSIMI ANGLIAE REGNI METROPOLIS
THE TOWRE
The Spital fielde
Lamberth
Cum Priuilegio.
est regia illa totius Angliæ ciuitas LONDINUM, ad flu
Thamesim sita. Cæsari, vt plures exis timat, Trinobantum
pata, multarum gentium cōmertio nobilitata, exculta domib. ornata te
excelsa arcibus, claris ingenijs, viris omnium artium doctrinarumq. gene
estantibus, percelebris. Denig. omnium rerum copia, atque opum excelletia
ilis. Inuehit in eam totius orbis opes ipse Thamasis, onerarijs nauibus per
inta millia paßuum, ad vrbem præalto alueo nauigabilis
STILLIARDS) Hansa, Gothica dictio, conuentum, vel congregationem sona
tarum ciuitatum est confœderata Societas, tum, ob præstita Regibus, ac Ducib.
cia: tum, ob securam terra, marique, mercaturæ tractationem, tum denique
quillam Rerumpub. pacem, & ad modestam adolescentum institutionem co
dam, instituta: plurimor. Regum, ac Principum, maxime Angliæ, Galliæ, D
Magnæ Moscouiæ, nec non Flandriæ, ac Brabantiæ Du cum priuilegijs, ac in
tatib. exornata fuit. Habet ea quatuor Emporia, Cuntores quidam vocant, in
ciuitatum negotiatores resident, suosque mercatus exercent. Hor. alterum h
ni, domestica oeconomia nitet, habens domum Gildehallā Teutonicā, quā vulgo Stilia

A map of Cornwall from the first detailed county survey of England and Wales which became the basis for all English maps for the next two hundred years.

Useful and Ornamental

BY THE 1570s the political and strategic climate in England called for reliable maps for civil control and national defence. The choice of surveyor fell on an obscure Yorkshireman, Christopher Saxton, who completed his county surveys between 1574 and 1579. In the latter year his pioneering atlas of all the English and Welsh counties with one general map was published. His map of Cornwall is one of the finest.

How Saxton went about his work is still disputed. Possibly he had access to predecessor material by John Rudd, Reyner Wolfe or Laurence Nowell. The number of Saxton's assistants is also unknown but he seems to have worked quickly, ascending church towers and prominent hills to fix directions and distances, perhaps by a simple form of triangulation. For the south-western counties, including Cornwall, he may have adopted a grid based on the Elizabethan beacon system set up to transmit news of any intended invasion from one high point to another.

His map shows many villages and towns designated by conventional symbols; also parks, woods, bridges and 'sugar loaf' hills. However, no roads are shown – these rarely appeared on county maps until the 1670s. The boundaries of the hundreds are delineated and Saxton uses the old English mile in his scale, roughly equal to 2,065 metres. Set alongside a modern map, Saxton's Cornwall is out of position; for instance, Land's End is shown north of Plymouth whereas its true alignment is almost 20 angular degrees south. This irregularity was not corrected until the maps of John Bill in 1626.

In 1576 Saxton's manuscript was engraved by the accomplished Flemish artist Lenaert Terwoort, one of several émigrés from Flanders. Both Saxton's name and Terwoort's appear on the map. Much of the artistic distinction of the map lies in the elegance of the engraving in typical Flemish style with neat italic lettering. Sea monsters and galleons enliven the seas as befits maritime Cornwall. Birds, flowers and strapwork adorn the cartouches.

Right Saxton's map of Cornwall. **Above** Frontispiece to Saxton's atlas of the counties of England and Wales showing Elizabeth I flanked by the figures of Geography and Astronomy.

The influence of Saxton's maps was long-lasting and his topography was adopted by Gerard Mercator for his great atlas published in 1595. Most of the county maps drawn by John Speed for his 1612 *Theatre of the Empire of Great Briataine* rely on Saxton's surveys although in a few instances – Cornwall being one – he drew on the later surveys of John Norden. The groundwork created by Saxton remained the basis for all English maps for nearly two hundred years.

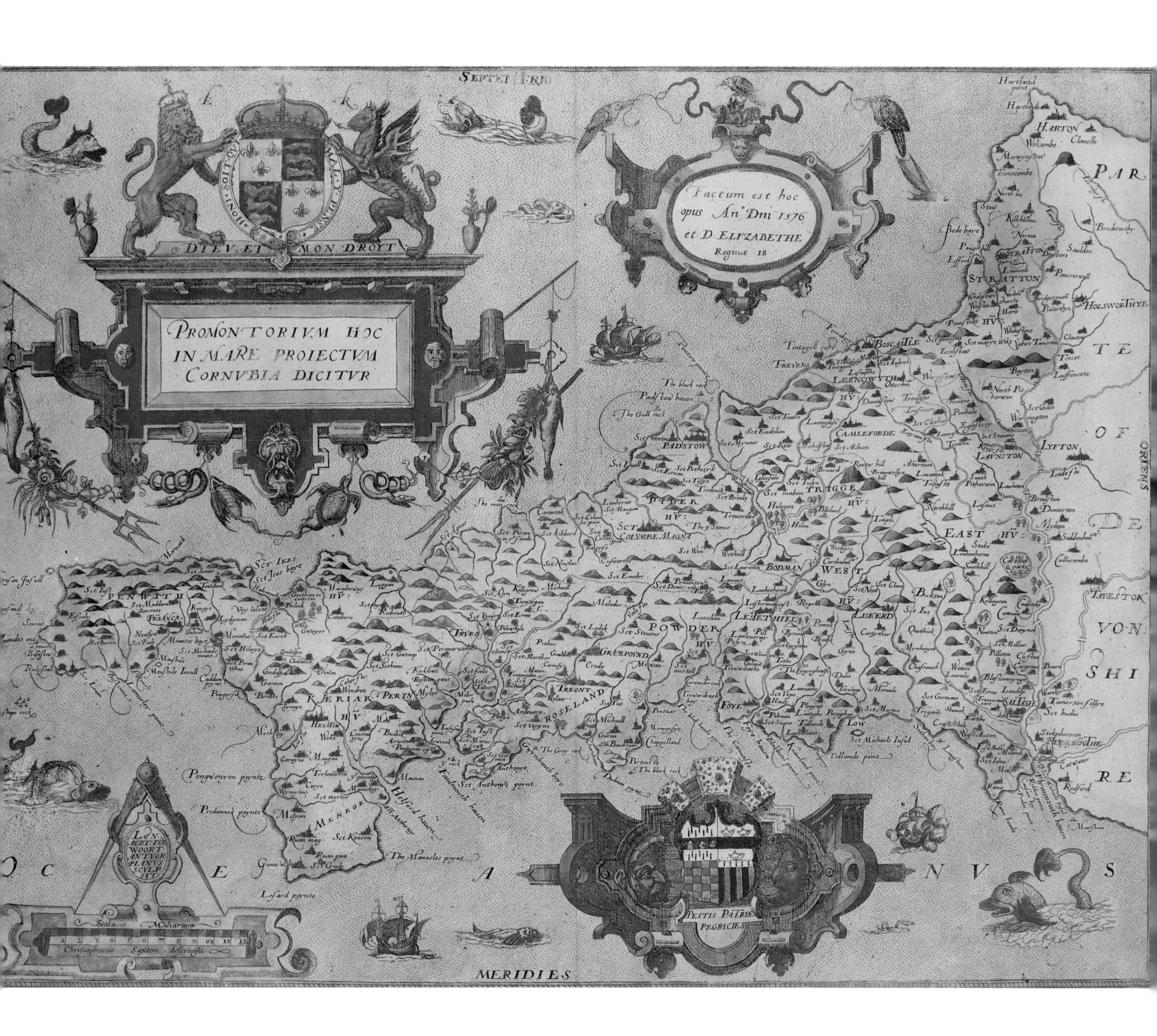
SEPTENTRIO
DIEV ET MON DROYT
HONI SOIT QVI MAL Y PENSE
Factum est hoc opus An° Dni 1576 et D ELIZABETHE Reginæ 18
PROMONTORIUM HOC IN MARE PROIECTUM CORNUBIA DICITUR
PAR
TE
OF
DE
VON
SHI
RE
ORIENS
OC
E
A
NVS
PESTIS PATRIÆ PEGRICIES
LEN AERT TER WOORT ANTVERPIANVS SCVLPSIT
Scala Miliarium
Christophorus Saxton descripsit
MERIDIES
PENWITH
KERIAR
PERYN
POWDER
TRIGGE
LESNOWYTH
STRATTON
EAST HV
WEST
ROSELAND
MENEGE
PADSTOW
BODMAN
LISKERD
FOYE
TRURO
GRAMPOVND
TREGONY
LAVNSTON
BOSCASTLE
CAMLEFORDE

An Amerindian map from Mexico created in response to a questionnaire issued by King Philip II of Spain.

Aztec but Christian

1579

Right Map of Cholula, Mexico, 1579.

THE MAP OF CHOLULA, made by an Amerindian painter, is the inheritor of the great pre-Hispanic tradition of mapmaking. It portrays a colonial city that was once an important pilgrimage centre before the conquest of the Aztec Empire by the Spanish in 1521. Its maker was likely born too late to have been trained in a pre-Hispanic painting academy; instead, he was almost certainly the product of Franciscan schooling. Across sixteenth-century Mexico, mendicant friars established schools for Amerindian boys from elite families and fostered a hybrid (and Christian) indigenous style of art. The Franciscan school in Cholula would have been housed in its monastery complex, which dominates the centre of this map.

The Cholula map reveals the survival and transformation of the native mapmaking tradition under Spanish colonial rule. In the map, the Amerindian preference for pictographic writing survives in the symbol for 'hill' that is seen at upper right. The pre-Hispanic pyramid is next to it. Also present is an Amerindian view of territory, understood less as a collection of geographic features than as a space defined by human action and social structure. For instance, the Cholula map shows the town as a grid plan, a design newly imposed by Spanish colonizers, but in the organization and arrangement of its neighborhoods, it reinstates the social layout of the pre-Hispanic city. Each of the six parishes seen on the map replaced one of Cholula's pre-Hispanic parts, their counterclockwise order on this map echoing the traditional hierarchy of Cholula's polities.

While many sixteenth-century maps made by Amerindians for Amerindian audiences exist, Amerindian mapmakers were frequently employed by the Spanish colonial government. Many maps were made as part of law suits and land grants. The Cholula map was one of the dozens of known maps created in response to a questionnaire issued by King Philip II in the final years of the 1570s.

The same questionnaire was sent to the Viceroyalty of Peru, but here it seems to have elicited few maps. This lack may be attributed to the traditional preference of native Andeans to encode ideas in elaborate knotted string bundles called *khipus*, rather than using graphic marks on paper or cloth.

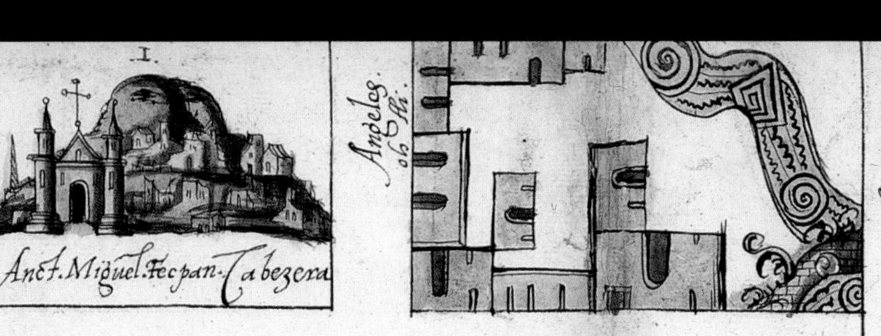

.1.
S. Anct. Miguel Tecpan Tabezera
Angeles ohtli

TOLLAN. CHOLULA.

S. Anct. Andres Cabezeras

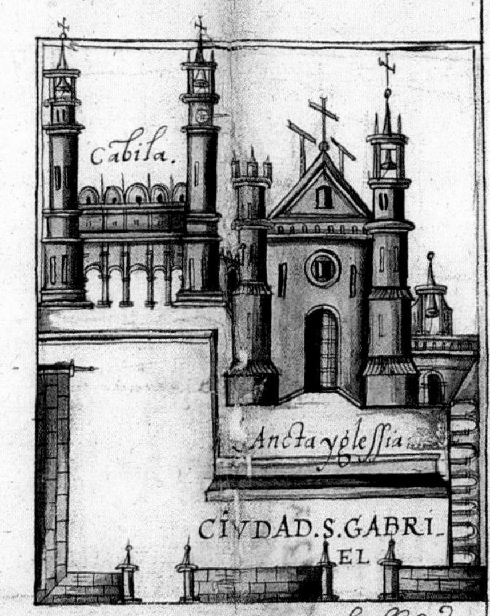

Cabila.
Ancta yglessia
CIUDAD. S. GABRIEL

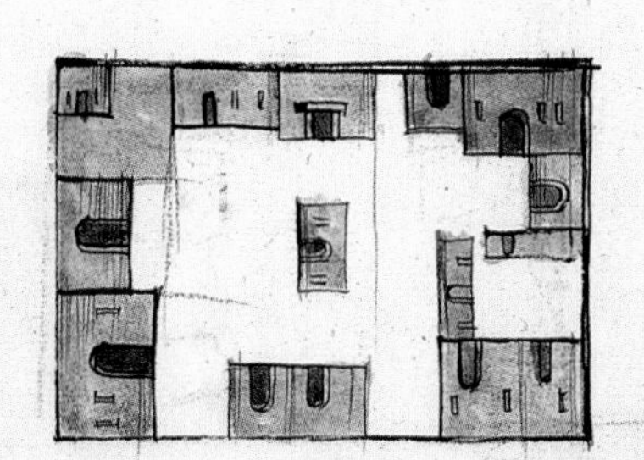

.2.
S. Anctiago. Tabezera

.5.
Sanct. Pablo. Tabezera

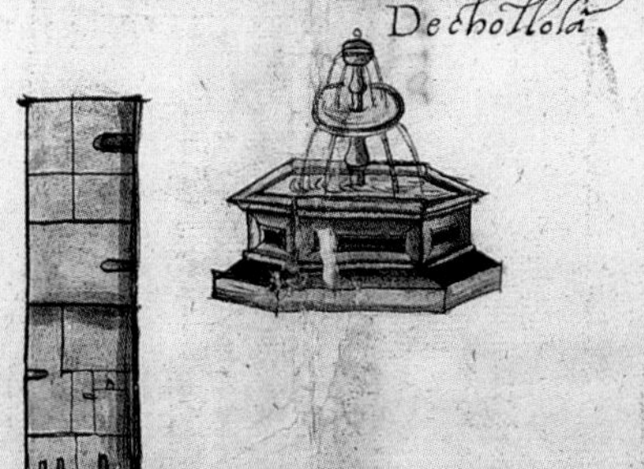

De chollola

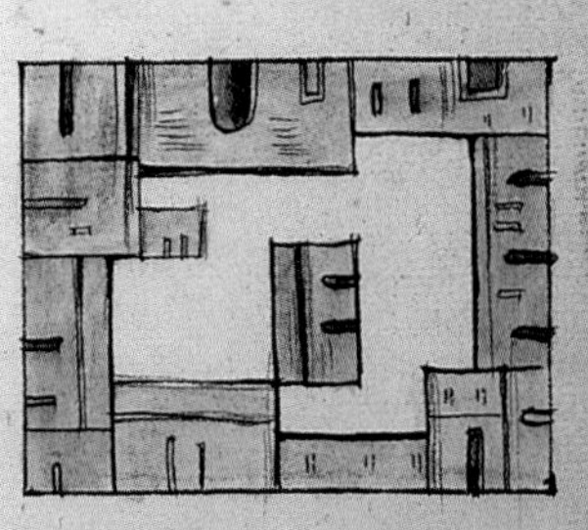

Tlaxcatlan ohtli
Tianquizco.
acapetlavahcan. ohtli.

.3.

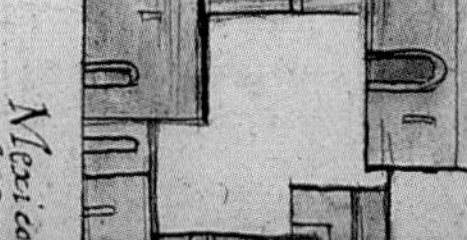
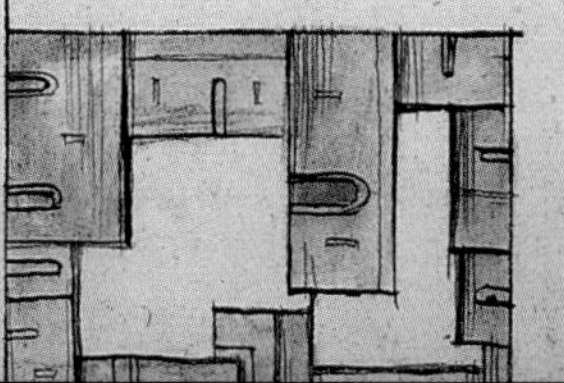

Mexico. ohtli.

.4.

A cartographic mural of the Italian Renaissance displayed in a gallery of maps in the Vatican Palace, Rome.

Papal Power

1580

IN THE RENAISSANCE painted maps decorated the audience halls of royal, papal and republican palaces, the libraries of monasteries and the loggias of suburban villas. Undoubtedly, this predilection for painted cartographic images relates to the increased uses of maps for such diverse activities as the administration of the state, the calculation of the freight of merchandise and the reading of the Bible and the classics. But it also relates to Renaissance patrons' desire to emulate their ancient predecessors who, as Livy, Pliny and Aelian reported, had displayed monumental maps of durable materials in public spaces. Transforming the ancient, sporadic display of individual maps in a veritable fashion, Renaissance patrons used painted maps as primary vehicles to convey their political legitimacy, religious supremacy or universal knowledge. These painted maps, which were carefully selected in relation to each other, were displayed alongside other non-cartographic images, such as allegories, religious scenes and historical events. The earliest and most spectacular painted maps are documented in Italy (Florence, Rome, Naples and Caprarola), but later they also appeared in Fontainebleau, Madrid and Salzburg.

The most accomplished cartographic mural among the surviving examples is the Gallery of Maps in the Vatican Palace, a long corridor decorated with forty-four monumental maps of Italian regions and cities. Commissioned by Pope Gregory XIII, designed by the Dominican polymath Egnazio Danti and painted in around 1580 by an army of artists, the Gallery of Maps was conceived as an imaginary walk along the Apennine spine of the Italian peninsula: the maps are displayed on the walls, while fifty-one events of church history that happened in the lands mapped below are depicted on the vault of the gallery. This unprecedented atlas of Italy did not represent the ancient geographical unit described by Strabo and Pliny or any future political partition.
It represented, instead, the boundaries of the Western Church that had readily submitted to the ecclesiastical authority of the popes after the Council of Trent (1563). In these cartographic representations of the post-tridentine primacy of Italy, the mapping of episcopal and archiepiscopal seats, recent acquisitions to the apostolic chamber and major sites for the collection of ecclesiastical revenues was as defining an element as the detailed geographical representation of places, local history and landscape.

Right One of the regional maps: Bologna. **Following pages** The Gallery of Maps, designed by Egnazio Danti, in the Vatican Palace, Rome. **See also** 200, 300.

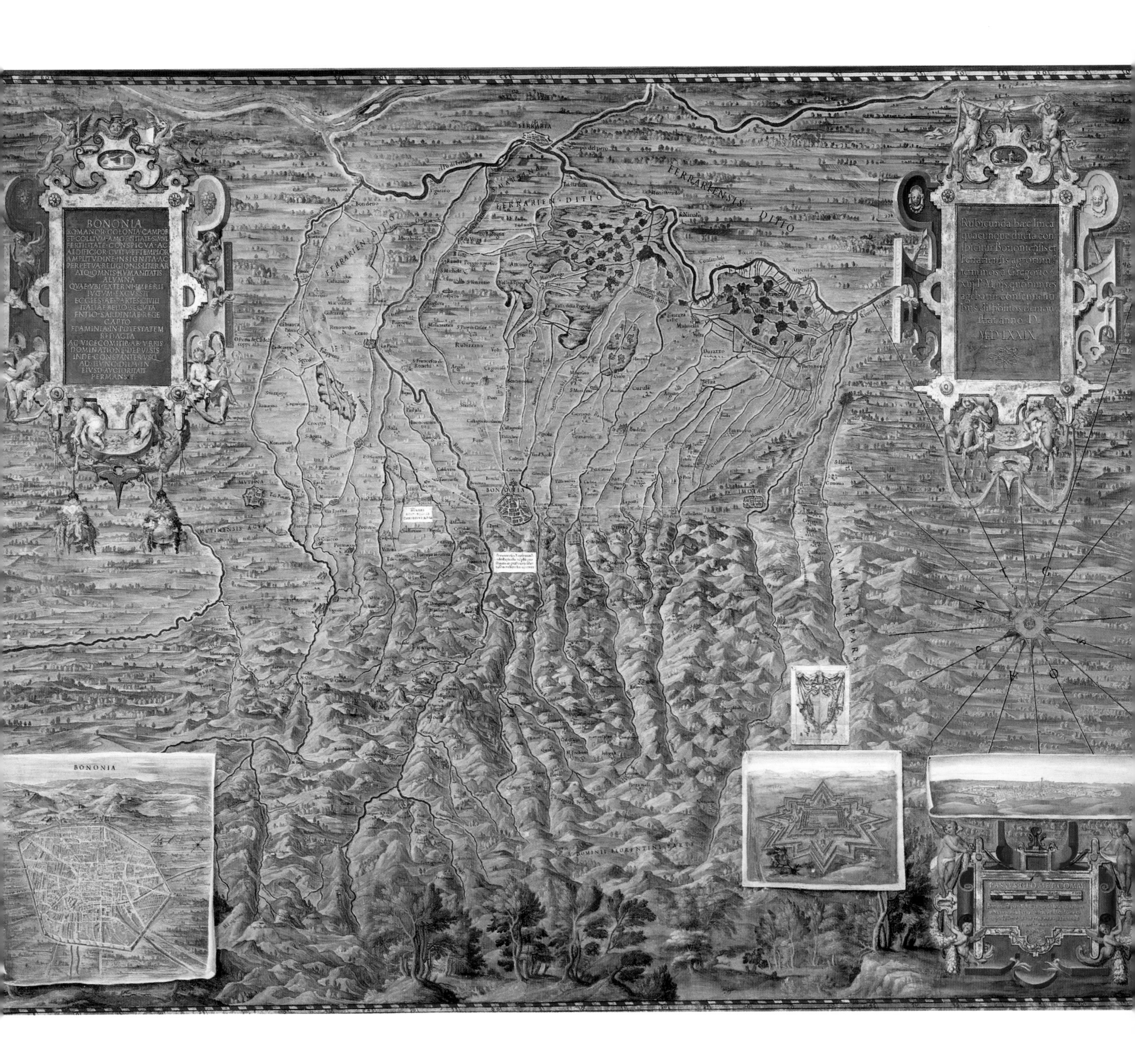
BONONIA
ROMANOR·COLONIA·CAMPOR
ET·COLLIVM·AMOENITATE·SIMVL
FERTILITATE·CONSPICVA·AC
AEDIFICIOR·CVLTV·ET·TEMPLOR
AMPLITVDINE·INSIGNITA·AC
PERPETVA·RELIGIONIS·LITERAR
ATQ·OMNIS·HVMANITATIS
ALVMNA
QVAE·VBI·EXTERNI·IMPERII
IVGVM·EXCVSSIT
ECCLESIAE·PARTES·CIVILI
ITALIAE·BELLO·SECVTA
ENTIO·SARDINIAE·REGE
CAPTO
FLAMINIA·IN·POTESTATEM
REDACTA
AC·VICECOMITIB·AB·VRBIS
DOMINATIONE·DEPVLSIS
INDE·CONSTANTER·VSQ
AD·HVNC·DIEM·IN
EIVSD·AVCTORITATE
PERMANSIT
Rubicunda haec linea
quacunque ducta con
spicitur Bononiensis et
Ferrariensis agrorum
terminos a Gregorio
xiij P.M. iis quorum res
agebantur consentienti
bus dispositos demon
strat anno D.
MDLXXIX
FERRARIA
FERRARIEN·DITIO
FERRARIENSIS DITIO
BONONIA
MVTINA
IMOLA
FLAMINIAE PARS
DOMINII FLORENTINI PARS
BONONIA

ITALIA ANTIQVA

ITALIA NOVA
ITALIA
ARTIVM
STVDIORVMQVE
PLENA SEMPER
EST HABITA

An early printed town plan of Exeter celebrates the fact that in the sixteenth century it was one of the wealthiest towns in England.

Urban Pride

1587

Right John Hooker's map of Exeter. Engraved by Remigius Hogenberg. **See also** 1521, 1574.

IT IS NOT SURPRISING that Exeter finds itself in the company of London, Cambridge, Oxford and Norwich as one of the earliest English towns to be surveyed and mapped; these were amongst the wealthiest towns in early modern England. Hooker's map is a bird's-eye view of the city taken from an imaginary viewpoint from the west with the River Exe running across the map from left to right. Suburbs outside the walls are shown, buildings are portrayed with details of their elevations, the principal buildings and streets are named, and there is much detail picturing the life of the city: a crane on the quay, fishermen with their nets, ships on the river, people walking the streets, market stalls on the cathedral green. The original engraved map survives in three states and the research of William Ravenhill and Margery Rowe suggests that they are unique copies. Hooker's map was much reproduced, adapted, and copied by contemporaries and near contemporaries (for example, by John Speed in 1611 and in Braun and Hogenberg's *Civitates Orbis Terrarum*) and even as decoration on a leather screen.

Hooker's portrayal of the city in which he was Chamberlain from 1555 has been studied from many angles and analyzed for what it can tell us about Exeter's architecture and townscape, but important questions remain – why was this map made at all? Why was a considerable sum of money expended on it? What was its intended purpose? There are no straightforward answers to these questions but we can be certain that Hooker's map is more than a factual, topographical description of the walls, streets and buildings of a cathedral city. It tells us something of the economic importance of the place and its iconography conveys much deeper messages. The arms of the Queen and the City of Exeter pictured in the upper corners of the map doubtless acknowledge Queen Elizabeth I's patronage. In this sense, the map is a celebration of the wealth and power of this important town under her government. Reflecting as they do the political significance of a concentration of inhabitants, their activities and their wealth, maps of towns, such as John Hooker's Exeter, serve many interests beyond the strictly practical.

Isca Damnoniorum britanice Kaier penhuelgorte Saxonice
Monketon Latine Exonia Anglice Exeancestre vel Exestre
at nunc Vulgo Exeter: urbs peratiqua, et Emporium celeberrimum
The Castle
Churche Yarde
S. Paule
Goldsmithe Streete
North gate Streete
Archer Lane
Southgate streete
South Gate
Pallace gate
Exe River
Post Mortem vita
Easte
Weste
Southe

A sketch map from the first detailed survey of Scotland depicts, amongst much other detail, the waning influence of the old Gaelic world.

Sketching out Scotland's Bounds

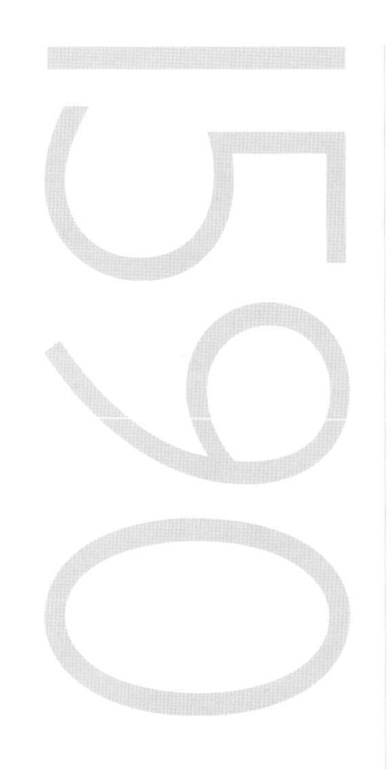

FOR JAMES VI and his ministers, seeking greater control over their '*alluterlie barbares*' Highland clans and unruly nobles, defining the bounds of their kingdom through maps was a crucial means of state integration, prestige and defence. In a predominantly agricultural society, poor by comparison with her southern neighbour, Scotland's merchants and magnates were also keen on the exploration and discovery of new sources of wealth. In addition, following the Scottish Reformation of 1560, a newly emerging body of ministers, lawyers, teachers and lairds were coming to appreciate the value of detailed, historically grounded descriptions of their country in words and pictures.

In this context, Timothy Pont (*c.*1565–1614) heralds the beginning of Scottish mapmaking, delineating a uniquely important snapshot of late sixteenth-century Scotland. His father Robert was one of the leading figures in the Scottish Reformation, and Timothy himself became a minister in the last decade of his life. Without employing trigonometry or accurate measurement, his surviving sketch maps are perhaps better regarded as qualitative representations of what was where rather than accurate geometric depictions.

This detailed map of the Loch Tay environs shows a distinctive Scottish Highland landscape: a dispersed rural settlement including tower houses, farms, mills and kirks, is set against an impressive backdrop of mountains and wooded glens around the Loch itself. Pont's 'landscape of power' is reflected at several levels. Finlarig in the west, and the newly constructed castle of Balloch (today Taymouth) in the east symbolized the extending Campbell control over this area, whilst the predominantly Gaelic place names, appearing here for the first time on maps, were steadily being superseded and altered by the more powerful linguistic forces of Scots and English. Pont took detailed descriptive notes on his travels, and frequently these spilled over onto his maps, with glowing references to the fish and pearls in Loch Tay, as well as the seventh-century King Domnall, who drowned in the Loch whilst fishing for salmon, and the 'steep and glennish' burns nearby.

Right Pont's map of the Loch Tay environs, Scotland, c.1583–1600.
Above Detail showing Elgin.
See also 1576

Pont's maps, largely unpublished in his own lifetime, were the major source material for Scotland's first atlas, published by Blaeu as part of his *Atlas Novus* in 1654. To assist with this, Robert Gordon of Straloch edited and revised many of Pont's maps in the 1630s and 1640s, and his darker, blacker ink can be seen here emboldening many of Pont's original details.

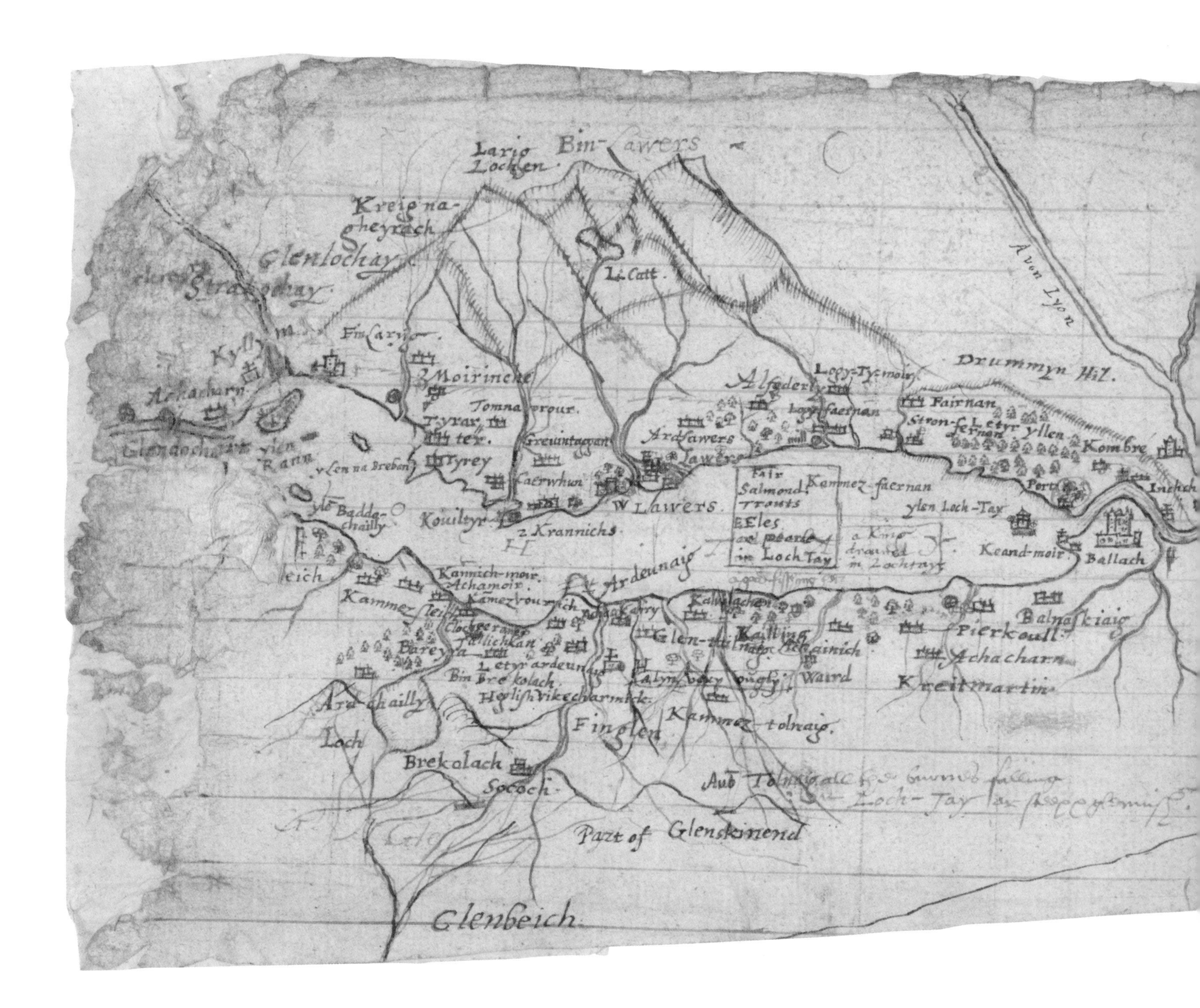

Lario Lochen
Bin-Lawers
Kreig na gheyrach
Glenlochay
Li.Catt
Avon Lyon
Drummyn Hil.
Achacharn
Tomna vrour
Fairnan
Kombre
Lawers
Fair Salmond Trouts Eles and pearle in Loch Tay
ylen Loch-Tay
Keand-moir
Ballach
2 Krannichs
Kouiltyr
Ardeunaig
Balnaskiaig
Pierkoull
Achacharn
Kreitmartin
Waird
Finglen
Loch
Brekolach
Sococh
Part of Glenskinend
Glenbeich

Tapestry maps commissioned to decorate a new home, to impress visitors and to demostrate a Catholic family's loyalty to Elizabeth I.

Pride and Patriotism Mapped in Wool

IN ENGLAND EXQUISITE and enormously expensive tapestries often played the same role that wall paintings did in warmer climes. In the 1590s, borrowing the idea of the wall map to achieve a unique decorative scheme for his new house, Ralph Sheldon commissioned four woven tapestry maps showing the counties in which he lived, owned land and had friends and family – Warwickshire, Worcestershire, Oxfordshire and Gloucestershire. Each tapestry measured approximately 4 metres (20 feet) high and 6 metres (13 feet) wide; together, they formed a panoramic view across England from London to the Bristol Channel.

The unknown designer composed his picture from information on the recent, and first, attempt at country-wide mapping, Christopher Saxton's county surveys. In this detailed view of Sheldon's home ground on the Warwickshire/Oxfordshire border, he may also have had local help. Separate colours demarcated the counties, four of which meet here. Hills, always emphasized with a green edge, became gently rounded or exaggeratedly conical to indicate change of terrain. Trees in more open country, as here, were short and squat, but in heavily wooded areas might be taller than the church towers or spires, illustrated without exactitude, which dominated the houses of the named villages. Bridges, differentiated between stone and wood, crossed rivers, some of them omitted from Saxton's maps. Palings and parkland surrounded the larger houses of gentlemen, including Sheldon's own at Weston, and those of many of his friends, depicted with at least an attempt at accuracy. Towns too, often shown in the conventional bird's-eye view from the south, were based on reality not imagination. However, the tapestry designer improved on his model; he included roads which, in yellow, wander uncertainly across a landscape dotted with windmills – usually Sheldon's property – the occasional fire beacon, or special landmarks such as the boundary marker known as the Four-Shire Stone close to Morton Henmarsh (now Moreton in the Marsh), or the Rollright Stones by the windmill south of Long Compton.

The corners emphasized Sheldon's patriotic loyalty to Elizabeth I, despite his own Catholicism. They showed the royal arms, a scale and dividers, the family arms and a long descriptive passage paraphrasing William Camden's recently published *Britannia*, the verbal equivalent of Saxton's maps. The wide decorative borders, filled with architectural and allegorical motifs, also showed globes, named the points of the compass and contained doggerel verse celebrating the county's natural attractions.

Right Sheldon's tapestry map of Worcestershire (detail). **Above** Another area in detail.
See also 1576, 1580

DASCOT
ILMINGTON
LONGDON
TREDINGTON
IDLICOT
COMPTON
COMPTON
SCORFIN
HVNNINGTON
WINDERTON
CHARINGWORTH
SHIPSTON
BRAILES
BARCHESTON
STRETTON
TIDMINGTON
BVRMINGTON
SVTTON
STVER FLVD
TODNAM
STVRTON
CHERINGTON
DORNE
LEMMINGTON
WOLFORD
MAGNA
WESTON
WOLFORD
PAR
WHICHFORD
MORTON
HENMARSH
BARTON
ONTHEHEATH
LONGE
COMPTON
SHEER

An early thematic map of northern Lancashire fit for its purpose – as a blueprint for the defence of the realm in the event of a Spanish invasion.

In the Event of Invasion...

THERE CAN BE few maps that meet modern standards of cartographic excellence less than this map of northern Lancashire. Although it has a scale bar, its geographical relationships are extremely approximate. The shorelines are vague, the sandy beaches grossly exaggerated. Rivers not roads are shown and you have to search hard to find towns like Manchester or Liverpool among the profusion of family names, family seats and family parks. The map was made for Elizabeth I's great minister, Lord Burghley, when England was at war with Spain in the years following the Armada.

Yet it shows precisely what Burghley considered important – and understated or completely ignored everything else. His understandable preoccupation with invasion meant that he wanted to see immediately where the Spaniards might land. Given the bad conditions of most sixteenth-century roads, rivers were the equivalent of our motorways (the width of which is exaggerated on most modern maps). In Tudor England population was more evenly distributed between town and country and, with the exception of cities like London, Norwich and Bristol, it was the voice of the country and not the towns that counted. Elizabeth and Burghley ruled the country through landed magnates like the earls of Derby and through country squires, from whom justices of the peace were selected. This map told Burghley who they were and where they lived.

Right Lord Burghley's map of Lancashire, detail.
Above Lord Burghley.

The map is not a static mirror. It provides a blueprint for action in a crisis. The beacons, again shown unrealistically large, enabled news of invasion to reach court quickly. Burghley, whose spikey, sloping handwriting can be seen adding information all over the map, put crosses against the homes of the Catholic gentry. He could thus see at a glance which nearby Protestant families (names without a cross) could be called on to take action should any Catholic families join forces with Spanish invaders. Meanwhile, the deer parks, indicated as fenced enclosures, provided places of assembly – and meat and water – for the militia charged with fending off the invaders while more substantial help was on its way.

A good map is one that is fit for its purpose. For all its scientific weaknesses and seeming oddities, few are better than this. As Lord Burghley was to say, 'the practick part of wisdom is the best'.

Magna crosbie chappel
Formby chap
Anker chap
Leatham p.
Birkdaile chap
Halsaule
Meylis
Kirham p.
melton p.
Plumpton p.
Broughton p.
Garston chap.
Chiltnall p.
Heaton p.
Ormschurch p.
west darby chappel
Latham
Hay parke
Lathom pk.
Hesket
Ostlande
west Darby Hundreth
Holland pch.
Farnworth p.
Wiggan
standish p.
Ashall.
charley p.
Leylond p.
Penkforth
Lay
Warington p.
Leigh p.
Deane p.
dolton p.
Ribchest p.
Brookhall
Blagburne Hundred
Blagburne p.
Turton water
Irk wat.
Salforth Hundred
Flixton p.
Manchester p.
Birrie p.
Ratchdall p.
Haslingden
Harwood p.
Padiam p.
Brunley p.
Penell beacon
new church in Pendill
new church Rostindall
Townley
Colne p.
colne water
Oldam p.
Mersey water
Midleton water

An early map of Utopia which by its very design, the details and typography, is shown to be a fantasy.

What's in a Name?

1596

ORTELIUS'S MAP OF Utopia was only known from a letter which Ortelius wrote to his beloved nephew Jacob Cools, living in London in 1596, in which he promised to send him twelve copies. No example was known until one copy emerged at a British auction. No others have been found since. The name *utopia* in Greek means 'a place that does not exist' though a second, incorrect derivation is *eu-topos*, that is 'good place'. Ortelius says that he made this map just to please his friends by providing an illustration for their copies of Thomas More's book *Utopia*. There is no text on the back.

The map is easily recognizable as an Ortelius map, not only because his name occurs twice on the map but also because of its characteristic cartouches, style of engraving and decorations. The number and nature of its toponyms (place names) make this map unique. Normally, Ortelius' maps have between 300 and 1,500 topographical names, but here we find only 111, about half of them referring to rivers. The bottom right cartouche says: 'To the spectator: Behold the joys of the world. See the fortunate Kingdom. What could be better? This is Utopia, Fortress of Peace, centre of Love and Justice, safe Haven and trusted Coast. Praised elsewhere, Venerated by you who knows why. This Land, more than Any Other, offers you a Happy Life. Dedicated to Johannes Wacker von Wackerfels, as told by Raphael, recorded by More, and drawn by Abraham Ortelius. Joy and Prosperity to You.'

Right Abraham Ortelius's map of Utopia. **See also** 1846, 1867, 1998.

The languages of the place and river names are not scattered over the map but form a pattern: apart from a few anagrams of his friends' names, starting clockwise at ten o'clock we find place names in Latin, then Greek, Spanish, Dutch, Slavonic, Turkish, French, Utopic, at seven o'clock Italian, and lastly German. The names all have an internal inconsistency or impossibility: river without water, city without people, etc. All these contradictory names essentially explain that Utopia does not exist, never did and never will, except in the contemplations of Hythlodaeus, More's narrator, who is a dreamer. The world would have been converted to Utopia were it not that man is inclined to pride and evil. Thus, Utopia becomes a contradiction in terms, having rivers without fish and cities without people. Ortelius's hesitation to make this map may have been sincere after all.

VTOPIAE
TYPVS, EX
Narratione Raphaelis Hythlodaei,
Descriptione D. Thomae Mori,
Delineatione Abrahami Ortelij.
MERIDIES.
ORIENS.
OCCIDENS.
SEPTEMTRIO.
Zapoleti
Achorij.
Macarenses.
Anemolij.
Idelbeck
Laerdael
Sottenbroeck
Sombriosa
Malogoscz
Golienfdorp
Niebylowna
Gugertzinluc
Tzorlichifar
Alaschezer
Caraoiluc
Nulleville
Manobia
Amaurotus metropolis
Rotiliana
Horsdumonde
Iamais
Saxirupia siue Tisvilaget
Barzanea
Artemolia
Tranboria
Trapemeria
Sysograntia
Belsogno
Ariauento
Cuccagnola
Nonnata
Casteluoto
Pointneys
Keinstadt
Onhausen
Larburg
Niemanzingen
Ventadinada
Villauazia
Aunno
Almagro
Madiapolis
Midepolis
Nusquamia
Nullibia
Nondumia
Scilicetia
Favolia
Quippeia
Vdepolis
Amyndus flu.
Anydrus fluuius
Nulliflus flu.
NOBILISS. VIRO: IO: MATTHÆO WACKHERIO À WACKENFELS SAC. CÆS. Mtis CONSILIARIO: ET EPI WRATISLAV. CANCELLARIO.
Amico optatissimo.
Ab. Ortelius dedicabat, L.M.
AD SPECTATOREM.
En tibi delicias mundi: regna ecce beata!
Queis melius, queis nil pulchrius orbis habet.
Hæc illa Vtopia est; arx pacis; nidus Amoris,
Justitiæ, ac summi portus et ora boni.
Lauda alias terras: istanc cole qui sapis. Isto
Vel nullo fixa est Vita beata loco.
I. M. W. à W. f.
Lustravit Raphael: Descripsit Morus: Abrahamus
Edidit Ortelius. Tu fruere atq; vale.

An early printed moral-political map on a very early Mercator projection is both up-to-date geographically as well as an allegory of contemporary events.

The Knight, the King and the World

ILLUSTRATION DOMINATES THE lower portion of this world map by the Netherlandish cartographer Jodocus Hondius the Elder. Entitled 'An image of the lands of the whole globe on which the struggles on earth of the Christian Knight are ... graphically depicted', the map contains an allegorical message alongside cartographic information. Hondius was one of the foremost cartographers of his day and this map, one of the first on the Mercator projection, is an up-to-date representation of the state of geographic knowledge at the time. He was also a Protestant refugee from Ghent who had fled to England in the 1580s to escape persecution during the Dutch revolt against the Spanish crown. Hondius's religious background provides insights with which we may better understand how he intended the allegorical elements of this map to be read.

The centre of the illustration shows a knight being threatened from all sides. He battles a figure entitled 'Peccatum' or Sin, which is attacking the knight with a snake. The knight stands on another figure, 'Caro' or Lust. 'Diabolus' and 'Mors', the Devil and Death, prepare to attack, and 'Mundus', the world personified, offers the knight the poisoned chalice of the whore of Babylon which was often used as a symbol of the Catholic Church.

Right Jodocus Hondius, the 'Christian Knight' World Map.
See also 1450, 1500, 1507, 1525, 1544 (p. 102), 1599.

The purpose of the allegory appears to be bound up with key political and religious events of the period. In 1589, the leader of the French Protestants, Henri, King of Navarre, became Henri IV of France. Spanish armies invaded the country to support the Catholic majority. Contemporary medals and portraits of Henri show that the Christian knight on the map, originally based on an engraving, has been subtly recast to look like the French King. In 1593 Henri converted to Catholicism and at the head of a reunited country expelled the Spaniards. Despite his conversion, he remained a hero to many Protestants for defeating Catholic Spain.

The map's illustration, therefore, shows the plight of Henri IV, the 'Christian Knight' beset on all sides by trials and tribulations. Hondius was inspired by existing allegories of the struggle of the knights and angels who attempted to save the world from evil and temptation. By introducing Henri into the scene, Hondius has given the moral image a contemporary relevance and a political message, underlined by the depiction of the real world.

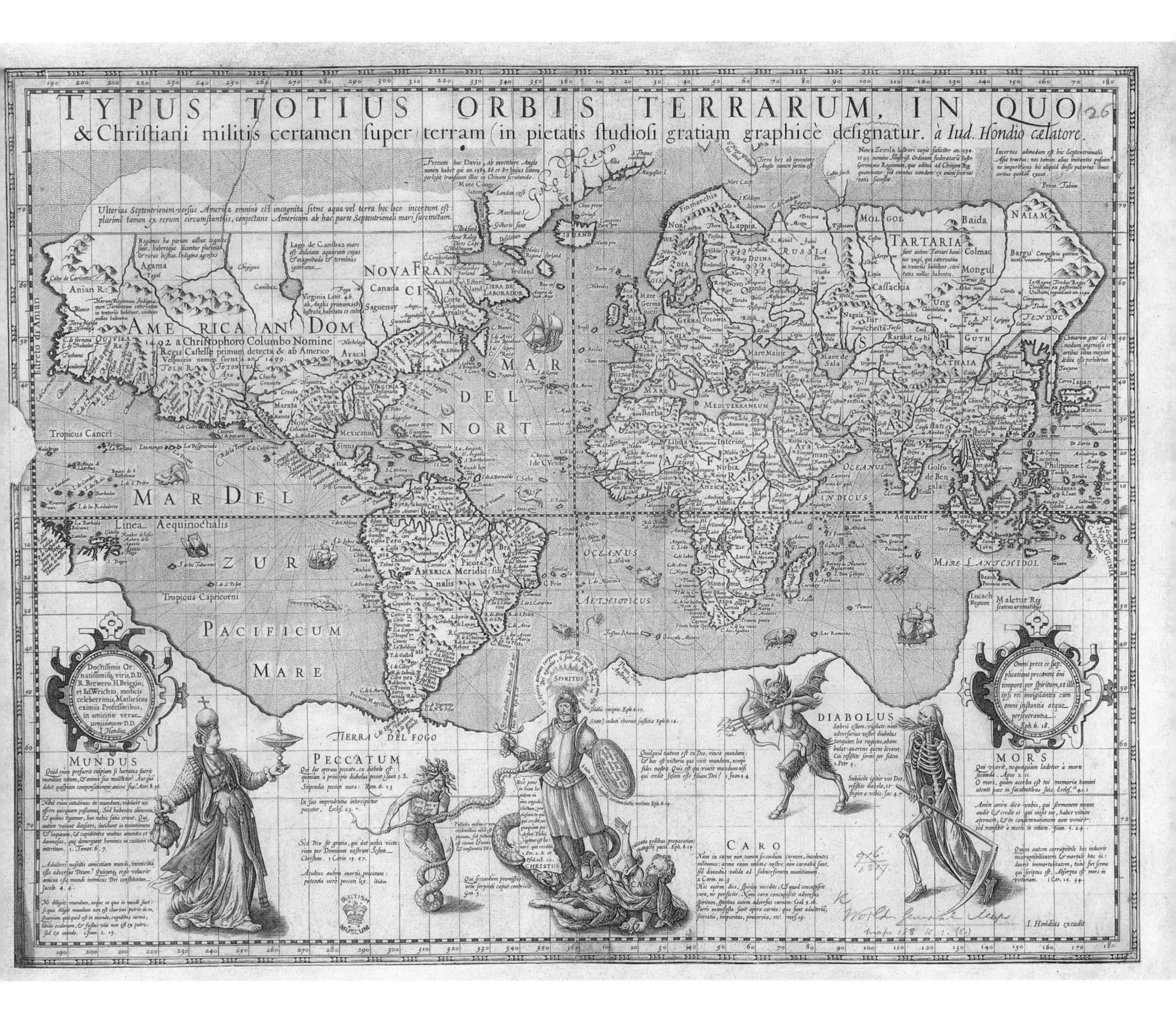
TYPUS TOTIUS ORBIS TERRARUM, IN QUO
& Christiani militis certamen super terram (in pietatis studiosi gratiam graphicè designatur. à Iud. Hondio cælatore.
AMERICA AN DOM
1492 a Christophoro Columbo Nomine Regis Castellæ primum detecta & ab Americo Vespucio nomen sortita an. 1499.
NOVA FRANCIA
MAR DEL NORT
MAR DEL ZUR
PACIFICUM MARE
Tropicus Cancri
Tropicus Capricorni
Linea Aequinoctialis
AMERICA Meridionalis
TIERRA DEL FOGO
MEDITERRANEUM
TARTARIA
RUSSIA
CATHAIA
OCEANUS INDICUS
OCEANUS AETHIOPICUS
MARE LANTCHIDOL
MUNDUS
PECCATUM
SPIRITUS
CHRISTUS
CARO
DIABOLUS
MORS
I. Hondius excudit

An early estate map shows a moated house, lands and parkland typical of the gentry of late sixteenth-centry England.

Ambition on the Map

THE EARLIEST SURVIVING English local maps drawn to a consistent scale date from the mid-1570s. Particularly suited to indicating boundaries with precision and durability, revealing a whole estate at a glance for estate management purposes, and often works of art in themselves, wealthy landowners were soon commissioning them in considerable quantities.

It is not known why Dudley Fortescue, the uncle by marriage and guardian of the young Robert Crane, commissioned this now rather worn map of his ward's lands near Sudbury on the Essex/Suffolk border. It may have been the final act of his guardianship. The map presents an evocative picture of a landscape in transition. There is a thumbnail sketch of Chilton Hall, a moated manor house, typical of Suffolk, with outhouses south of the moat. The orchard and the 'kytchyn corner' are named to its north. Some distance to the south is a sketch of the church, which still survives. It stands surrounded by fields, but names to the north suggest that within living memory it had been surrounded by Chilton village. There is 'Church House Meadow' – the church house was the social centre of medieval villages where feasts and fairs were held to raise money for the church. Interestingly, the lanes (coloured red) meet there and not at the church. The meadow to its left is 'Dye House Meadow'. The field to its right is 'Tower Meadow'. It seems that Chilton was one of the villages that was moved from the vicinity of the great house and given over to the sheep for which Suffolk was famous. To the right of the hall, the social aspirations of the Crane family are represented by that almost essential status symbol of large Elizabethan landlords: the park, distinctively outlined in green.

Young Robert Crane became a baronet, or hereditary knight, in 1605. His second wife was a great-granddaughter of Elizabeth I's minister, Lord Burghley, and was to be the great-grandmother of Sir Robert Walpole, traditionally considered to have been Britain's first prime minister. Crane commissioned a grand tomb to himself. But he left no male heirs. Chilton Hall decayed, the surviving east wing becoming a farm house. The park – if it ever existed (for the plan could represent proposals as well as the reality) – did not last long. But the map remains as a monument to the ambitions of the Crane family.

Right Survey of the manor of Chilton, Suffolk, 1597.
Above Detail showing the church and its surroundings.
See also 350, 1608, 1719, 1720.

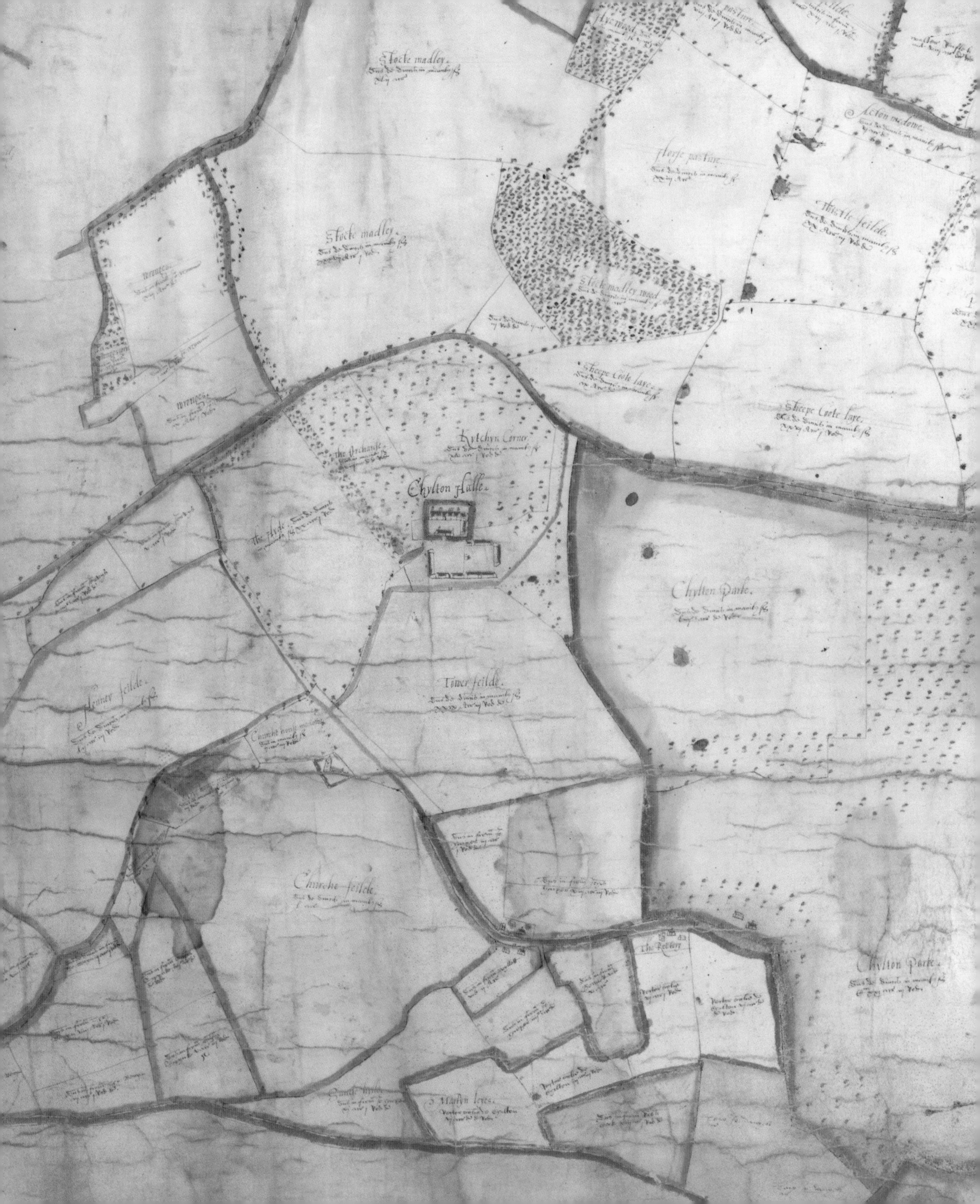

Stocke madley
Horse pasture
Acton medowe
Stocke madley
Thistle feilde
Stocke madley wood
Sheepe Cote laye
Sheepe Cote laye
Kytchyn Corner
the Orcharde
Chylton Halle
the Hyde
Chylton Parke
Tower feilde
Church feilde
Chylton Parke

An early Dutch printed world map with allegorical surrounds acts as a visual encyclopedia of the known world.

The World Personified

1599

RENAISSANCE MAPS OFTEN functioned as visual encyclopedias, integrating information about the ethnography, natural history and geography of the world. This map by the Netherlandish cartographer Petrus Plancius arranges much of this information around six allegorical illustrations. These divide the world into six parts: Europe, Asia and Africa in the Old World, and 'Peruana', 'Mexicana' and 'Magellanica' in the New World. Each part of the world is represented by an image containing a female personification of the continent and scenes from daily life in the region. Each personification sits atop an animal, with the exception of Europe, whose dignity is not compromised by being mounted on her bull, which lurks in a docile manner nearby. Europe is dressed in a rich gown and crown. She holds a sceptre and has a foot on a globe topped with a cross, which signals her place as a ruler of the world by the grace of God. The objects of civility around her feet may offer further justification of her right to rule: a book and an armillary sphere represent knowledge, with perhaps an emphasis on cosmographical learning; her arts are illustrated by the lute at her feet; a helmet, a gun and a hunting horn advertise her military prowess. Her right arm encircles a cornucopia containing vines – seen by the ancients as evidence of the highest level of civility. The personifications of Africa, Asia and America emphasize the commodities that these continents could provide for Europe. Asia, Mexicana and Peruana have a box of jewels, coins and foodstuffs at their feet. Magallanica holds spice plants – cloves and nutmeg – in each hand.

The map's illustrations reflect Plancius's professional concerns. An official cartographer to the Dutch East India Company, Plancius was concerned with finding routes to Asia, particularly to the Spice Islands. He produced maps, sailing instructions and navigational instruments for the Dutch fleets in the late sixeenth and early seventeenth centuries. He supported Dutch attempts to carve out a niche the Moluccan spice trade, which had been dominated by the Portuguese.

Plancius's map was widely disseminated through its inclusion in the 1599 and later editions of Jan Huyghen van Linschoten's travel account, *Navigatio ac itinerarium*, a copy of which was for many years supplied to all ships sailing from the Netherlands to India.

Right Petrus Plancius's world map
See also 1450, 1500, 1507, 1525, 1544 (p. 102), 1597 (p. 134).

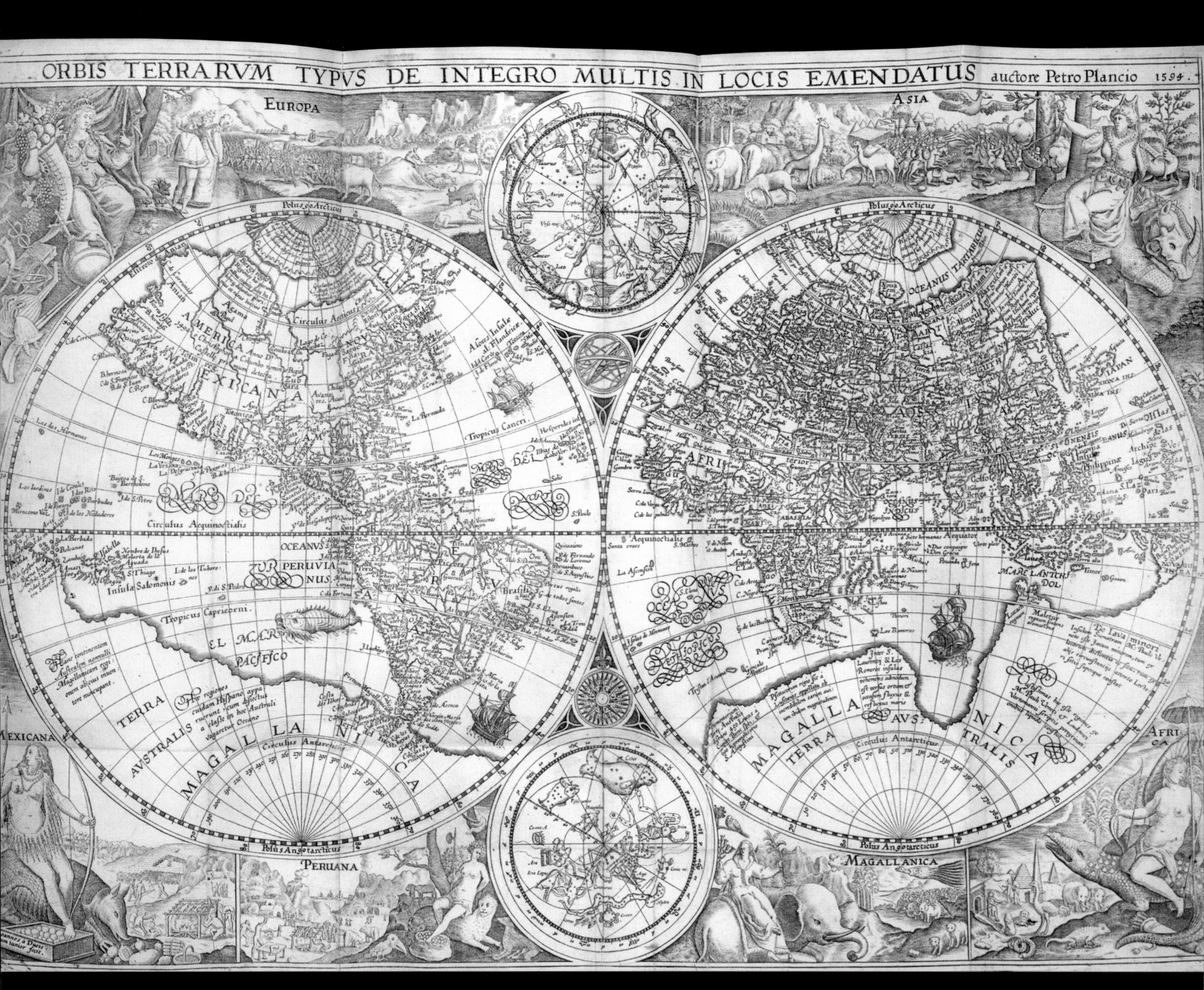
ORBIS TERRARVM TYPVS DE INTEGRO MULTIS IN LOCIS EMENDATUS auctore Petro Plancio 1594.
EUROPA
ASIA
Polus Arcticus
Polus Arcticus
Circulus Arcticus
AMERICA
MEXICANA
Tropicus Cancri
Circulus Aequinoctialis
OCEANUS
PERUVIANUS
Tropicus Capricorni
EL MAR PACIFICO
TERRA AUSTRALIS
MAGAL LA NICA
Circulus Antarcticus
Polus Antarcticus
OCEANUS TARTARICUS
AFRI
ASIA
Aequinoctialis
Aequator
ETHIOPICUS
MAGALLA NICA
TERRA AUSTRALIS
Circulus Antarcticus
Polus Antarcticus
MEXICANA
PERUANA
MAGALLANICA
AFRICA

A detailed map of one of France's border regions, one of the first to be created, compiled by a military engineer as an aid to defensive planning.

Defence and Mountains

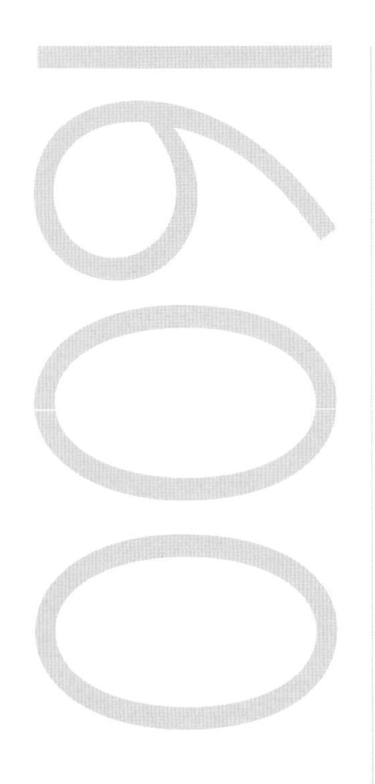

DURING THE REIGN of Henri IV of France (1589–1610), five or six military engineers (*ingénieurs du roi*) made detailed manuscript maps of all the great frontier provinces – Brittany, Picardie, Champagne, Dauphiné and Provence. These maps were primarily designed for organizing a defensive border, but they soon passed into printed form, and were widely published in atlases like the ones sold by Christophe Tassin; this was how French people came to have an idea of what their country looked like.

The engineer for Dauphiné at this time was Jean de Beins, who had come originally from Paris, and had fought in the wars of religion as a mounted carabinier. When the wars ended, about 1598, he was sent to Dauphiné to prepare the fortifications against the most probable enemy, the Duke of Savoy. Jean de Beins remained there for many years, and worked on a number of mountain-forts, several of which remain more or less as he left them. But he also found that in order to make sense of his task, he needed to draw maps of the province.

Dauphiné is very mountainous, and the existing maps were rather crude. Jean de Beins worked out a system of surveying the province by mapping one valley after another, and then putting the sketches together into a master-map. Indeed, he was so successful at this that the maps which he drew about 1600 were often still in use at the end of the century.

Right Jean de Beins, *Paisage de Grenoble*. **See also** 1597 (p. 134), 1684.

Like many cartographers of his time, he was also a considerable artist, who used the 'view' to give a convincing impression of the lie of the land (this was long before the general use of contour-lines). Our plate shows his view of the countryside around Grenoble, which lies on the River Drac, and is partly surrounded by mountains. Jean de Beins shows not only the fortified town (on which he worked a great deal), but also the appearance of the countryside, with its villages and vineyards.

This image was no doubt prepared for Henri IV, and was kept at Paris in the engineers' drawing-office. But at some period – perhaps at the time of the French Revolution – it was liberated from this office, and eventually found its way to London, and to the British Library. It is now part of a magnificent atlas of maps and plans prepared in the early seventeenth century by the engineers of Henri IV.

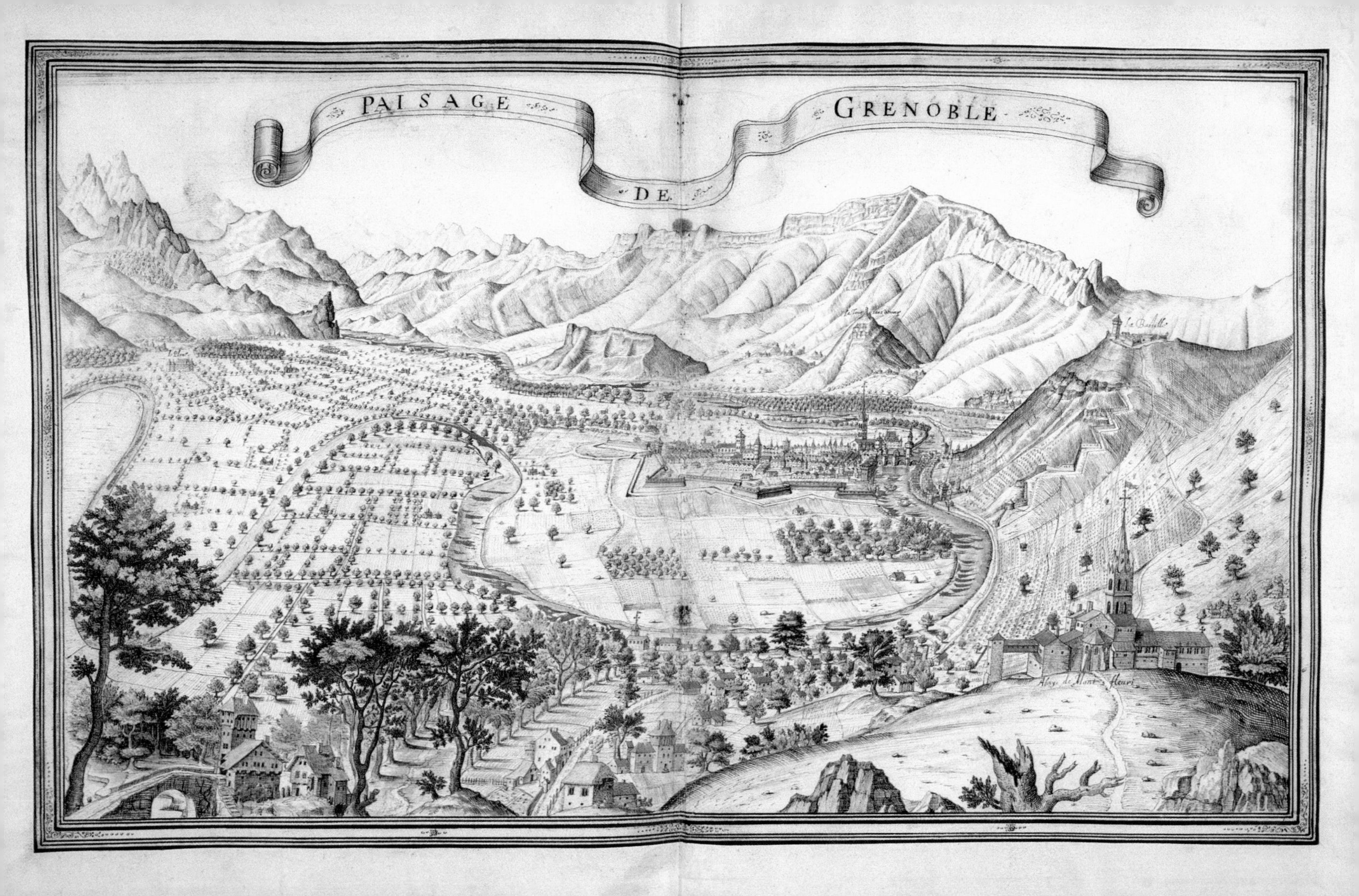
PAISAGE DE GRENOBLE

The first large map of the kingdom of Great Britain for over fifty years, prepared for presentation to the new King, James I.

A Royal Genealogical Map

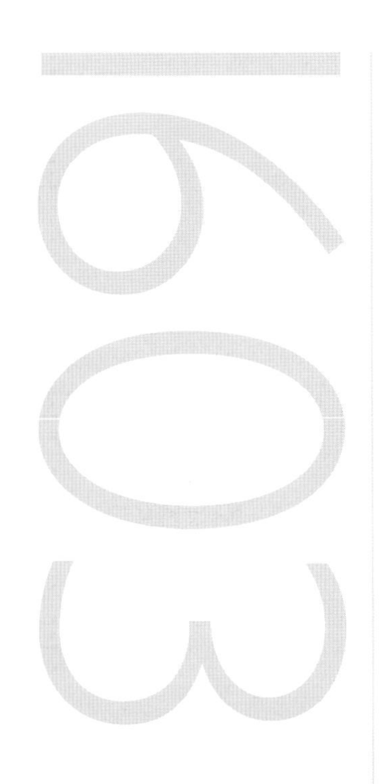

A DUTCH ÉMIGRÉ in England, Hans (John) Woutneel, originated this large map of the British Isles which is dated 1603 and is dedicated to the newly enthroned King James I who as James VI of Scotland succeeded Queen Elizabeth I in March of that year. It seems likely that Woutneel seized an early opportunity to present to its supreme head a complete picture of his recently united realm. As such, it was the first large map of the British Isles since Mercator's pioneering wall map of 1564 and is contemporary with a large general map by Jodocus Hondius also from around 1603.

Several features of Woutneel's map add gloss to his patriotic motive. To the right is a large genealogical tree with portraits of all the monarchs from William the Conqueror down to James I and his Queen, Anne of Denmark. Vignettes and text provide a proud reminder of the recent defeat of the Spanish Armada in 1588. The Spanish fleet is shown proceeding up the Channel, being scattered into the North Sea by fire ships off Calais, and being wrecked on the coasts and islands of Scotland and Ireland. In its title Woutneel's map also purports to show by a somewhat ungrammatical announcement: 'Al the Battails that have bin fought, Both in those kingdoms, & in france, Since the Norman Conquest Anno 1066, Expressed by the figures of Tents in their proper places, with a breif relation of the events, Sett Downe in the margent.' Text may have been printed separately for attachment to the map's margins, but in the unique surviving example of Woutneel's map no such text is present. The engraver of the map was another Dutch émigré William Kip, and both he and Woutneel are believed to have become British citizens.

Right Hans Woutneel, map of the British Isles. **Above** Detail showing James I and Anne of Denmark.

In compiling his map, Woutneel used a combination of sources. For England and Wales he seems to have relied principally on a rare two-sheet map of 1592 by his compatriot Jodocus Hondius who was probably the first to present a royal genealogical tree. The table of shires comes from an anonymous map of 1594 while the Armada story may have been taken from Speed's single-sheet map of *c*.1601. Scotland is based on Mercator and Ireland seems derived partly from Hondius's map of 1592 and partly from the map of Ireland printed only a year beforehand (1591) by yet another Dutch émigré, Pieter van den Keere.

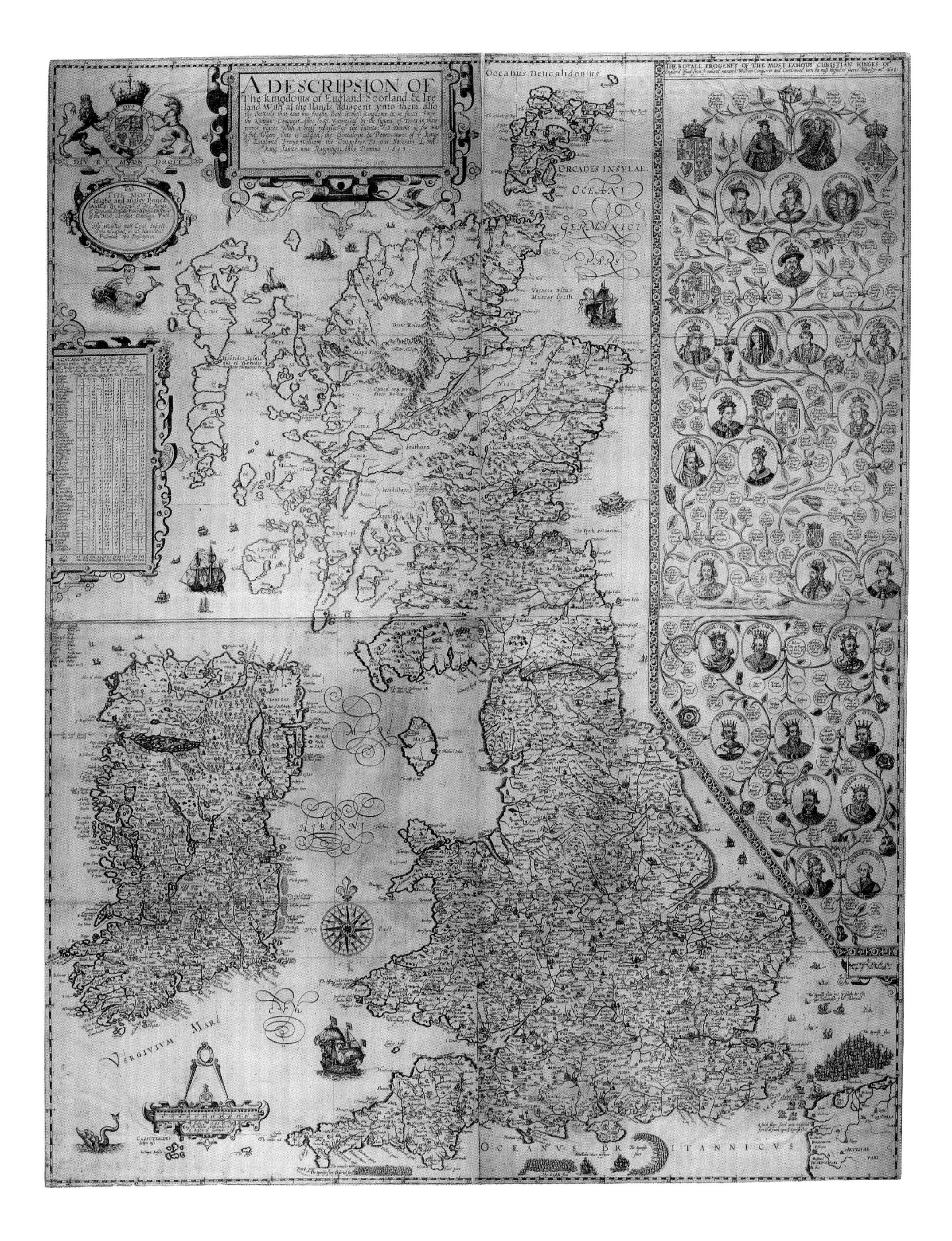
A DESCRIPSION OF
The kingdoms of England Scotland & Ireland With al the Ilands adiacent vnto them. also the Battails that haue bin fought. Both in these kingdoms & in france Since the Norman Conquest Anno 1066. Expressed by the figures of Tents in their proper places. With a breif relation of the euents. Set Downe in the margent. Where Vnto is added the Genealogie & Pourtratures of ye kings of England Frome William the Conqueror To our Soverain Lord King James now Raigning. Anno Domino 1603.
DIEU ET MON DROIT
THE ROYALL PROGENEY OF THE MOST FAMOUS CHRISTIAN KINGES OF England
Oceanus Deucalidonius
ORCADES INSULAE.
OCEANI GERMANICI
HIBERNI
MAN
VERGIVIUM MARE
OCEANVS BRITANNICVS

An exquisitely detailed local map drawn to aid legal deliberations in the case of a man charged with trespass in a royal rabbit warren.

A Jacobean Landscape

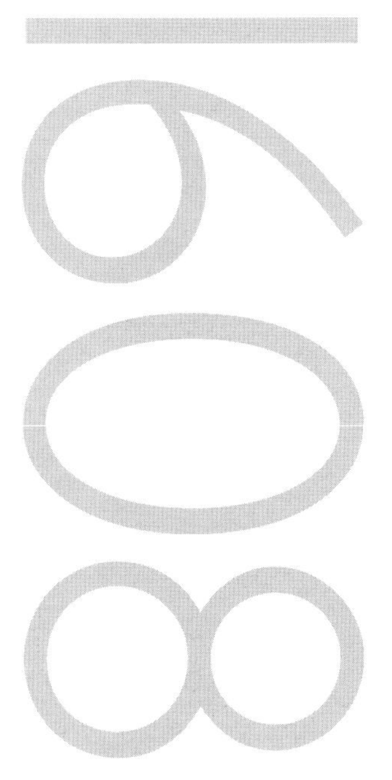

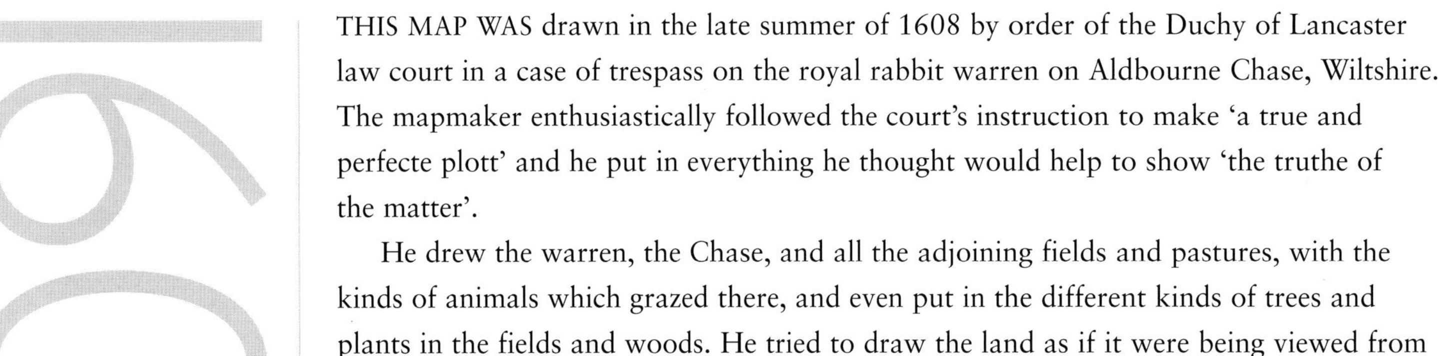

THIS MAP WAS drawn in the late summer of 1608 by order of the Duchy of Lancaster law court in a case of trespass on the royal rabbit warren on Aldbourne Chase, Wiltshire. The mapmaker enthusiastically followed the court's instruction to make 'a true and perfecte plott' and he put in everything he thought would help to show 'the truthe of the matter'.

He drew the warren, the Chase, and all the adjoining fields and pastures, with the kinds of animals which grazed there, and even put in the different kinds of trees and plants in the fields and woods. He tried to draw the land as if it were being viewed from the sky, an early attempt at balanced cartographic perspective.

In addition to all these drawings, the mapmaker also wrote notes on the map to convey information. He noted 'much uneven grounde like unto auncient furrowes'; evidence of the ridge and furrow effect created by the medieval ploughing of the communal open fields before enclosure. The newer fields are named and the artist was careful to show their boundaries, since these were a prominent feature of the dispute.

By the early seventeenth century it was common practice to include a scale on local maps. This mapmaker used a scale bar to indicate that he had made the map to scale, but did not state what unit he used. The common measure was the perch, suggesting a scale of 1 inch to 16 perches (1:3168). Cardinal points placed in the margins tell us that this map is oriented with south at the top; the modern convention of orientation to the north had not yet arrived. This was an important consideration when the map came to be used, as the finished map was shown to witnesses in the case, who knew the area, and were asked to comment on whether the map was an accurate depiction of it.

While this map had a practical purpose, the mapmaker took care to embellish it. The scale bar and title are enclosed in decorative cartouches. A particularly handsome border of fruit and flowers includes acorns, pinks, sweet pea pods and grapes, reminiscent of motifs in contemporary Jacobean embroidery.

Right Detail showing the very careful delineation of crops and animals as clear as the all-important field boundaries.
Following pages Map of Aldbourne Chase, Wiltshire.
See also 350, 1597 (p. 136), 1719, 1720.

is Copies is Called Hastes hoock and is claymed to
bee waldrons inheritannce Butt the kinges Tennaunts
haue the pasture therof and the king hathe y Connies
FEELDE

Southe
A true plott of the Lye plaine Ewens hill and the Stockes beinge the grounds in Controversie wth the landes neare adioyninge
This is called Southwood lodge and standeth in Fowles Close within the mannor of
This is called Stubbes Coppice and is clayméd to bee waldrons inheritannce Butt the kinges Tennantes haue the pasture therof and the kinge hathe the Connies
In this and other partes of Ewens hill the ground is uneven like unto auncient furrowes
Est
S O V
THE
COM
MON
E

105
West
FEELDE
ABLE
North

Maps, Mankind & Morality

The earliest maps in this book (1500 BC) have been interpreted as appeals to the divine for protection. Another early map (600 BC) is an attempt to place its creator and his society in relationship to the known environment and to the wider world. From the very beginning of human time, one of the principal functions of maps has been to place man in his spiritual as well as his physical and geographical context.

In many cases this has involved mapping the universe as well as the world. This is as true of Indian and Aztec societies as of European society. Because of the similar need to show one's place in the scheme of things, these maps have a broadly similar structure despite vast differences in the cultures that have produced them. Man's holiest place – such as Jerusalem for Christians (see 1265, 1300), or for Aztecs, Tenochtitlan (1524), China for the peoples of eastern Asia (1800) or, for Jains, Mount Meru (see 1750 p. 206) – is at the centre, with zones radiating outwards from that point taking the viewer from the known and familiar via unfamiliar regions inhabited by mankind and the marginal (usually containing strange peoples) to the totally unknown and divine. At this point, differences are usually to be found, depending on the philosophical beliefs held by the mapmaker. The ultimate purpose of the Jain world map is to demonstrate the reality of reincarnation, that of the Christian *mappaemundi* to remind the viewer of the ultimate reality of the Last Judgement and/or the possibility of redemption for sins.

This use of mapping is not unlike the modern satirical maps showing the world as viewed by New Yorkers or Londoners, where geographical features are ranked according to their supposed importance to the typical New Yorker or Londoner regardless of physical size or importance according to accepted value systems. The first map to be printed in North America (1677) suggested that God was punishing early colonists for their moral and theological shortcomings. In the early fourteenth century an Italian monk, Opicinus de Canistris, moulded the shapes of countries as shown on portolan charts (see 1325, 1465) into profiles resembling people in order to make moral points about his contemporaries, in a tradition which led, via patriotic curiosities such as the Belgic Lion (a map of the Netherlands manipulated into the form of a lion) of the sixteenth and seventeenth centuries, and George III as Great Britain (see 1762), to the cartoon maps of Frederick Rose (1877); though occasionally, as in the maps in the Gallery of Maps in the Vatican (1580), the map depicted in Vermeer's paintings (1658) or the cartoon relating to the partition of Poland (1772), the map itself can act as a moral symbol within a building, a painting or a cartoon. A subsidiary tradition has been the map of Utopia, where the mapping of the impossible acts as a symbolic representation of the impossibility for man to live up to his own moral codes. It is represented here by two examples, one from the sixteenth century created by one of the giants of the history of mapmaking, Abraham Ortelius (1596), and the other by the Japanese artist Satomi Matoba (1998). In a related vein didactic spiritual maps have warned good Catholics of the dangers of Jansenism (1660 p.162); or through maps, modelled on route maps (see 1675, 1790) illustrating John Bunyan's moral fable *The Pilgrim's Progress*, of the dangers that beset mankind; and, through mock late nineteenth-century Admiralty Charts (see 1903) of the dangers to North Sea fishermen of godlessness. More light-heartedly, maps have warned innocent youth of the dangers to be met with in the pursuit of love and marriage.

The superficially mundane and utilitarian Peters Projection World Map also serves a primarily moral and admonitory purpose. In itself, this projection, known for over a century before being popularized by the German cartographer Arno Peters, enables the viewer to see the true sizes of continents relative to one another, though at the price of losing their true shapes. As a result of clever marketing leading to its adoption by UNESCO and other international charitable organizations, it has become a moral icon, acclaimed as a 'new' and 'true' map because it shows the true size of the Third World as compared to the developed world, whose size is exaggerated on the more familiar Mercator projection. Named after one of its earliest exponents, the sixteenth-century geographer Gerard Mercator, this has consequently and most unfairly (as it was intended to serve a quite different purpose) been demonized as Euro-centred and colonialist.

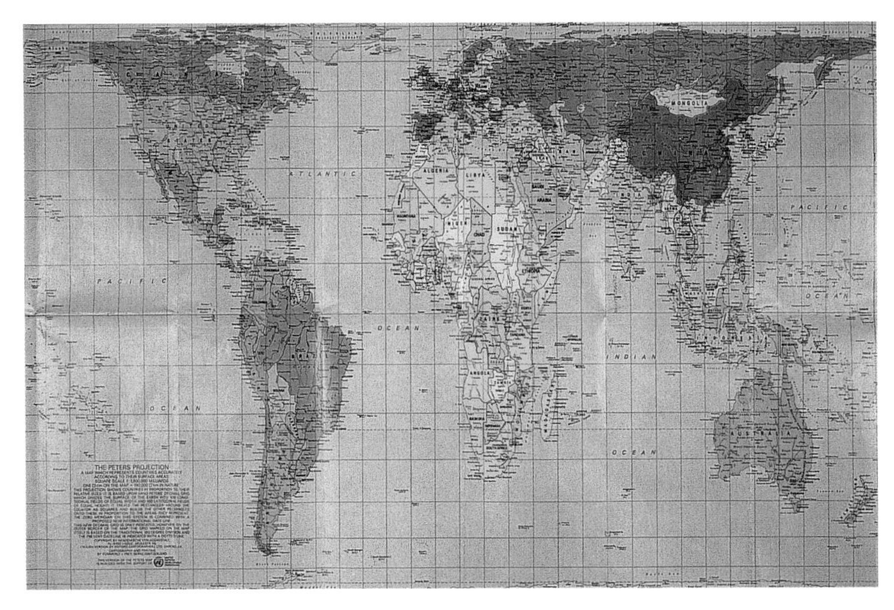
A Peters Projection World Map which shows the true sizes of continents relative to one another.

Rather than stirring the world's conscience, maps have more often been used to evoke patriotic sentiments. Often there is nothing controversial about it. The maps of countries such as France or Cyprus are often to be found on their coins and stamps. In 1603 Hans Woutneel used a map to associate James VI of Scotland, who had just acceded to the thrones of England, Wales and Ireland, with a new, British, patriotism that embraced all his dominions. More controversially, following the execution of Charles I in 1649, the map of England and Wales without Scotland replaced the depiction of the monarch on the Great Seal which was and still is appended to all major documents of state. In the 1831 (p. 262) map of Poland we see how a map can be used as an eloquent patriotic appeal against the injustices of the international political system. In 1914 (p. 312) an official Indian government official poster somewhat cynically used a map of India as a means of encouraging military recruitment.

No map can be totally value-free because of the inevitably subjective process of selection that goes into its compilation. Over the millennia, however, many mapmakers have sought to map man's soul through their depiction of his physical and spiritual environment.

An influential early European map of India drawn by two Englishmen resident at the court of the Emperor Jehangir Khan.

The Mughal Empire in India

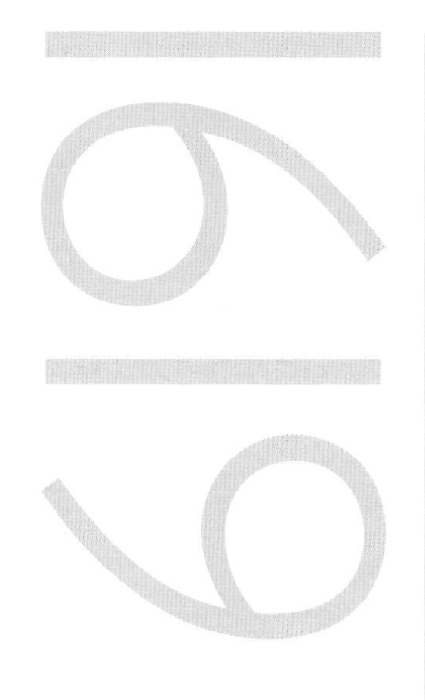

Right William Baffin, Thomas Roe, *A Description of East India.*
See also 1782.

THE PORTUGUESE WERE the first European nation to build ships strong enough to sail around Africa and into the Indian Ocean, and they held a monopoly on the rich spice trade from their colonies in India and South-East Asia for over a hundred years. However in 1596 the Dutchman Jan Huyghen van Linschoten published an account of his seven years in Goa as assistant to the Archbishop, in which he exposed the decadence that had corrupted the rich Portuguese Empire, and so other European nations ventured to trespass into these lands.

When a Portuguese ship was captured and towed to Dartmouth harbour, English people saw the rich plunder of spices and silks, carpets and calicoes, jewels and drugs, hides and ivory. A group of merchants approached Queen Elizabeth for a charter and the English East India Company was founded on 31 December 1600. After facing opposition from the Portuguese, they requested King James I to send an ambassador to the Mughal court, to obtain official decrees allowing trading posts to be established.

Thomas Roe was chosen to represent England at Agra, the seat of the Emperor Jehangir, and he spent four years there, learning about the customs of this rich oriental court and gaining information about the country. On the ship in which he returned to England, the first mate was William Baffin (later to map large parts of Canada), and between them they drew the first detailed map of the Mughal Empire. The various provinces are placed and named and major rivers are shown. The road from Agra to the former capital of Lahore had been planted with trees to provide shade in the hot summers, and the Ganges was shown as the main route to the rich province of Bengal in the east. This holy river was reputed by the Hindus to pour forth from the sacred mountains through the mouth of a cow, and so Roe depicted it on his map. Cartographers of later versions found this hard to believe so they changed the cow's head to a small lake.

This map was influential among European cartographers and was copied many times with very little change despite the growing number of European travellers to India.

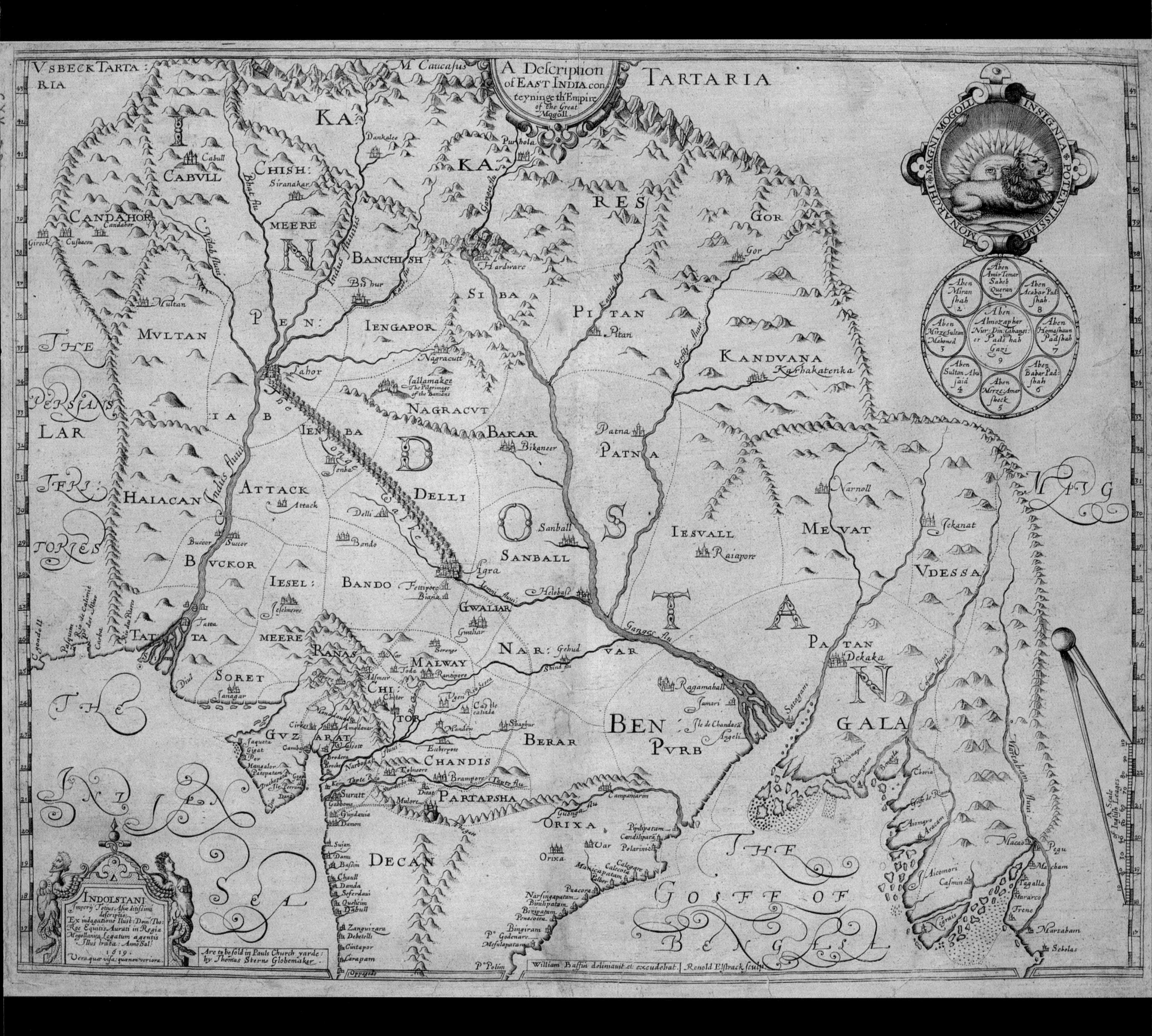
A Description of EAST INDIA conteyninge th'Empire of the Great Mogoll
TARTARIA
VSBECK TARTARIA
M. Caucasus
MONARCHI MAGNI MOGOLL INSIGNIA POTENTISSIMI
Aben Almozaphar Nur Din Gehangir Padshah Gazi
Aben Amir Temar Saheb Queran
Aben Miran shah
Aben Acabor Padshah
Aben Mirze Sultan Mahomed
Aben Homashaun Padshah
Aben Sultan Abusaid
Aben Babar Padshah
Aben Mirze Amar sheck
CABVLL
CANDAHOR
MVLTAN
HAIACAN
ATTACK
BVCKOR
IESEL:
MEERE
SORET
GVZARAT
CHISH: MEERE
BANCHISH
PEN: IAB
IENGAPOR
NAGRACVT
IENBA
DELLI
SANBALL
BANDO
GWALIAR
MALWAY
CHITOR
CHANDIS
PARTAPSHA
BERAR
DECAN
ORIXA
BEN PVRB
PATNA
PITAN
SIBA
KAKARES
GOR
KANDVANA
IESVALL
MEVAT
VDESSA
PATAN
BENGALA
MAVG
INDOSTAN
Lahor
Agra
Indus fluvius
Ganges flu
THE PERSIANS
LAR
IERI TORIES
THE INDIAN SEA
THE GOLFE OF BENGALA
INDOLSTANI Imperii Totius Asiae ditissimi descriptio: Ex indagatione Illust: Dom: Tho: Roe Equitis Aurati in Regia Mogollanica Legatum agentis Illustrata: Anno Sal: 1619.
Vera quae visa; quae non veriora.
Are to be sold in Pauls Church yarde: by Thomas Sterne Globemaker
William Baffin delineavit et excudebat.
Renold Elstrack sculp.
A Scale of English Leagues

A rare example of a professional chart created in pursuit of Anglo-Dutch interests in the Chinese silk trade.

Anglo-Dutch Cartographic Collaboration

Right Chart of Celebes and Java by Gabriel Tatton, 1621.

THIS MEDIUM- TO LARGE-SCALE collection of charts drawn by Gabriel Tatton (d. 1621) is the sole surviving example of original marine surveying in the East Indies by a professional English chartmaker in the second decade of the seventeenth century.

The significance of Tatton's charts rests on their immediacy as a record on a particular English East India Company voyage, 1619–21. Although the English and Dutch were flexing their maritime muscles against each other in the East Indies, this was the moment when they joined together briefly, as the euphemistically termed 'Fleet of Defence', against the Spaniards and the Portuguese in the Philippines and Manilas to seize the Chinese coastal silk trade. Tatton drew these highly detailed and beautiful charts on the voyage of the *Elizabeth*, one of the ships in the 'Fleet' before he died in Hirado (i.e. Firando), Japan on 12 September 1621. They show the course of the *Elizabeth* and, probably, those of other ships in the Anglo-Dutch fleet.

The rapprochement between the Dutch and English was short lived and the resumption of hostilities resulted in the so-called Massacre of Amboina in 1623, which confirmed the wisdom of the English in retreating from east of Celebes to Sumatra and Java. There they continued to trade successfully for spices until the 1670s.

In cartographical terms the two countries were more evenly matched and often worked together. Born of their joint cause against the Spaniards, the two countries' sailors and cartographers had a long tradition of crewing on each others' ships and, at this period, of exchanging and copying charts. Tatton had worked in Holland at the turn of the century and had honed his professional skills there. Here we see him at work in the East Indies in 1619–21 at a critical juncture for both English and Dutch East India companies and for their respective cartographical organizations.

Hessel Gerritsz, hydrographer to the Dutch East India Company between 1617 and 1632 had prepared a set of experimental charts from Bantam to Japan in 1618 and then a more definitive official set in 1619. Neither survives but this set of charts by Tatton in 1620–21 may be a further revision of that work based on actual observation. If so, the charts are of importance to the development of mapping in England and the Netherlands, and reveal what the charts of 1618 and 1619 by Gerritsz may have looked like.

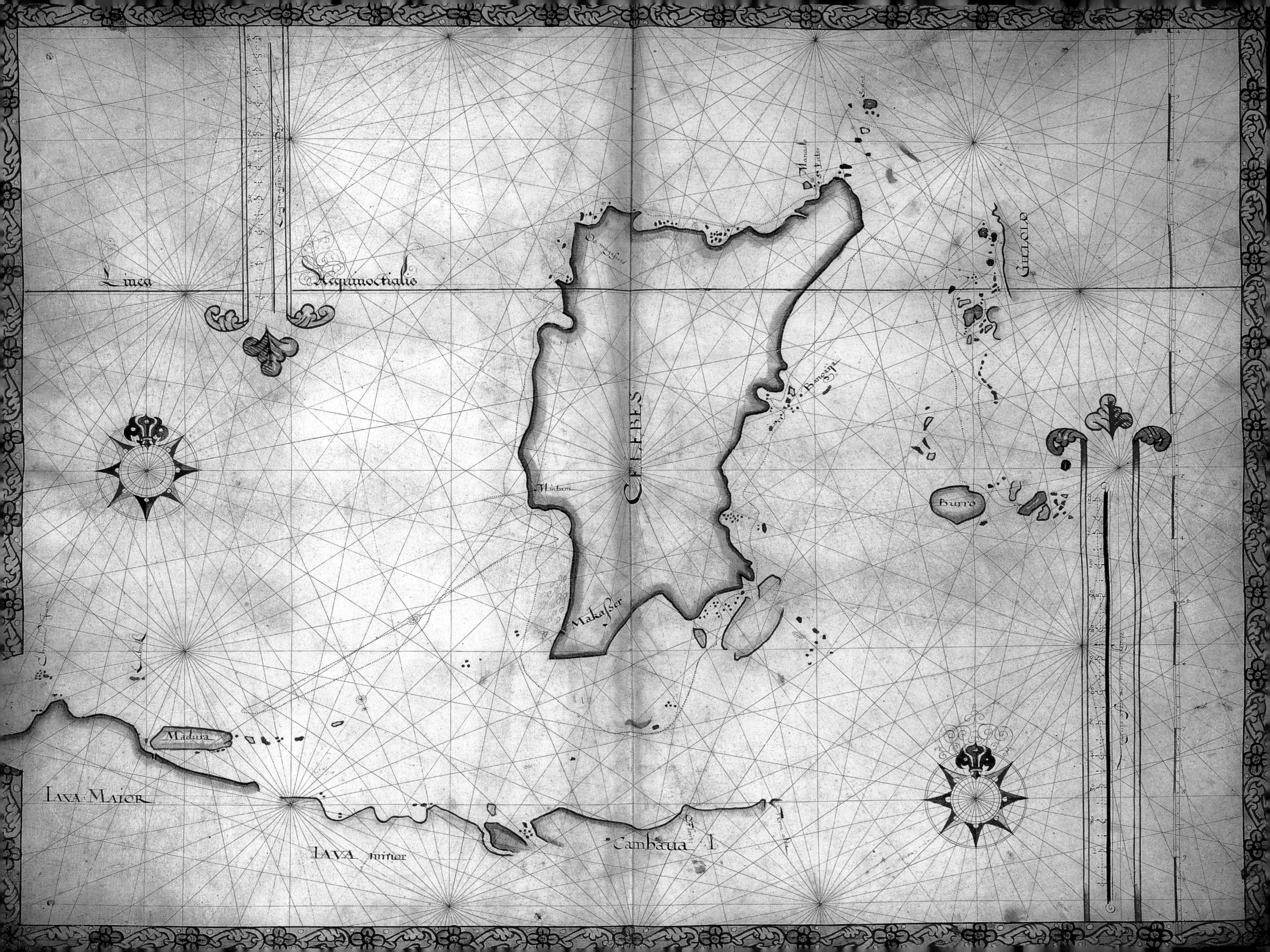

Linea Aequinoctialis
CELEBES
Makasser
Burro
Madura
IAVA MAIOR
IAVA minor
Cambaua I

A cornucopia of figures and symbols used in this Dutch title-page anticipate the history and geography of the world contained in the atlas maps that follow.

A Title-page Display of the Whole World

1623

THE DUTCH MAPMAKER and publisher Jodocus Hondius acquired the rights to the atlas of the great Gerard Mercator a few years after Mercator's death in 1594. Hondius expanded the contents and devised a new main title-page to the work, retaining the central figure of Atlas with additional representations of the continents placed around a grand architectural structure. The Hondius-Mercator title-page was used for editions from 1606 until the mid-1630s, the later editions incorporating the name of Jodocus's son Henricus.

The central figure shows the mythological Atlas, wise King of Mauretania who is shown measuring a globe of the earth. Mercator chose him to provide the appellation for his original book of maps, rather than the other 'Atlas', who for his misdeeds was punished by having to support the world on his shoulders.

Starting clockwise to the left of Atlas, 'Europa' alone is crowned and she holds a sceptre as further evidence of her ruling authority. Her floral cornucopia suggests peace and plenty. At top left is 'Mexicana', a generalized warrior with bow and arrow, accompanied by an animal which may be a tapir or an armadillo. 'Africa' at top right is black and is shaded from the burning sun by a primitive umbrella. She also holds arrows and is seated on a large crocodile. Below her is 'Asia', dressed in flowing oriental robes. In her right hand she holds a censer of myrrh or similar spices and in her left a sprig of balsam which (with the cat at her foot) was associated with the lands of Egypt. 'Magalanica' represents the as yet unknown continent south of the tip of South America. In spite of the coldness her female figure is unclothed, but holds a flaming torch signifying the large fires along the Patagonian coast seen by Magellan and which gave rise to the name Tierra del Fuego. By her is a misplaced diminutive elephant. The sixth figure is headed 'P[e]ruana' and shows a native of the Spanish Viceroyalty of Peru. She has an Inca-style headdress and holds a symbolical battle axe, perhaps of gold as it is often coloured as such. Behind her is a puma or jaguar.

Readers opening up the atlas would engage with these figures and symbols which anticipate, through the maps and text that follow, the history and geography of the world.

Right Title-page of the Hondius-Mercator atlas. **Above** Title-page of Abraham Ortelius's atlas of 1570.

MEXICANA
AFRICA.
GERARDI MERCATORIS
EVROPA
ASIA.
PRVANA.
MAGALA NICA.
ATLAS
SIVE
COSMOGRAPHICÆ
MEDITATIONES
DE
FABRICA MVNDI ET
FABRICATI FIGVRA.
Denuò auctus
DESCRIBO, ORNO, VITUPERO ET OPTO FIDELITER
EXCUSUM SUB CANE VIGILANTI
EDITIO QVINTA
Sumptibus & typis æneis Henrici Hondij, Amsterodami An. D. 1623

A scholarly map using linguistic evidence and the Book of Genesis to plot the dispersion of peoples all over the globe.

Illustrating the Phoenician Link Between Noah and Us

IN THE EARLY MODERN PERIOD many scholars and cartographers gave serious attention to sacred geography. Normally this would have meant the description of the Holy Land, but often, following a practice which goes back to late antiquity, sacred geography explained the dispersion of peoples all over the globe.

This is the main focus of the *Geographia Sacra* (1646), a monument of erudition by Samuel Bochart (1599–1667). As a young scholar Bochart wandered between Protestant capitals of learning, where he studied theology and mastered the oriental languages. He then settled in Caen, Normandy, as a pastor to the Huguenot community, and led there a life of preaching and scholarship. Indeed, the *Geographia Sacra* emerged directly from Bochart's sermons on Genesis and was based on his conviction that the Bible was the only credible source about human prehistory, and that only a careful historical analysis could recover this knowledge. This he did with great patience, and when the *Geographia Sacra* was finally published, having overcome the hurdles mounted by the Catholic authorities, the provincial pastor reached the summit of the Republic of Letters.

The first part of Bochart's work, 'Phaleg', is a close exposition of Genesis 10, the chapter describing the progeny of Noah. Working with complex etymology – the study of the origin of words, which in the period was valued as today we value DNA testing – Bochart mapped out the geographical spread of Noah's descendants, from Spain in the west to Persia in the east. Bochart also proved that pagan myths were in fact a corrupt version of biblical truths. In 'Chanaan', the second part, Bochart turned to the only other source that in his view could complement the Bible in the recovering of ancient geography – Phoenician navigation. Having been expelled from Canaan by Joshua, Phoenicians established colonies all over the Mediterranean and beyond. For Bochart they therefore formed a crucial link between biblical and profane geography. To trace this process Bochart had to work backwards from classical geography, using philological hints in related languages, which could point to Phoenician and hence Hebrew origins. His linguistic researches were summarized on this map of Phoenician navigation 'from Thule to Taprobana', one of the few early modern maps to make consistent use of Hebrew lettering.

Right Bochart's map of Phoenician navigation from Thule to Taprobana. **Above** Samuel Bochart.

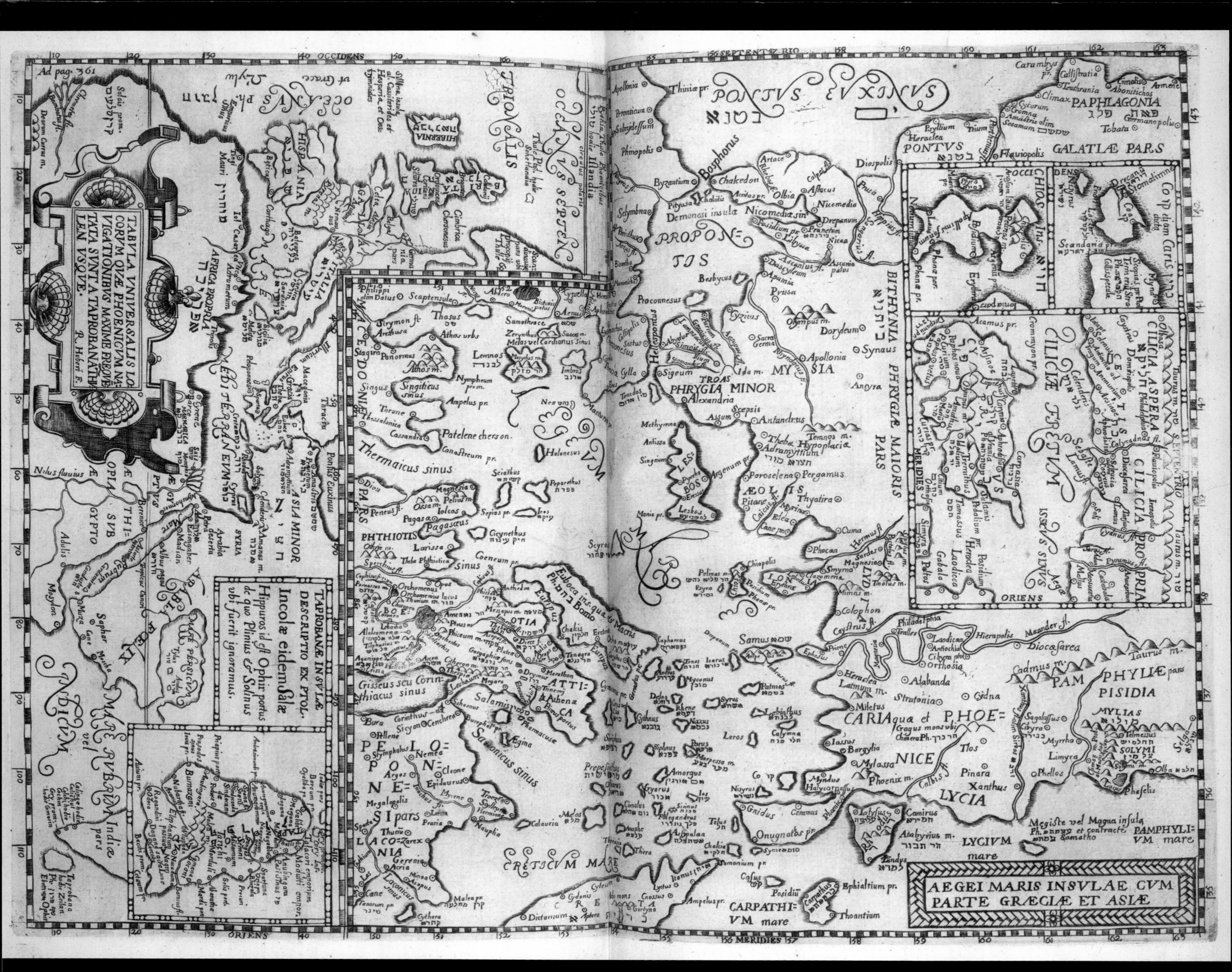

AEGEI MARIS INSVLAE CVM PARTE GRÆCIÆ ET ASIÆ
PONTVS EVXINVS
PAPHLAGONIA
GALATIÆ PARS
PONTVS
BITHYNIA PHRYGIÆ MAIORIS PARS
PROPONTIS
Bosphorus
MYSIA
TROAS
PHRYGIA MINOR
LESBOS
ÆOLIS
IONIA
LYDIA
CARIA qua et PHOENICE
LYCIA
PAMPHYLIÆ pars
PISIDIA
CILICIA ASPERA
CILICIA PROPRIA
CILICIÆ FRETVM
ISSICVS SINVS
CHIOS Ins.
MACEDONIA
Thermaicus sinus
PHTHIOTIS
Euboea ins. qua et Macris
BOEOTIA
ATTICA
Saronicus sinus
PELOPONNESI pars
LACONIA
CRETICVM MARE
CARPATHIVM mare
LYCIVM mare
PAMPHYLIVM mare
TABVLA VNIVERSALIS LOCORVM QVAE PHOENICVM NAVIGATIONIBVS MAXIME FREQVENTATA SVNT A TAPROBANA THVLEN VSQVE. R. Huberi F.
OCEANVS SEPTENTRIONALIS
HISPANIA
ITALIA
MEDITERRANEVM
AFRICA PROPRIA
ASIA MINOR
SYRIA
Arabia deserta
MARE PERSICVM
ARABIA FOELIX
MARE RVBRVM vel INDICVM
ÆTHIOPIA SVB ÆGYPTO
TAPROBANÆ INSVLÆ DESCRIPTIO EX PTOL.
Incolæ eidem Salæ
Hippuros id est Ophir portus de quo Plinius & Solinus ubi fuerit ignoramus.
Indiæ pars
Taprobana hodie Zeilan
Ad pag. 361

A map from the survey of Ireland undertaken to tighten the administrative grip on the country following Cromwell's brutal conquest.

Mapping the Spoils of War

OLIVER CROMWELL'S BRUTAL conquest of Ireland after 1649 was followed by confiscation of the lands of Catholic Irish 'rebels' and their reallocation to loyal Protestants. A large part of the island was then surveyed.

The relatively detailed and uniform mapping between 1655 and 1659 was the achievement of William Petty, a multi-talented economist and organizer of genius. Remarkable for the extent of its coverage and its clear, logical organization of information for the purposes of government, it was far ahead of anything that had hitherto been attempted in Europe. It demonstrated that colonization, like warfare, was often a laboratory for mapmaking. Petty's surveys were carried out by mainly English surveyors but drawn, often in vigorous style, by local draughtsmen. They went down to parish level, though the original parish maps were destroyed in 1922 during the Irish Civil War. What are left are the 'Barony' maps, compiled from the parish maps and covering groups of parishes.

This map comes from the set retained by Petty and owned by his descendants until recently. The clarity of organization is reflected in the way in which the map folds out so that text and map can be consulted simultaneously. The information is similar to that found on English estate maps. The text describes the landmarks of each parish, lists the owners and gives the plot names and the acreages of the profitable and unprofitable land, notably bogs. The map, at a scale of two inches to the mile (5 centimetres to 1.6 kilometres), distinguishes the individual landholdings and the parish boundaries, notes arable land and contains thumbnail sketches of important buildings and a couple of views.

The survey earned its name, the 'Down Survey', from being put 'down' on paper. It became cursed by Irish Catholics because it was followed by centuries of British Protestant domination. But Petty would have had reason to object that the condemnation was unjust. The survey shows that loyal Catholics, like Lord Dempsey, and 'two sisters of William Soare, the one married to an English Protestant, the other to an Irish Papist' continued to be the majority landowners in several of the parishes shown on this map and by so doing confirmed their continuing right to the land so long as they remained loyal to their British overlord.

Right *The greatest part contiguous of the Barony of Philipstown in the Kings County [Leinster] by Patricke Ragett. E. Lucas deliniavit.* From the 'Down' Survey. **Above** William Petty. **See also** 1827.

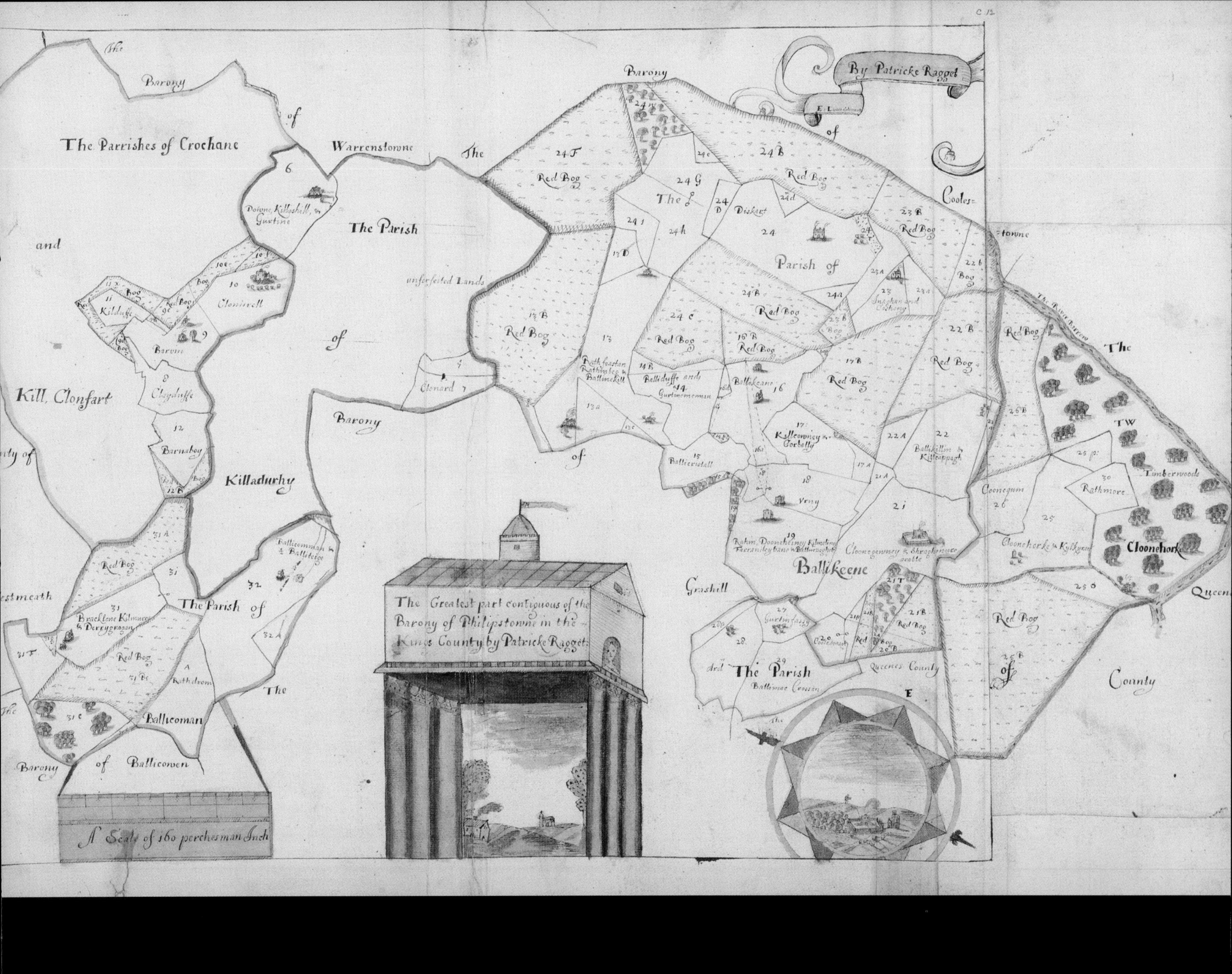

By Patricke Ragget
The Parrishes of Crochane
and
Kill, Clonfart
The Barony of Warrenstowne
The Parish of Killadurhy
unforfeited Lands
Barony
The Parish of
The Barony of Coolestowne
Red Bog
Parish of
Ballikeene
The Parish
Queenes County
The TW
Timberwoods
Cloonehorke
Rathmore
Grashill
Balliconman
Barony of Balliconen
The Greatest part contiguous of the Barony of Philipstowne in the Kings County by Patricke Ragget.
A Scale of 160 perches in an Inch

A Dutch painting shows how wall maps were originally displayed but also suggests their allegorical role.

Profitable and Useful, Allegorical and Ornamental

JOHANNES VERMEER FREQUENTLY included wall maps in his paintings. Since it is to be seen in two other paintings, he probably owned this map of the provinces of Holland and West Friesland by Floris van Berkenrode and his sons, published in 1621 by Willem Blaeu, the most important map publisher of the 'Golden Age' of Dutch mapmaking.

At one level, map and painting seem severely realistic. The Dutch did decorate their living spaces with maps. Girls did, doubtless, chat to officers. Dutch map publishers needed to make their maps as pretty as well as informative as they could so that they would sell: the original of this map contains an extensive statement emphasizing Blaeu's sole right to produce and sell the map. The map itself had been commissioned by the States of Holland for practical, administrative purposes since its was largely based on a detailed survey undertaken between 1611 and 1615 of the Dutch polders, essential for protection against flooding. A close look adds a further touch of reality. The map has been discoloured by over thirty years of exposure to light. The original green of the land has faded to blue, the map is peeling and the paper is becoming brittle with oxidization (with the result that only one complete original, now in awful condition, survives).

There is another, allegorical level however. The map is accompanied by a patriotic historical text and the handsome title cartouche is surmounted by the allegorical figure of Holland holding a figure of Victory symbolizing the recent victory of the Dutch Republic – of which Holland was the richest and most important province – in its fight for independence from Spain. Flanking the title beneath are the figures of a soldier and Neptune – representing the sea and trade – and at the bottom are symbols of music and science representing the province's cultural attainments. The painting also seems to have a patriotic message. Vermeer's hometown of Delft is shown on the map and, as well as depicting a private moral dilemma, as innocent girl talks to a soldier who has presumably had a string of female conquests, the painting may also refer to the dilemma facing the Dutch as peace returned after more than ninety years of almost continual war, mainly against the Spanish, but more recently also against the English.

Maps served many roles in seventeenth-century Holland.

Right Johannes Vermeer, *Soldier and Laughing Girl*. **Above** The title cartouche of van Berkenrode's map.

A celebrated satirical map created as part of a propaganda campaign to mock the views of a discredited Bishop.

A Map of Heresy

Right Nicolas Cochin, chief propaganda artist to Cardinal Mazarin, *The Country of Jansenia*.
See also 1684.

IN 1653, POPE Innocent X issued a bull condemning five propositions published by the deceased Flemish Bishop Cornelius Jansen (1585–1638). The Vatican's verdict set off a propaganda campaign which sought to win the support of the French cultural elite against the Bishop's partisans, the so-called 'Jansenists'. As part of this process, visual media became increasingly important. In 1660 an imaginary travel account was published. The author of the *Relation du Pays de Jansénie* was an aged capuchin friar who upon entering the order had taken the spiritual name of Zacharie de Lisieux (1596–1661). During his career as a court preacher, he had gained a reputation as a political satirist.

The folded engraving that accompanies the book displays characteristics common to both pictures and maps. The pictorial frame shows wooden shoes, wicker baskets and ice skates, and thus suggests Flanders as the hot spot of the Jansenist reform movement. Following the principle of guilt by association, the land of 'Jansenia' is bordered by three provinces: 'Libertinie' (Libertarianism) 'Calvinie' (Calvinism) and 'Désesperie' (Desperation). The influx of heretic thought from these provinces leads to a constant rise in level of the 'River of Heresy' which winds its way – analogous to the theologians' argument – through the country, until nothing can oppose it. The river metaphor points to a dreadful end in the near future and thus is meant to be an unmistakable warning.

In his *Notes manuscrites sur les peintres et les graveurs* of 1775, the artist and historian Jean-Pierre Mariette identified the folding map as the work of the etcher Nicolas Cochin (1610–86). Cochin was made chief propaganda artist by the young Louis XIV's first minister, Cardinal Mazarin, and was commissioned to document the successes of the French Army in the war against Spain and the Spanish Netherlands. For him, the publishing campaign against the Jansenists was simply a spiritual continuation of the war in Flanders.

The map of 'Jansenia' stirred up a lot of public indignation. Prominent members of both parties commented on it. A mob gathered before the publisher's premises in order to seize hold of the remaining copies. Jansenists were particularly offended by the fact that Jesuits used later editions of Zacharie de Lisieux's satire as a schoolbook for Latin lessons, to the amusement of their students.

Septentrion
Occident
Orient
Tombeau foudroié
LAC
Midy

The largest atlas in the world presented to the newly restored Charles II by Dutch merchants eager for preferment.

Bribery not War

IN BAROQUE EUROPE they thought big, whether it was palaces, gardens, globes… or atlases. This monster is said to be the largest atlas in the world. It was presented to Charles II on his restoration to the British throne in May 1660 by a consortium of Dutch merchants led by Johannes Klencke. Two similar, but slightly smaller atlases are known. One was presented to the Elector of Brandenburg and the other was purchased for the Duke of Pomerania. These are now in Berlin and Rostock.

The merchants would have had plenty of reasons for wanting to court Charles. Personal vanity was one – Klencke was duly knighted. Much more important would have been the hope of winning royal support in their struggle for commercial advantages. Britain was emerging as the principal rival to the Dutch for trade outside Europe. There had already been one Anglo-Dutch war and there were to be more in the coming decades.

The calf leather binding is stamped with the heraldic symbols for England, Scotland, Ireland and – since the kings of England still included 'King of France' among their titles – the fleur-de-lys of France. Inside there are forty-two printed wall maps by the leading map publishers of the golden age of Dutch mapmaking. Unlike the examples that were mounted and exposed to the elements (as seen in the paintings of artists like Vermeer), these maps have remained in pristine condition. They cover all parts of the world. Though the atlas, which is now on wheels, is brought out for consultation a few times each year, for most of the time students consult more handily sized photographs of the maps.

Charles II, a map enthusiast, placed the atlas not in the Royal Library but in his 'Cabinet and Closset of rarities' at Whitehall. It was there that the diarist and polymath John Evelyn saw it in the 1680s, describing it as 'a vast book of Mapps in a Volume of neere 4 yards large', together with 'rare miniatures… a vast number of Achates, Onyxes & Intaglios… rare Cabinetts of Pietra Commessa… a curious Ship modell… & severall other rarities'. Since 1998 it has adorned the entrance lobby to the Maps reading room at the British Library in London, impressing passing readers and tourists as much as it once did John Evelyn.

Right The Klencke Atlas open at Willem Jansz [Blaeu]'s wall map of Italy of 1614–17. **Above** Portrait of Charles II by Thomas Simon on his Coronation medal of 1661.
See also 1621, 1664, 1924.

NOVA DESCRITTIONE D'ITALIA DI GIOANN. ANTONIO MAGINO.

A map of New Amsterdam, soon to be renamed New York, almost certainly presented to the Duke of York, later James II.

For New York's Godfather

NEW YORK BEGAN life as the Dutch colony of New Amsterdam and various place names such as Staten Island and the Bowry, derived from Dutch words for the States (of the Dutch Republic) and the (Dutch governor's) farm still recall those times. It thrived and in an attempt to keep track of its rapid growth from 1657 the town was repeatedly surveyed and mapped for the Dutch authorities by Jacques Cortelyou.

In late August 1664 the English captured New Amsterdam from the Dutch. An example of Cortelyou's map must have been sent to London not long afterwards. It provided the model for this map. Its colourful but rather naive decoration strongly suggests that it was created by a draughtsman specializing in decorative mapmaking who worked in the docklands east of the Tower of London, where chartmakers had been active since the 1590s. The map reflects the transitional status of the town. Although it is still called New Amsterdam, an English fleet is shown in the harbour and what had previously been the North River has already been renamed the Hudson River. Although there is no formal dedication, it was almost certainly intended (like other maps in the same collection) for Charles II's brother, James, Duke of York (later James II), hence its popular name the 'Duke's Plan'. Charles had appointed his brother proprietor of the town and surrounding country that March, in anticipation of the conquest.

No documentation relating to its presentation to the Duke has survived, but its splendour suggests that it was intended to draw James's attention to the new conquest, which had not yet been formally renamed, at a time when his attention was fixed on matters closer to home. The map gives a favourable but perhaps not excessively flattering impression of the town, which at that time was well known for the extent of its greenery. The town wall that was to give its name to Wall Street and the Broad Way leading from the fort on the site of today's Battery Park can be clearly seen, while the canal occupying the modern Broad Street adds a distinctively Dutch touch.

The map remained in the royal collection until the 1820s when it was presented to the British Museum.

Right The 'Duke's Plan' of New York. **Above** Part of the wall from which Wall Street takes its name. **See also** 1621.

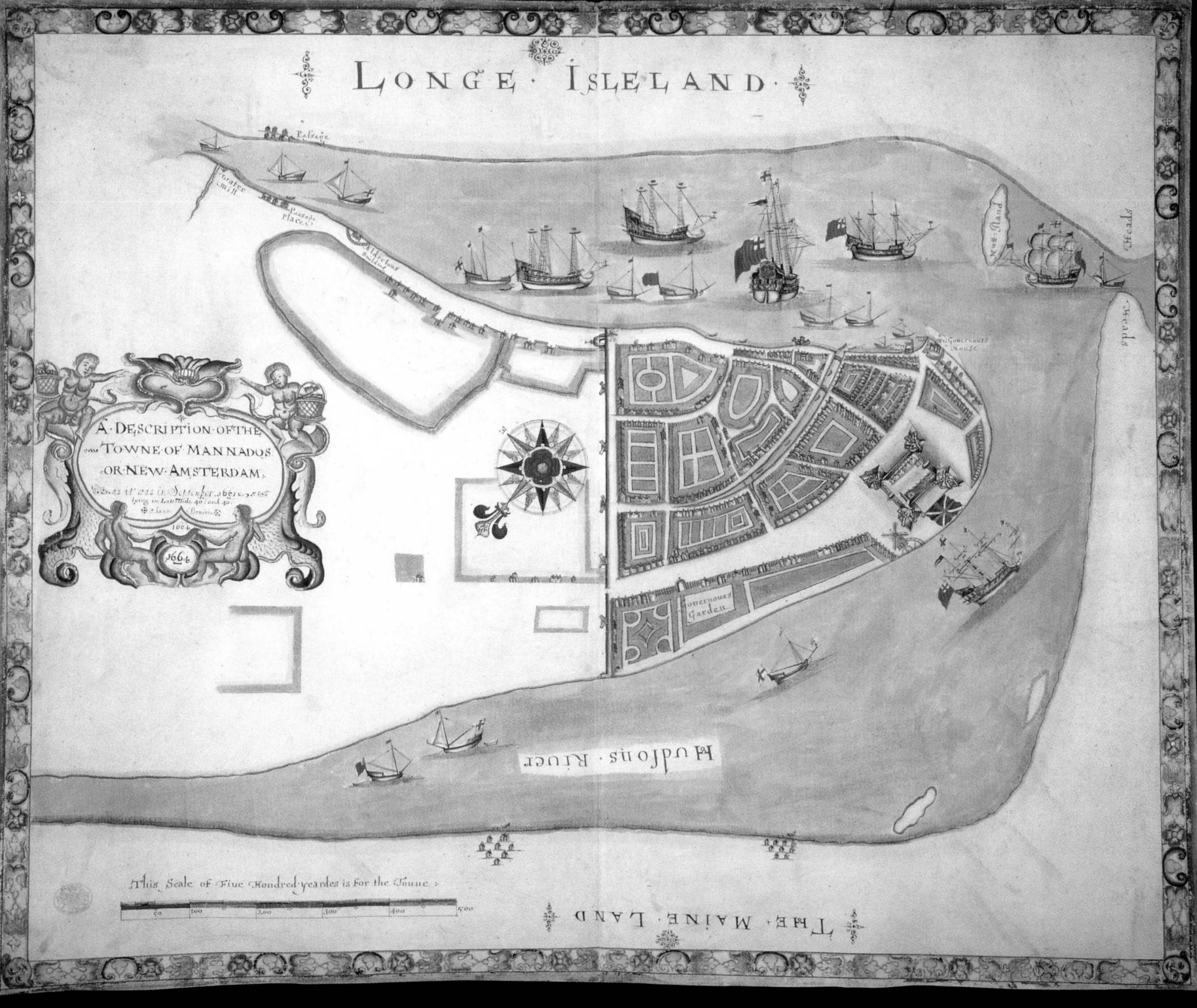

LONGE · ISLELAND ·

A · DESCRIPTION · OF · THE TOWNE · OF · MANNADOS · OR · NEW · AMSTERDAM

1664

1664

Passage

Water mill

Passage Place

Heads

Nutt Island

Heads

The Governours House

Governours Garden

Hudsons River

This Scale of Five Hondred yeardes is for the Towne

50 100 200 300 400 500

THE · MAINE · LAND

One of the regional maps from the first Japanese atlas describes rice output and the distances between locations in a stylized and symbolic form.

Traditional Japan

Right Map of the Kantō region, Japan, from *Nihon Bunkei-zu*.

PRIOR TO MODERN times, the Japanese cartographic imagination was shaped by many influences. These ranged from agricultural maps from ancient China to western European sea charts introduced to the Japanese in the late 1500s. The history of Japanese maps is one characterized by great variety and richness.

One important work is the *Nihon Bunkei-zu*, (日本分形図) the first Japanese atlas, by an unknown cartographer, published in 1666 in Kyoto. This printed atlas consists of sixteen maps with tables describing regional rice output and distances between locations on land as well as by sea. Interestingly, there is no one map featuring the entirety of the country. Rather, the atlas is composed of several regional maps each covering a number of provinces. The *Bunkei-zu* shows how, in the Japanese world-view, medium and reading practice are intimately related. The atlas, despite making contemporary references to both political and geographic features, does not betray its pictorial origins. At its heart lies an abstracted simplification. Bold animated lines stretch and deform the Japanese landscape. The origins for this map type can be traced directly to an eighth-century work known as the *Gyōki-zu*. In order to fully decipher the maps in the atlas, the reader must have a background in the principles of Asian ink painting and calligraphy. This background allows a symbolic understanding of the form that can be as effective as one based on Western science, precise measurement, and geographical verisimilitude.

In the image of the Kantō region, the location of Nikkō (日光) is given. In 1636, Nikkō was where Toshogu Shrine was built as the mausoleum of the Tokugawa Ieyasu (1543–1616), the great shogun and founder of the Edo period (1615–1868).

An intimate personal object, the *Bunkei-zu* was circulating during a time when notion of place in Japan was greatly affected by religious, literary and Japanese aesthetic ideas. Made when growing urban affluence and interest in domestic travel was on the rise, it marks the beginning of an emergence of a popular printed culture that would later peak with the widespread popularity of books, novels, woodblock *ukiyo-e* prints, as well as other maps of various subjects.

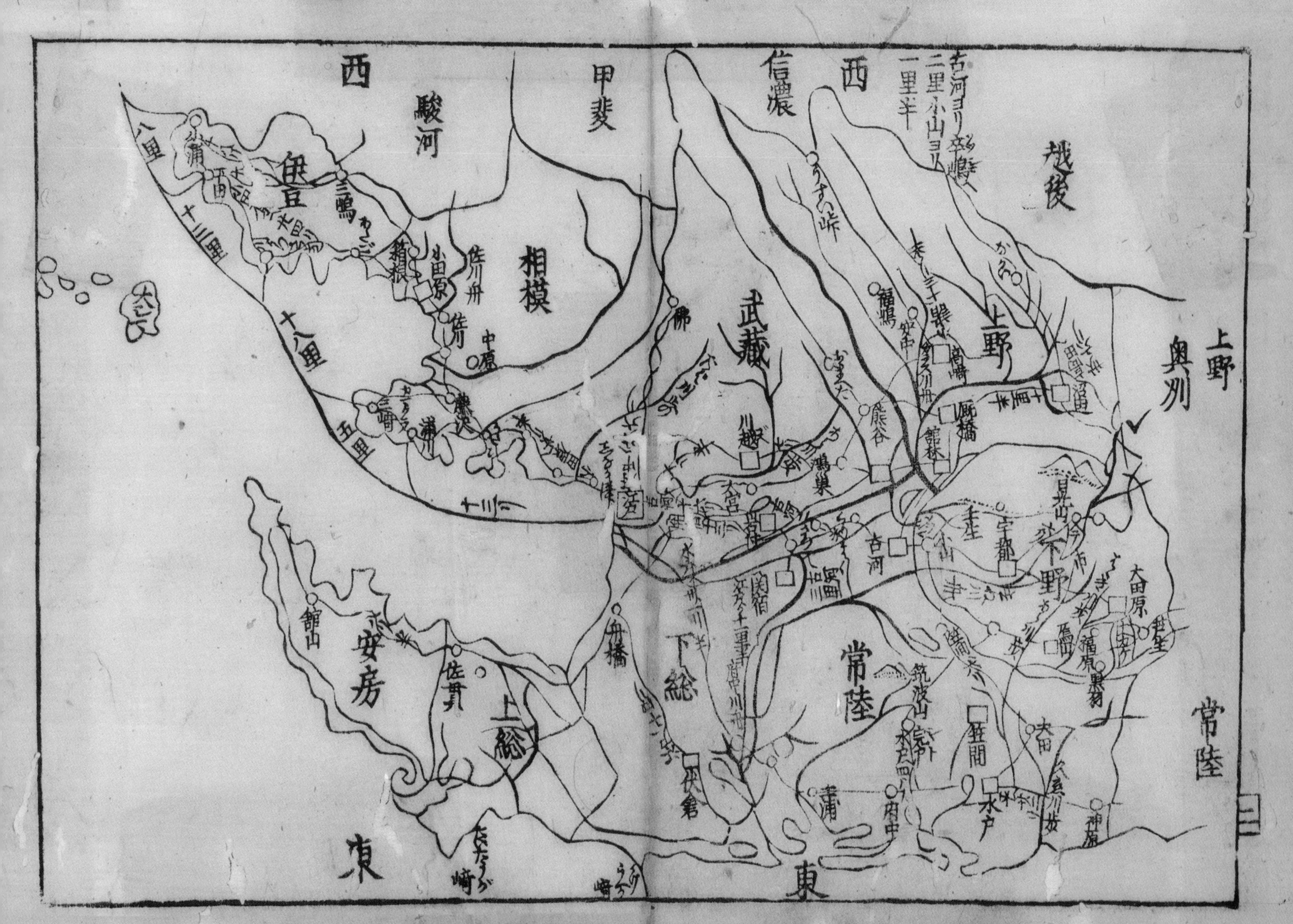

西
駿河
甲斐
信濃
西
越後
伊豆
相模
武蔵
上野
奥州
上野
八里
十三里
十八里
五里
十三里
三嶋
箱根
小田原
中原
藤沢
うすい峠
高崎
館林
熊谷
川越
鴻巣
大宮
古河
宇都宮
下野
大田原
黒羽
安房
館山
佐貫
上総
舟橋
下総
関宿
佐倉
常陸
筑波山
笠間
太田
水戸
府中
土浦
常陸
東
東

A map of the Empire of Germany, sponsored by Prince Rupert of the Rhine, from the first truly English world atlas, reveals a microscopic secret.

Who Made the Map?

WHO IS REALLY responsible for this map? It was published in London in the winter of 1669 to 1670 by Richard Blome (1635–1705) for inclusion in *A Geographical Description of the Four Parts of the World*, the first world atlas to be completely engraved, printed and published in England. In order to raise the money for the expensive copperplates Blome sought sponsorship for each of the maps from wealthy individuals. In this case Charles II's cousin, the dashing English Civil War hero Prince Rupert of the Rhine (1620–82), provided the money in return for a prominent dedication and the display of his arms. A German prince of the Wittelsbach family who had been born in Prague, which was shown on the map, Rupert also took a lively personal interest in the art and science of engraving.

Although the map was printed and published in England its content, as Blome openly acknowledged, was taken from the work of Guillaume Sanson. Sanson, the leading French cartographer of the time, had published the model in his atlas, *Cartes Générales de toutes les Parties du Monde* in 1658. Sanson, however, had not surveyed the region himself. He had compiled the information from a variety of printed and manuscript, official and private, graphic and printed sources, evaluating and selecting the information as he went. The overall design of Blome's maps, with its putti and coats of arms, was copied not from French but from Dutch maps, which then dominated the European map market.

Finally and easily overlooked was the man whom Blome employed to engrave the copperplate. The engraver had to possess a skilled and steady hand capable of engraving images and letters in reverse, and cutting clear but miniature lettering at various sizes. Blome employed the famous Bohemian engraver, Wenceslaus Hollar, for some maps in the atlas, but this plate is unsigned. Judging from its occasional awkwardness, it may be the work of an apprentice. His master would have decided on the overall design. At some point, though, whoever it was got bored, rebellious or made a bet. What appears to be a scratch under the map is a testimonial to his skill: they are the words beginning the Lord's Prayer 'Our Father', written in microscopic writing.

In truth Blome, Prince Rupert, Sanson, his sources, the bored engraver and his master all 'made' the map.

Right *A Generall Mapp of the Empire of Germany*… from Richard Blome, *A Geographical Description of the Four Parts of the World*. **Above** 'Our Father' in microscopic writing.

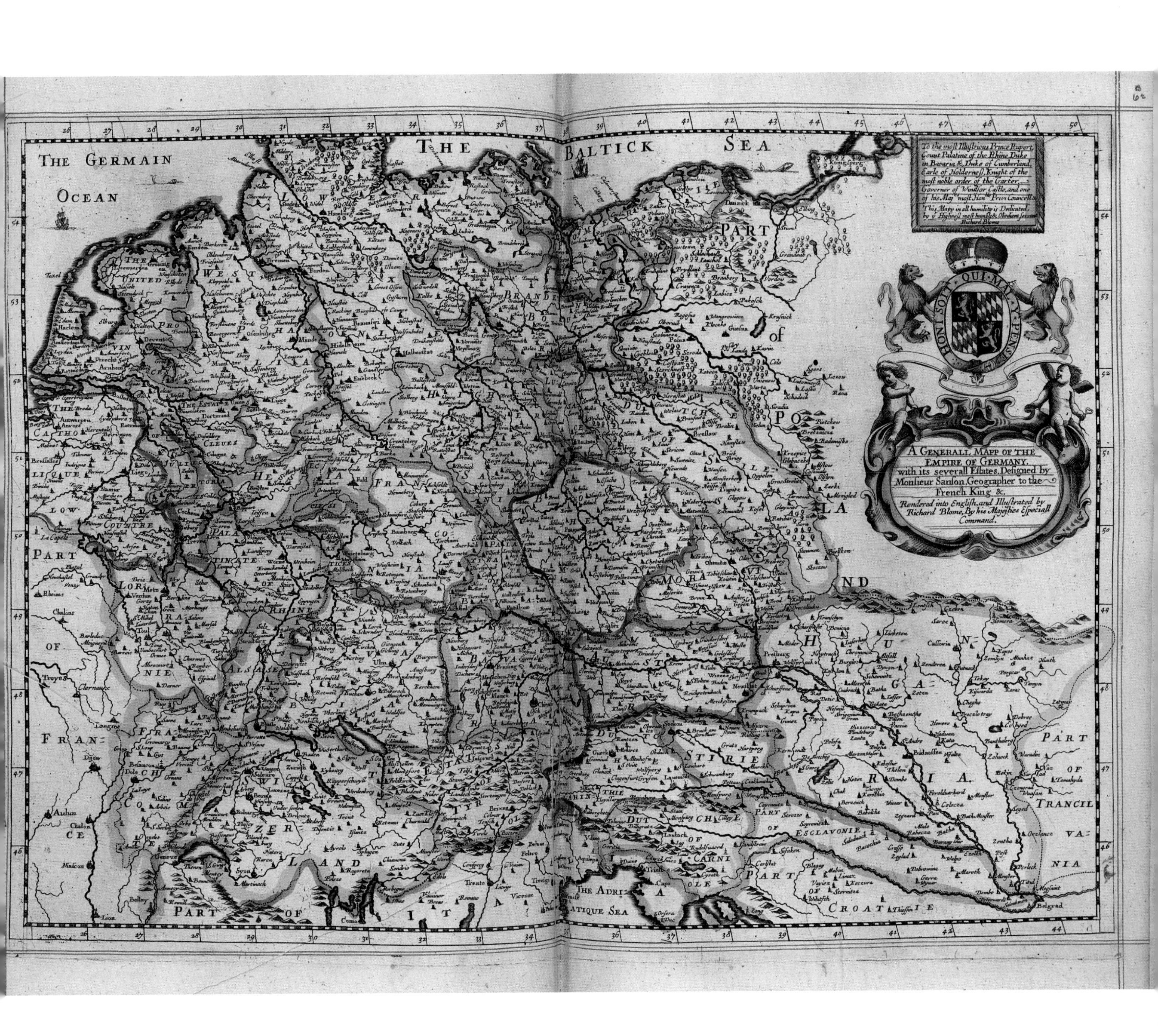
To the most Illustrious Prince Rupert,
Count Palatine of the Rhine Duke
in Bavaria & Duke of Cumberland,
Earle of Holderness, Knight of the
most noble order of the Garter,
Governer of Windsor Castle, and one
of his Maj^es most Honble Privi Councell
This Mapp in all humility is Dedicated,
by y^r Highness most humble & Obedient servant
Richard Blome
HONI SOIT QUI MAL Y PENSE
A GENERALL MAPP OF THE
EMPIRE OF GERMANY,
with its severall Estates, Designed by
Monsieur Sanson, Geographer to the
French King &.
Rendered into English, and Illustrated by
Richard Blome, By his Majesties Especiall
Command.
THE GERMAIN
OCEAN
THE BALTICK SEA
PART of POLAND
HUNGARIA
THE ADRIATIQUE SEA
PART OF TRANCILVANIA
PART OF ESCLAVONIE
CROATIE

A detailed map of Virginia and Maryland presented to the proprietor of Maryland, Lord Baltimore, and subsequently used as evidence in successive border disputes.

Bohemia in Maryland

1670

Right *Virginia and Maryland as it is Planted Inhabited this Present Year 1670 Surveyed and Exactly Drawne by the Only Labour and Endeavour of Augustine Herman Bohemiensis.* Engraved by W. Faithorne.

ONLY FIVE FURTHER examples of this map have come to light since the 1870s when this copy, then considered unique, was discovered by commissioners investigating the line of the Maryland–Virginia border. Hermann's map was the first improvement in the depiction of the region since John Smith's remarkable map of Virginia in 1612. By way of derivatives, it was to remain the definitive image until 1754 and was used in arguably the most famous of colonial border disputes, that between Maryland and Pennsylvania culminating in the celebrated Mason-Dixon Line of 1767.

Hermann was probably born in Prague in about 1623 or earlier. He may have learned his surveying skills when serving in the Thirty Years' War, after which he joined the Dutch West India Company as a foreign agent. He claimed to have been involved with the early trading of tobacco in Virginia. Settling in New Amsterdam (New York) in the early 1640s, he initially worked for the Dutch authorities. He was later involved in fur-trading, farming, land speculation and diplomacy – and in determining the Maryland–Virginia border on the Delaware peninsula in 1659.

In exchange for drawing a sketch map of the region Hermann was granted land in Maryland in 1660. It took ten years to survey and complete this detailed map, and the manuscript was sent to Lord Baltimore, the proprietor of Maryland, in England in 1670. He was delighted, declaring it 'the best mapp, that was Ever Drawn of any Country whatsoever with recommendation to the Press'. The King was so pleased that he granted Lord Baltimore £40 and Hermann, by a royal privilege dated 21 January 1673/4, the sole right to the printed map for fourteen years. The map itself was the largest produced to date of a region in North America. Numerous plantations are identified lining the rivers and creeks of the Bay. The Maryland–Virginia border on the Delaware peninsula, determined in Maryland's favour on 25 June 1668, is identified by a line of 'dubble Trees' on this map.

In 1930 the historian Lawrence Wroth wrote that 'it is an ironical reflection that, in spite of Lord Baltimore's pleasure and satisfaction ... the Hermann map should have been used successfully against the interests of the State of Maryland, in every boundary dispute from the day of its construction to the present generation'. Among Hermann's descendants were the wives of John Randolph and Benedict Arnold.

VIRGINIA
MARYLAND
DIEU ET MON DROIT
Chowann River
The Goulden or Brass Hill
RAPPAHANOCK
STAFFORT.C
CHARLES C
ANNE ARUNDEL C
BALTEMORE County
CALVERT C
WESTMERLAND C
NORT UMBERLAND
MIDDELSEX C
NEW KENT C
HENRICO
SURREY C
ISLE OF WIGHT C
NANTEMOND
LOWER NORFOLK
COROTUK C
Cape Henry
Cape Charles
The Great Bay of Chesepeake
DORCESTER C
TALBOT C
CAECIL C
THE NORTH SEA
Delawar Bay
NEW JARS
Christina C
at present Inhabited Only or most By Indians
VIRGINIA AND MARYLAND
As it is Planted and Inhabited this present Year 1670 Surveyed and Exactly Drawne by the Only Labour & Endeavour of Augustin Herrman Bohemiensis
A Scale of 8 English Leagues
AUGUSTINE HERRMAN BOHEMIAN
W: Faithorne Sculpt

John Ogilby's magnificent atlas *Britannia* of 1675, with over 100 folio-sized maps, was the first to contain strip road maps of England and Wales.

The First English Road Maps

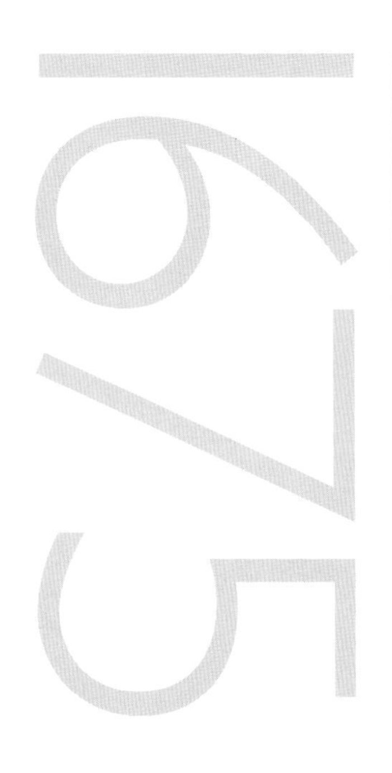

JOHN OGILBY'S REPUTATION is based on his final career, begun very late in life, as a publisher of maps. He lost his house in the Great Fire of London in 1666 and his first foray into mapmaking was as 'a sworn viewer' establishing the property boundaries of the burnt-out city. In concert with a number of professional surveyors he created an outstanding plan of London at a scale of 100 feet (30 metres) to the inch (2.5 centimetres). Armed with this experience he ventured into publishing and in a prospectus, dated 10 May 1669, he announced a six-volume description of the world. The final volume was to be the description of Britain, soon announced as a further six volumes at a cost of £34 per set. The total cost of the enterprise was the huge sum of £20,000.

Only one volume was ever completed. The *Britannia* was issued in 1675 and is a landmark in the mapping of England and Wales. It was the first national road-atlas of any country in western Europe. The maps, of seventy-three major roads and cross-roads, totalling 7,500 miles (12,500 kilometres), were presented in a continuous strip-form and, uniquely, on a uniform scale at 1 inch to a mile (1.6 kilometres). Of the hundred sheets of roads, most depicted a distance of about 70 miles (112 kilometres) on one sheet. The road is shown as a series of parallel strips and the surveyors noted whether the roads were enclosed by walls or hedges, or open, local landmarks, inns, bridges (with a note on the material of construction), fords and sometimes cultivation in the countryside on either side of the road. Hills were drawn to show the direction of their incline, and relative steepness. The frontispiece illustrates the surveying techniques employed. Distance, for example, was measured accurately by using a hand-pushed wheel. The changes in direction of the roads, necessary to overcome the schematic design, are shown by compass roses. The compass directions of significant local landmarks, established by triangulation bearings, are also given.

Among the many maps published before the nineteenth century depicting the English countryside, no other displays such a wealth of detail in such an immediate and comprehensible form. After its publication, no map in England could be published without incorporating the information it contained.

Right John Ogilby's survey of the road from London to Bristol from his *Britannia* published in London in 1675. **Above** Miniaturized plan of Reading: Ogilby's route may include the earliest plans of many small towns and villages. **See also** 1790.

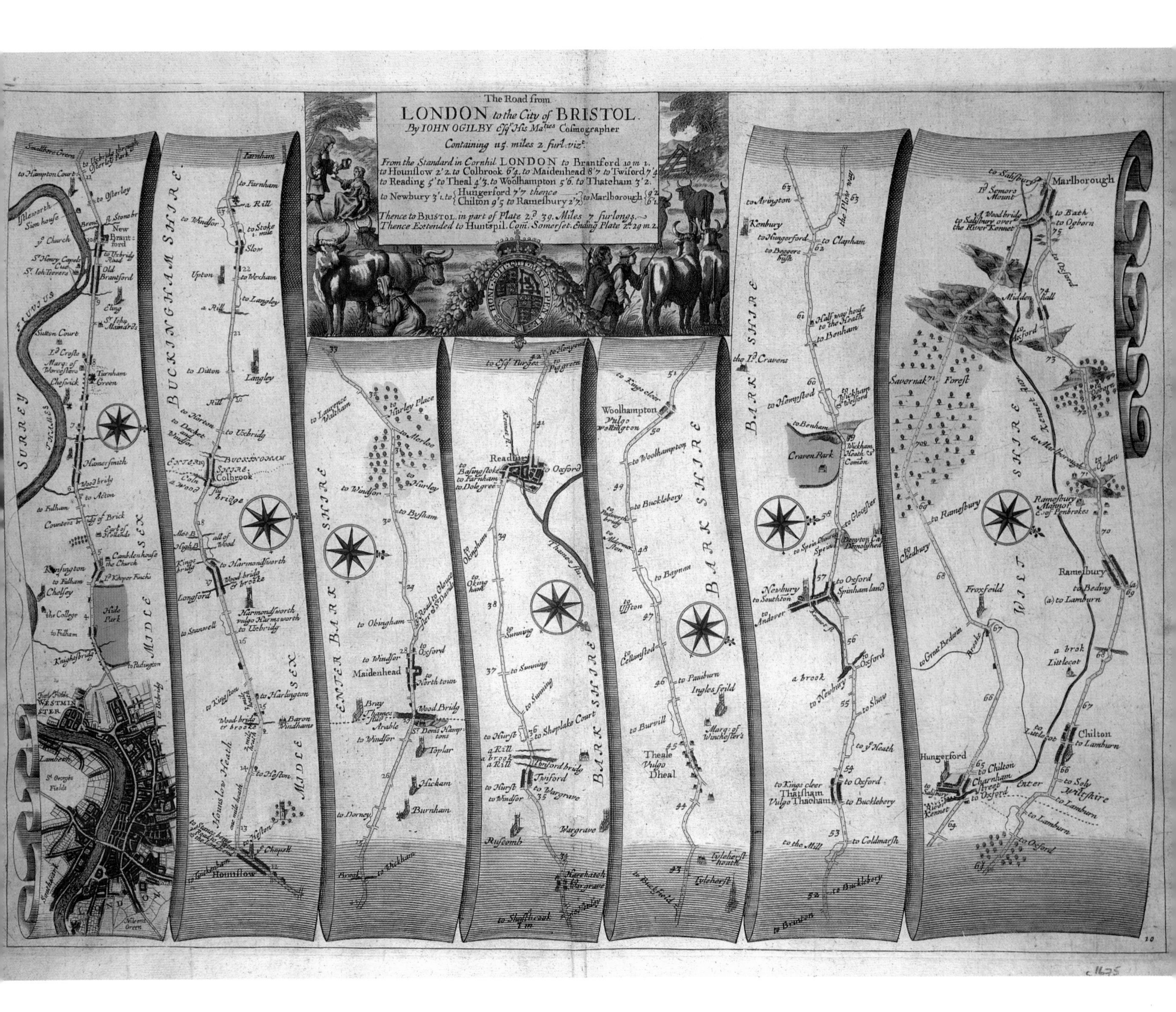

The Road from
LONDON to the City of BRISTOL.
By IOHN OGILBY Esq.r His Ma.ties Cosmographer
Containing 115. miles 2 furl: viz.t
From the Standard in Cornhil LONDON to Brantford 10 m 1.
to Hounslow 2' 2. to Colbrook 6' 4. to Maidenhead 8' 7 to Twiford 7' 4
to Reading 5' to Theal 4' 3. to Woolhampton 5' 6. to Thatcham 3' 2.
to Newbury 3' 1. to { Hungerford 7' 7 thence / Chilton 9' 5 to Ramesbury 2' 7 } to Marlborough { 9' 2 / 6' 1
Thence to BRISTOL. in part of Plate 2.d 39. Miles 7 furlongs.
Thence Extended to Huntspil. Com. Somerset. Ending Plate 2.d 29 m 2.
BUCKINGHAM SHIRE
MIDDLE SEX
SURREY
BARK SHIRE
WILT SHIRE
Hounslow
Maidenhead
Reading
Woolhampton
Newbury
Hungerford
Marlborough
Ramesbury
Chilton
Craven Park
Savernak Forest
Hide Park

The earliest map to be printed in North America was employed as evidence in a theological dispute and shows Indian incursions – an act of God to test the Pilgrims' faith.

Puritan New England's Precarious Perch on 'This Western Coast'

1677

Right William Hubbard, *A Map of New England.*

THIS MAP'S FORM, printed from a rough-hewn woodblock, exemplifies the material poverty of England's early colonies; its content, their cultural complexity. The map was not a general geographical guide; Hubbard knew his readers among New England's civil and religious elites were already familiar with the region's spatial organization. Instead, Hubbard designed it as a very selective geographical index to just some of the events – assaults by the Wampanoags and Narragansetts on English settlements, but not vice versa – that he narrated in his Anglocentric history of King Philip's War (1675–6).

The map was central to Hubbard's theological dispute with another Puritan minister over the war's significance. Increase Mather argued that God sent the Indians as a scourge to force the English to return to godliness and renew their covenant with God as His chosen people. Hubbard agreed that the Indian attacks had been sent by God but did not presume to know God's reasons. To discern those reasons, he argued, required the careful collection and evaluation of all evidence of possible divine interventions. In this context, he used the map to demonstrate symbolically the comprehensiveness of his analysis and the rigour of his conclusion that God sought to test the Puritans' faith, even to destruction.

The map itself was derived from a 1665 map of Massachusetts Bay's territorial claims. Other maps from the same source depicted New England as the eastern core of colonies extending indefinitely to the west. By contrast, Hubbard's map echoed his contention that the English were in fact 'hemm'd in on both sides and almost round about'. With west at the top, the map depicted a thin strip of English coastal settlement, untouched by Indian attacks, weighted down by a frontier of now devastated settlements, marked by indexical numbers, and the overarching zone of native peoples and their wild, wooded lands. (In the north, where conflict persisted, Indians emerge from forests to attack the English.) The map imagined New England to constitute a precarious Christian toehold on 'this Western Coast' of the Atlantic Ocean, as a prefatory poem put it. Hubbard's pessimistic map thus celebrated neither English colonialism nor an English victory in war, but a merciless God.

A MAP OF
NEW-ENGLAND,
Being the first that ever was here cut, and done by the best Pattern that could be had, which being in some places defective, it made the other less exact: yet doth it sufficiently shew the Scituation of the Country, and conveniently well the distance of Places.
The figures that are joyned with the Names of Places are to distinguish such as have been assaulted by the Indians from others.
A Scale of forty Miles.
10
20
30
40
Newhaven
Gilford
Seybrook
Newlondon
Mattabesick
Harfford
Winsor
Springfield
Northamton
Hadly
Deerfield
Squaheag
Pequid Country
Sqabaog
Nipmuk
Stoniton
Naraganset
Lancaster
Malborough
Sudbury
Groton
Water-Town
Concord
Chensford
Medfield
Cambridg
Woburn
Billerica
Haveril
Bradford
Roxbury
Dorchester
Boston
Braintry
Providence
Newport
Mount-Hope
RHODE ISLAND
Seaconk
Pocasset
Taunton
Weymoth
Hingam
Plimouth
Scituat
Hull
Lyn
Salem
Marblehead
Wenham
Ipswich
Rowly
Newbury
Salisbury
Hamton
Dover
Piscataqua R.
Winter-Harbor
C. Ann
Casco Bay
Kenebeck R.
Pemaquid
The White Hills
Sandwich
Martins Vineyard
Yarmoth
Cape Cod
Nantuket

A Baroque commemorative military map gives a detailed bird's-eye view of the seige of Spanish-ruled Luxembourg, eventually taken by the French.

A French Triumph

AFTER 1600 FRANCE's rulers sought to banish the spectre of foreign invasion that had plagued their country from the late 1580s by strengthening and expanding the kingdom's borders. The highwater mark came in 1684 with the successful siege of Spanish-ruled Luxembourg and the Twenty-Year Truce, forced by Louis XIV on his humbled enemies, which enshrined all the recent French gains.

Since the 1620s French military successes had been commemorated in superbly engraved and etched pictorial maps by outstanding artists such as Jacques Callot, Stefano della Bella and Nicolas Cochin, which were based on surveys taken on the field of battle by military engineers. From 1642 their work was coordinated by a retired military engineer, Sébastien de Pontault, Sieur de Beaulieu, for a projected atlas which was to immortalize the 'glorious conquests' of 'Louis the Great'.

Romeyn de Hooghe, perhaps the greatest etcher of the late seventeenth century, had been employed by Beaulieu. This bird's-eye view gives a detailed depiction of the siege of Luxembourg which lasted from 29 April to 4 June. It shows the French under the Duc de Crequi surrounding and bombarding Luxembourg from their camp at the top left, and the town's strong defences, which had been thought impregnable. Also to be seen are the munitions wagons, the devastated buildings inside the town, individual soldiers and even the gallows used by both sides to maintain discipline. To the right of the compass star in the foreground is a helmeted figure representing France with a shield bearing Louis XIV's symbol, the Sun, with his motto '*Nec pluribus Impar*' – 'Not unequal to many'. Brandishing a sword it chases away the figures of the Dutch Republic, with lion-skin and shield with bound arrows, and another helmeted figure bearing a shield with the double-headed shield of the Holy Roman Empire.

But despite first appearances, the map is no glorification of Louis XIV. By 1684 de Hooghe had become the principal propagandist of Louis XIV's most stubborn enemy, the Dutch leader, and later King of Great Britain, William III. Prominent among the figures being expelled by France is Justice bearing a pair of scales, while behind France is the demonic figure of War, on the point of firing a cannon. They express the sentiments of Louis's enemies. The following decades were not to be so kind to France. By 1710 foreign forces were once again deep inside French territory.

Right Romeyn de Hooghe, *Luxemburgum*. **See also** 1600, 1660 (p. 162), 1697, 1707.

LUXEMBURGUM,
DUCAT: COGNOM: METROPOL: Á CHIMACENS: PRINC: STREN: DEF: REG: GALLIÆ ARMIS VERO OCCUP: IV IUNII A°. MDCLXXXIIII
ex conatibus Romani De Hooghe et Formis Nicolai Visscher Amst: Bat: cum Privil: Ordin: General: Belgii Fœderat:
LUXEMBOURG
Ville Capitale du Duché du Mesme Nom. Apres avoir Soustenu, deux blocquades et esfuye les Violences terribles de bombes Violons etc: estant mis au sac comme une ruine, fut assiegé le 19 d'Avril 1684: de Monseig.r le Duc de Crequi par les François Vigoureusement defendu par Monseig.r le Prince de Chimay rendu le 4 de Juin 1684.
Eschelle de 160 Toises
Eschelle de 200 Roeden
TERRAIN PIERREUX
Place des Armes
Cour de Prince

A map by Vincenzo Coronelli charts the limits of the known landmasses of the 'Southern Sea' – the Pacific Ocean.

The Pacific Ocean: A Sea Without Storms

THIS MAP IS DIRECTLY derived, even down to the decoration of the title panel, from the great manuscript terrestrial globe made by Coronelli for Louis XIV of France in 1683, which was then printed in reduced form in Venice in 1689. It is part of the series of loose maps produced by Coronelli between 1688 and 1691, and bound from 1690 in various volumes of *Atlante veneto*.

The map represents the Pacific Ocean or 'Southern Sea' from 180 degrees to 320 degrees east of Hierro (Canaries), following French and Dutch cartography of the later seventeenth century. The coordinates are taken from maps by the leading French cartographer, Guillaume Sanson, while the design of the landmasses derives from Joan Blaeu's *Grand Atlas* (1664) and from M. Thevenot's *Recueil de voyages* (Paris, 1681). The text below the Equator notes the tranquillity of the Pacific waters, the constant eastern winds that blow between the Tropics, and the relative brevity of the crossing: only sixty days.

The ocean, of which only very few islands were then known, is marked by the paths of the ships of Jacques le Maire, the discoverer of Cape Horn (1615–17). In the north between 40 and 50 degrees north are the hypothetical coastlines of the Terra di Iesso (Asia) and Tartaria di Yupi: the first interpretations of the coasts of northern Japan – explored by the Dutch around 1643 (Hokkaido and Sea of Japan). The Strait of Anian still appears, albeit in a form that is less and less certain, between the 'Terra di Iesso' and America. It lies immediately to the north of California which, having been mapped as a peninsula in the sixteenth century, assumed the form of an island in the course of the seventeenth. It was only in the following century that this mistake was corrected.

Right Vincenzo Coronelli, *Mare del Sud.*

In the southern hemisphere, to the south-west of New Holland (Australia), the coasts of Van Diemen's Land (Tasmania) are beginning to take shape, along with those of New Zealand, also discovered by the Dutch (in the south an asterisk signals the Antipodes of Venice – Coronelli's homage to the city of his birth, but also the city that gave him a wage and the honorary title of 'Cosmographer of the Republic of Venice'). Further to the east, one sees the last remaining trace of a hope not totally abandoned, although by then mostly considered a fable: the undiscovered southern continent.

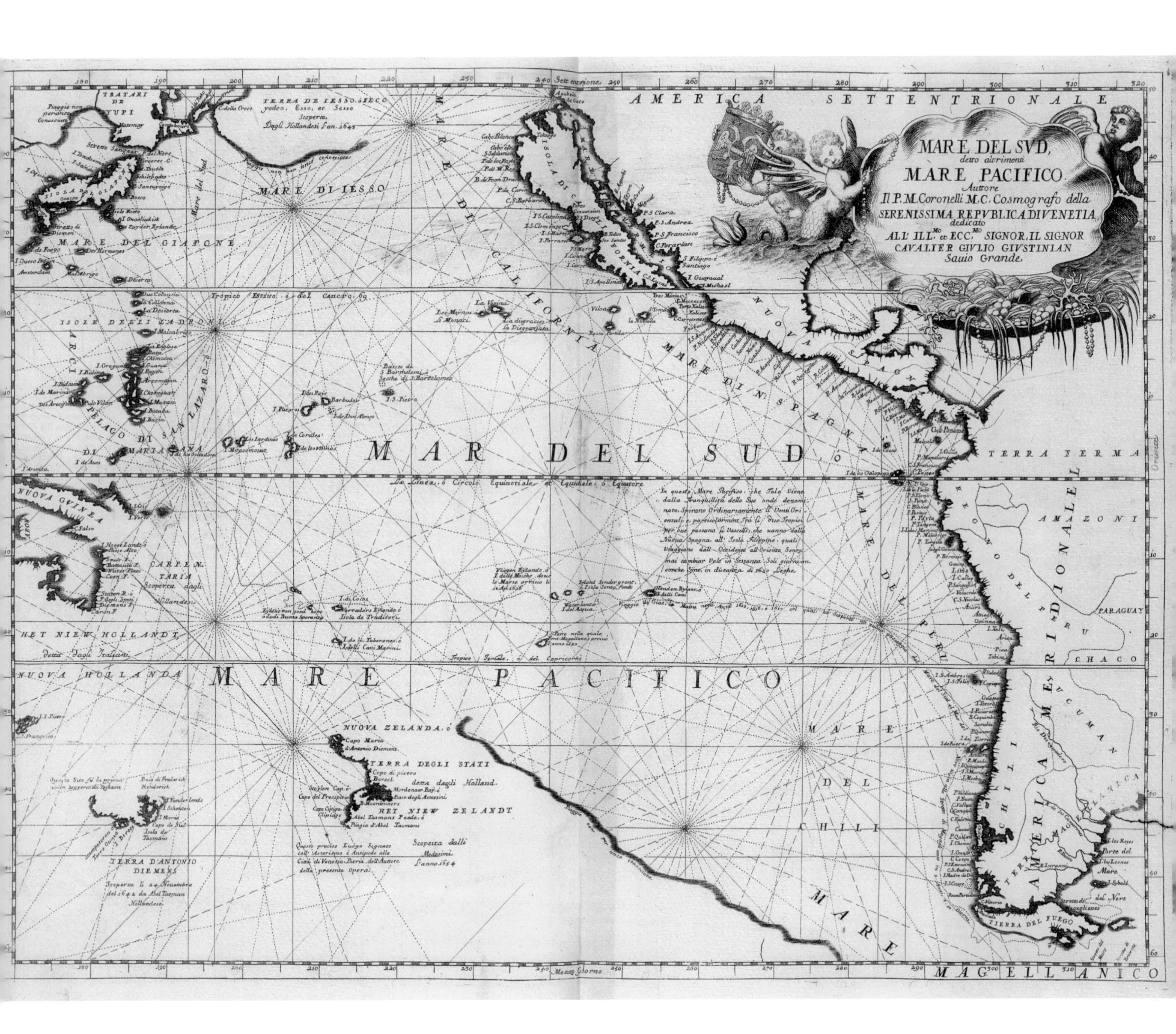
MARE DEL SVD, detto altrimenti MARE PACIFICO
Auttore Il P.M. Coronelli M.C. Cosmografo della SERENISSIMA REPVBLICA DI VENETIA dedicato ALL' ILL.MO et ECC.MO SIGNOR IL SIGNOR CAVALIER GIVLIO GIVSTINIAN Sauio Grande.
AMERICA SETTENTRIONALE
TERRA DE IESSO, ò ECCO yedeo, Esso, et Sesso Scoperta Dagli Hollandesi l'an. 1643
MARE DI IESSO
MARE DEL GIAPONE
MARE DI CALIFORNIA
ISOLA DI CALIFORNIA
NUOVA SPAGNA
MARE DI N. SPAGNA
ISOLE DELLI LADRONI
ARCIPELAGO DI SAN LAZARO
Tropico Estiuo, ò del Cancro
MAR DEL SUD
La Linea, ò Circolo Equinotiale, et Equidiale, ò Equatore
In questo Mare Pacifico, che Tale viene dalla Tranquillità delle Sue onde denominato, Spirano Ordinariamente li Venti Orientali, e particolarmente Trà li Due Tropici per cui passano li Vascelli, che uanno dalla Nuoua Spagna all' Isole Filippine, quali Viaggiano dall' Occidente all' Oriente Senza mai cambiar Vele in Sessanta Soli giorni, ancorche Sijne in distanza di 1620 Leghe.
NUOVA GUINEA
CARPENTARIA Scoperta dagli Hollandesi
HET NIEW HOLLANDT, detta dagli Italiani
NUOVA HOLLANDA
Tropico Iemale, ò del Capricorno
MARE PACIFICO
NUOVA ZELANDA, ò TERRA DEGLI STATI detta dagli Holland
HET NIEW ZELANDT
TERRA D'ANTONIO DIEMENS
MARE DEL PERU
MARE DEL CHILI
TERRA FERMA
AMAZONI
PARAGUAY
CHACO
TUCUMAN
AMERICA MERIDIONALE
TERRA DEL FUEGO
MARE MAGELLANICO
Settentrione
Mezzo Giorno

An Indian large-scale map combining different perspectives of the royal palace of Kotah, Rajasthan, together with the Rajah, his court and subjects.

Diwali in India

IN RECENT DECADES scores of pre-modern maps of Indian cities, towns, forts and places of pilgrimage have come to light and received scholarly attention. Dating from the seventeenth to the twentieth centuries, these maps, painted on either paper or cloth, originate from widely scattered regions and are rendered in a great diversity of styles. The richest finds, however, are from the princely states of Rajasthan, several of whose rajahs – especially those of Mewar and Jaipur – commissioned artists to provide them with maps for decorative, commemorative and utilitarian purposes. In the city palaces, now museums, of both those states one finds rooms whose walls were, and remain, largely covered with maps. The small paper map illustrated here, though relating to the capital city of Kotah state, is believed, on stylistic grounds, to come from Udaipur, the capital of Mewar.

Most maps of urban localities are essentially planimetric, though many features, such as town walls and major buildings, are shown as if seen from a frontal, generally exterior, ground perspective. Many others, like the early example illustrated here, combine one or more oblique perspectives. The artist's choice in this regard would be determined largely by his desire to highlight architectural details of interest to his patron and to those whom the patron sought to impress. Many also showed the general nature of the surrounding terrain. Within the city, one should assume neither completeness of detail nor great fidelity in the rendering of what was actually on the ground.

Many maps were vividly embellished to illustrate important events, such as battles, durbars (ceremonial royal audiences) or, as in this example, religious festivals. Within its small compass it manages to show not only the King and his harem, but scores of other happy celebrants, music, fireworks, acrobatics, and animal fights. Such maps, unlike those of a more utilitarian nature, had little need for explanatory text. The action spoke for itself.

Right Diwali Celebrations at the Royal Palace at Kotah, Rajasthan.
Above Detail of the courtyard.
See also 65, 1580, 1697.

An early printed Hebrew map charts the story of the Israelites' deliverance from Egypt with many illustrative references to biblical texts.

The Land of Milk and Honey

IN 1695 THE AMSTERDAM printers Asher Anshel & Partners issued a fine Passover Haggadah sponsored by Moses Wiesel. The Haggadah is the service book used in Jewish households on Passover Eve to celebrate the Israelites' deliverance from Egyptian bondage. This was the first-ever Hebrew book to be illustrated with copperplate engravings and with a map. The map and engravings were created by Avraham bar Ya'akov, a proselyte who had previously been a Christian pastor in the Rhineland.

In designing the map, Avraham bar Ya'akov intended to show Passover celebrants the journey made by the freed Israelites from Egypt through the Sinai desert into Canaan, the places where they stopped on the way (minute tents positioned along a meandering route), and the division of the Land amongst the Twelve Tribes.

Christian van Andrichom's map of 1590 influenced the design which has east at the top. The herd of cows labelled *halav* – milk, and the beehive labelled *devash* – honey, are clear references to 'A land flowing with milk and honey' (Leviticus 20:24). The inscription above the eagle spreading its wings reads: 'You have seen what I did to Egypt and how I bore you on eagles' wings, and brought you to myself' (Exodus 19:4). The barges carrying cedar wood heading from Tyre towards Jaffa, illustrate the verse: 'And we will cut wood out of the Lebanon, as much as thou shall need; and we will bring to thee on rafts by sea to Jaffa; and thou shall carry it up to Jerusalem [to build the Temple]' (Chronicles II 2:15). The ship about to overturn in the Mediterranean and the person being swallowed by a whale allude to Jonah's story, an element often encountered in contemporary maps. In the lower right-hand corner, Egypt is represented by a woman sheltering under a parasol, astride a crocodile.

Right Avraham bar Ya'akov, *Map of the Holy Land with the Route of the Exodus.* **Above** Jonah and the whale with (below) the name 'Avraham bar Ya'akov'.

The legend within a scrolled frame lists the forty-one stations where the Israelites camped on their long journey to the Promised Land. The River Jordan rises in the Lebanon mountains from two sources: the Yeor and the Dan, hence its name which complies with the legend found in the literature of medieval travellers. The printers Proops reissued Avraham bar Ya'akov's map in the 1712 and 1810 editions of the Amsterdam Haggadah.

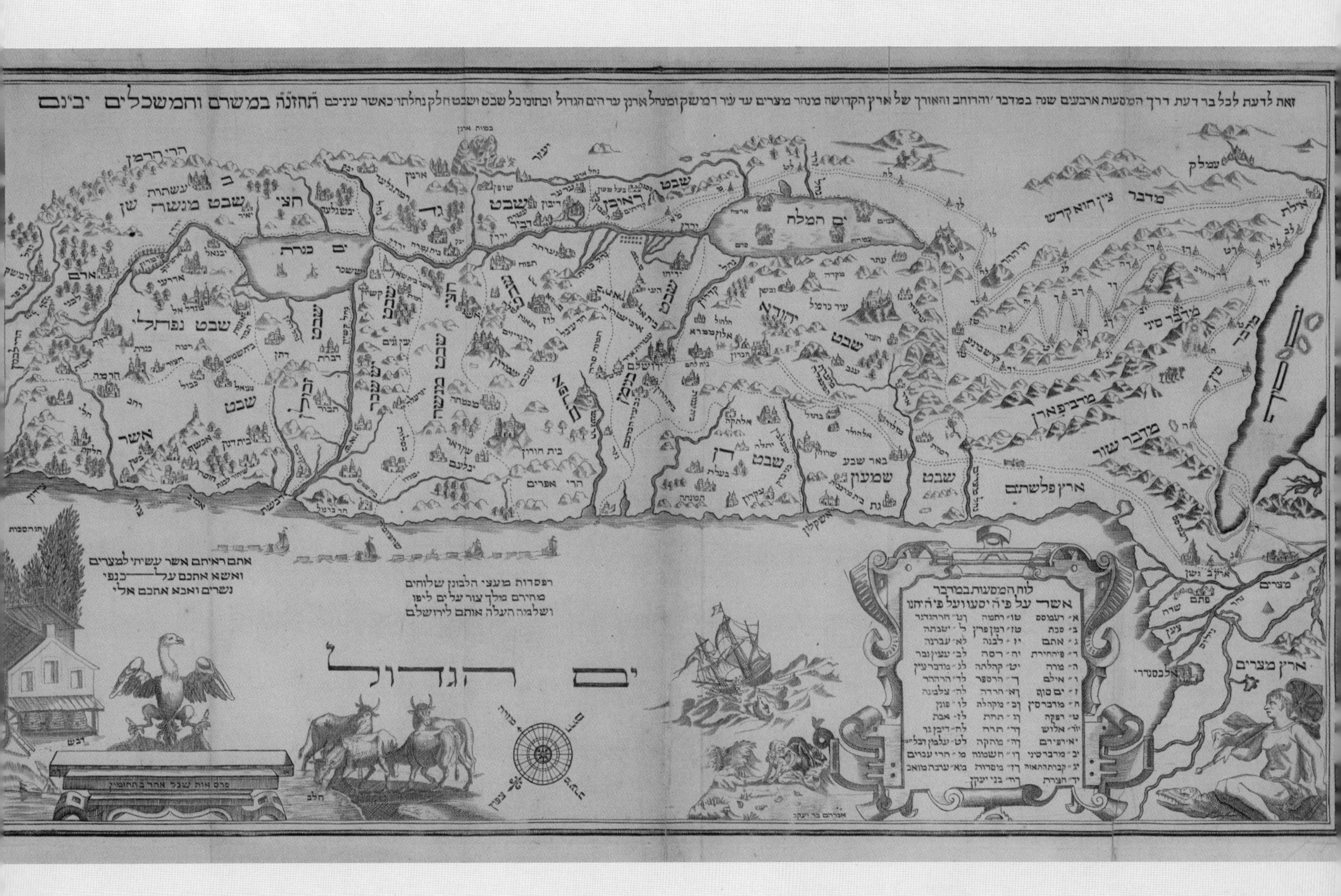
זאת לדעת לכל בר דעת דרך המסעות ארבעים שנה במדבר ׳והרוחב והאורך של ארץ הקדושה מנהר מצרים עד עיר דמשק ומנחל ארנן עד הים הגדול ובתומי כל שבט ושבט חלק נחלתו׳ כאשר עיניכם תחזנה במשרם והמשכלים יבינם
ים הגדול
ים כנרת
ים המלח
ארץ פלשתים
ארץ מצרים
מדבר שור
מדבר פארן
מדבר סיני
מדבר צין הוא קדש
עמלק
הרי חרמן
שבט נפתלי
שבט אשר
שבט מנשה
שבט דן
שבט שמעון
הרי אפרים
הר כרמל
אתם ראיתם אשר עשיתי למצרים
ואשא אתכם על כנפי
נשרים ואבא אתכם אלי
רפסדות מעצי הלבונן שלוחים
מחירם מלך צור על ים ליפו
ושלמה העלה אותם לירושלם
לוח המסעות במדבר
אשר על פי ה׳ יסעו ועל פי ה׳ יחנו
א׳ רעמסס טו׳ רתמה כט׳ חר הגדגד
ב׳ סכת טז׳ רמן פרץ ל׳ יטבתה
ג׳ אתם יז׳ לבנה לא׳ עברנה
ד׳ פיהחירת יח׳ רסה לב׳ עצין גבר
ה׳ מרה יט׳ קהלתה לג׳ מדבר צין
ו׳ אילם כ׳ הרספר לד׳ הרההר
ז׳ ים סוף כא׳ חרדה לה׳ צלמנה
ח׳ מדבר סין כב׳ מקהלת לו׳ פונן
ט׳ דפקה כג׳ תחת לז׳ אבת
י׳ אלוש כד׳ תרח לח׳ דיבן גד
יא׳ רפידים כה׳ מתקה לט׳ עלמן דבלתים
יב׳ מדבר סיני כו׳ חשמנה מ׳ הרי עברים
יג׳ קברות התאוה כז׳ מסרות מא׳ ערבת מואב
יד׳ חצרת כח׳ בני יעקן

A Dutch bird's-eye view of William III's villa and its immediate gardens used as the venue for the peace conference between France and her assorted enemies.

Mapping Peace Negotiations

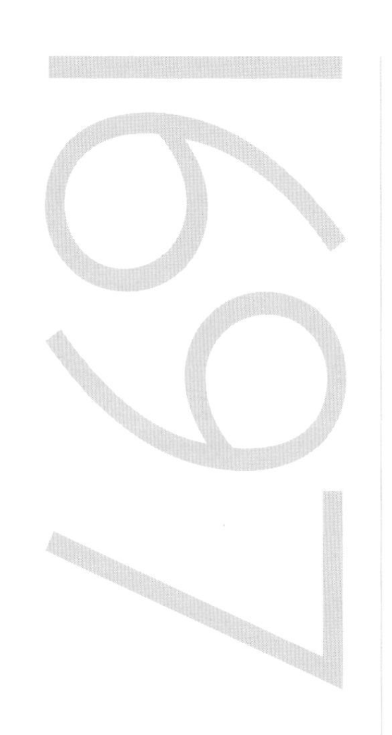

IN 1697, AFTER more than nine years of war, peace was formally concluded between Louis XIV's France and a large Allied consortium including the Holy Roman Empire, Great Britain, the Dutch and Spain. It marked the first French diplomatic setback in decades.

The wartime alliance against France had been led by William III, Stadholder General of the Dutch Republic and, since 1689, King of Great Britain, and the formal negotiations had been preceded by secret negotiations in Versailles between William's friend, the Earl of Portland, and French ministers. In the course of them the French had been forced to agree that the peace conference should take place not on neutral ground but in Rijswijk in the Dutch Republic, one of William's country houses. This conferred enormous prestige and the Dutch made the most of the opportunity, through prints such as this.

The house and its garden, depicted both in elevation and in bird's-eye view, are seen at their most immaculate and William III's ownership is underlined by his coat of arms (bottom right). The splendour of this copy is enhanced through having been coloured by a professional map colourist, or *afsetter*. William III was particularly interested in gardening and this is reflected in the formal gardens with their parterres, formal ponds, topiary hedges, green tunnels, pavilions, dovecote and fountains, in which the delegates presumably were free to relax or engage in informal discussions. The Swedish Mediator's coach is shown approaching the central entrance following the arrival of the French ambassadors' coach which had entered the front courtyard by a side entrance. The small plan of the first floor indicates the Mediator's quarters (d), overlooking the garden, and those of the French (c, f) and their enemies (b, e), separated by a central Saloon (a) where the Mediator passed messages to the different sides and where peace was finally signed.

Maps had been used increasingly to clarify negotiations since the middle of the seventeenth century. Though borders continued to be described and legitimated in terms of legal rights, the Treaty of Rijswijk was one of the first in which a physical feature – in this case the Alps between France and the Duchy of Savoy – which could unambiguously be followed on a map, was used to draw an international boundary.

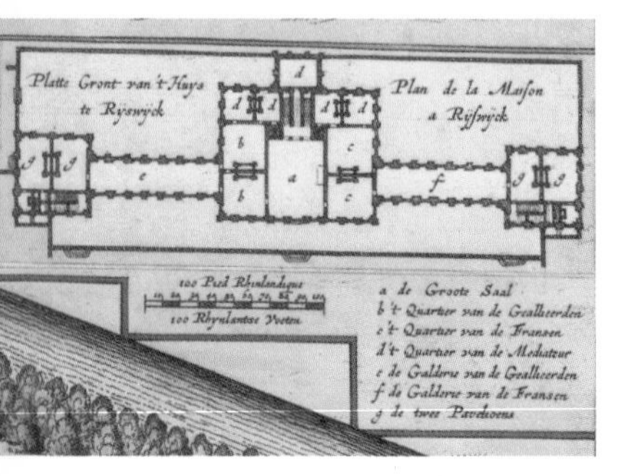

Right Frederik de Wit, *'t Konincklyck Huys te Ryswyck.* Plan of the ground floor showing quarters of the various negotiators. **Above** The villa and gardens of Rijswijk. **See also** 1600, 1684, 1690.

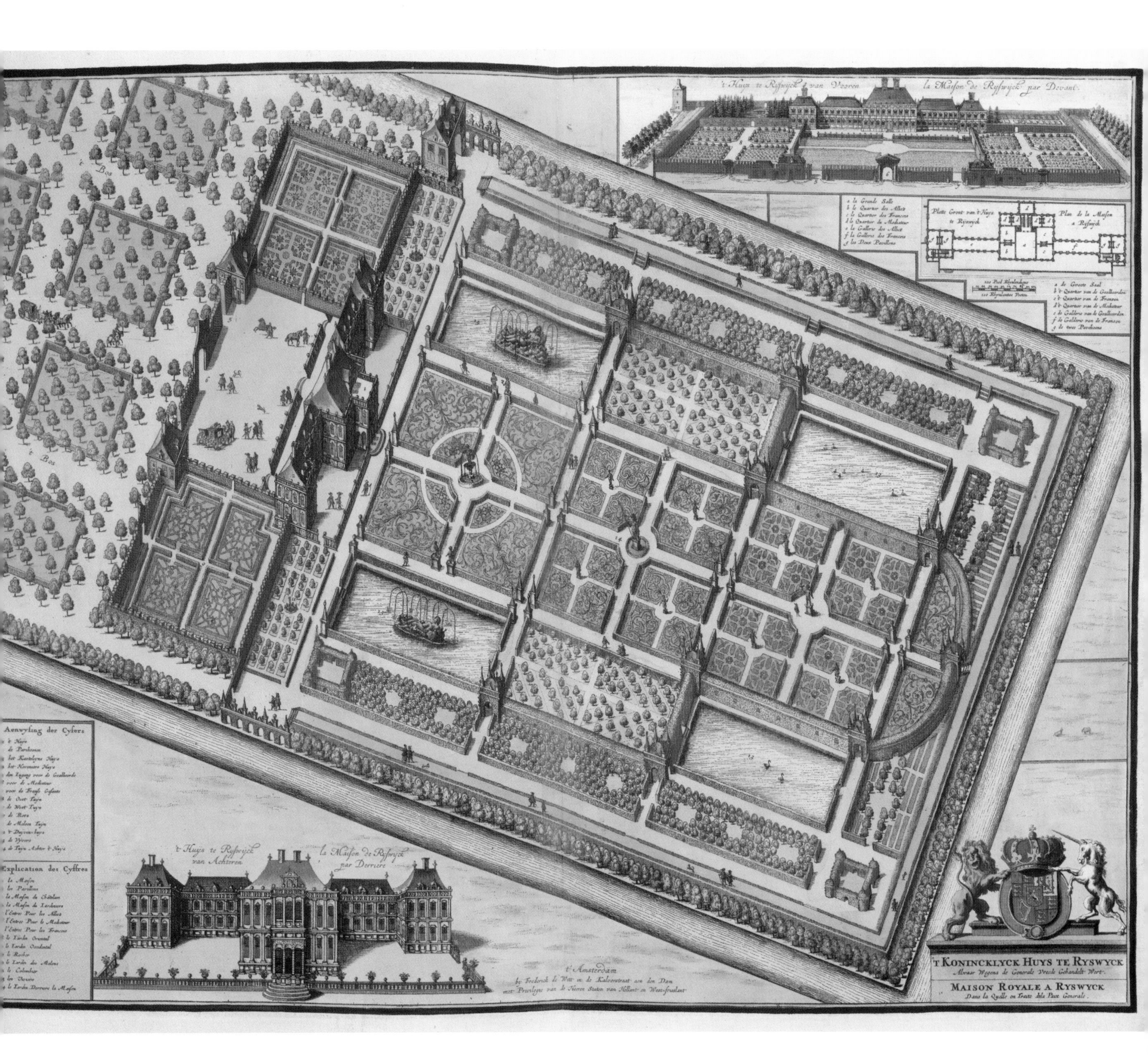
't Huys te Ryswyck van Vooren
la Maison de Ryswyck par Devant
'T Koninckliyck Huys te Ryswyck
Alwaar Wegens de Generale Vreede Gehandelt Wert.
Maison Royale a Ryswyck
Dans la Quelle on Traite dela Paix Generale.
't Huys te Ryswyck van Achteren
la Maison de Ryswyck par Derriere
t'Amsterdam
by Frederick de Witt in de Kalverstraat aen den Dam
met Privilegie van de Heeren Staten van Hollant en West-frieslant
Aenwysing der Cyfers
Explication des Cyffres

An English commemorative map depicting the British and allied triumphs over the French was given fresh impetus by another victory.

War and the Map Trade

IN THE CLOSING days of July 1708 news arrived in London of another stunning victory of the Duke of Marlborough over the French, this time at Oudenarde. Christopher Brown, a Devonshire man who had come to London to train as a mapseller in the 1680s, saw the general elation as an opportunity to get rid of some old stock. He advertised this map in the *London Gazette* of 29 July.

The basic map was a standard image, enhanced by marginal notes (left) on the climates, on the prime meridian of Teneriffe (bottom left), and with a grid of latitude and longitude. These scientific attributes were played on in the title, but it was fevered patriotic pride that was really expected to sell the map. It is surrounded by views and plans of British and Allied triumphs from 1701 to 1706, such as the capture of Gibraltar and victory at Blenheim, by a little-known Dutch-born engraver, De Ruyter, suggesting that it had first been issued late in 1706 or 1707. It has an attractive title panel showing Fame crowning a bust of Queen Anne, with beneath, an imperially crowned figure, possibly intended to be Minerva, with a horn of plenty and implements of the arts and sciences at her feet. We know from inventories that maps like this adorned the apartments of Queen Anne's consort, Prince George of Denmark, and presumably of many of her subjects.

Brown retired to a life of comfort in Devon in 1712. The copperplate for this map was purchased by another mapseller, Thomas Bowles. At some point in the troubled years between 1715 and 1724 which were disturbed by domestic unrest and foreign wars he in his turn saw an opportunity to earn some extra money by reissuing the map. He changed the dedication to George I who had succeeded Anne in 1714 but nothing else – except, and typically, for an advert in the bottom right corner for, among others, 'a new map of the world in which is contain'd the Theory of Tydes from Sr Isaac Newton', two eclipse maps and '2 sheet maps Adorned with Curious Compartment and Observations'. Wars have always been good for the map trade.

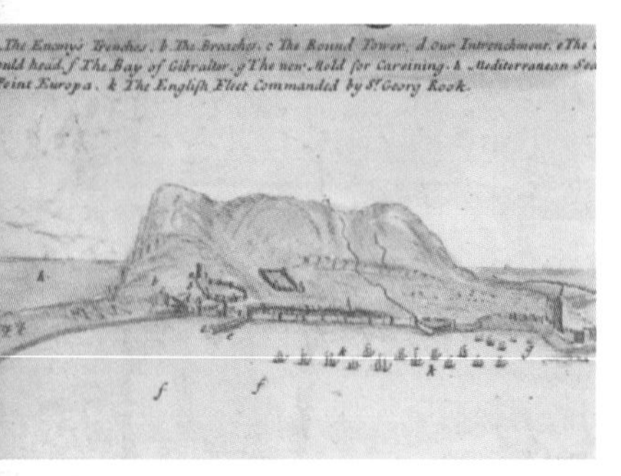

Right John Harris, *A New Map of Europe Done from the most Accurate Observations Communicated by the Royal Societies at London & Paris Illustrated with Plans &c Views of the Battles, Seiges and other Advantages Obtained by her Majesties Forces and those of Her Allies over the French.*
Above Detail showing Gibraltar.

A NEW MAP OF EUROPE
THE NORTHERN OCEAN
THE NORTH SEA OR GERMAN OCEAN
THE MEDITERRANEAN SEA
BLACK SEA
THE CASPIAN SEA
Tho. Bowles

A German emigrant on the fringes of London intellectual society produces a map of the King's dominions in North America as a means of self-advancement.

Patriotism and Self-Promotion

HERMANN MOLL'S MAP of the King's dominions in America is more commonly termed the 'Beaver Map', because of the large scene of beavers making a dam. It is the classic delineation of the British possessions along the eastern seaboard of the modern United States from the first years of the new Hanoverian dynasty. A compilation of contemporary printed and manuscript maps, it is of particular value for the depiction of the southern colonies which are highlighted in the insets.

Moll is an interesting figure in British cartographic history; he was an emigrant from Germany, and a talented 'geographer' as he termed himself. He was not a wealthy man but aspired to greater things: comfortable in the company of the great navigators of his day who would often visit his shop, he also sought acceptance in the inner political and intellectual circles of London society. He was known to men of the calibre of Robert Hooke, Daniel Defoe and Jonathan Swift (who made him Lemuel Gulliver's preferred mapmaker), but was always seemingly on the fringes of their group, often attempting to collect debts.

Right Hermann Moll, *A New and Exact Map of the Dominions of the King of Great Britain on ye Continent of North America.* **Above** Detail showing the border between the French colony of Louisiana and the British colony of Carolina. **See also** 1747.

Moll advanced himself through his maps: he hoped to gain social acceptance by promoting the values and interests of the circles he aspired to. He promoted the British Empire through cartographic metaphor; his 'Beaver Map' is as much about opportunity as geography. His maps were also about self-promotion with the title and dedication intended to bring him to the attention of the new King and of Lewis Dowglass, a powerful figure in colonial government.

Moll was also only too happy to use his maps to settle scores with rivals, reserving his most bitter attacks for the French. It is surprising, therefore, that this map tacitly accepts French boundaries for North America, granting the lands watered by the Mississippi and its tributaries to the French, rather than the British, whose charters defined the southern colonies as extending from the Atlantic to Pacific coasts. These conflicting claims would soon burst into cartographic, and then into actual, war. By the time Moll's next important map of eastern North America appeared in 1720, he had adopted an aggressive tone, railing against the French claims as unjustified 'Incroachments'.

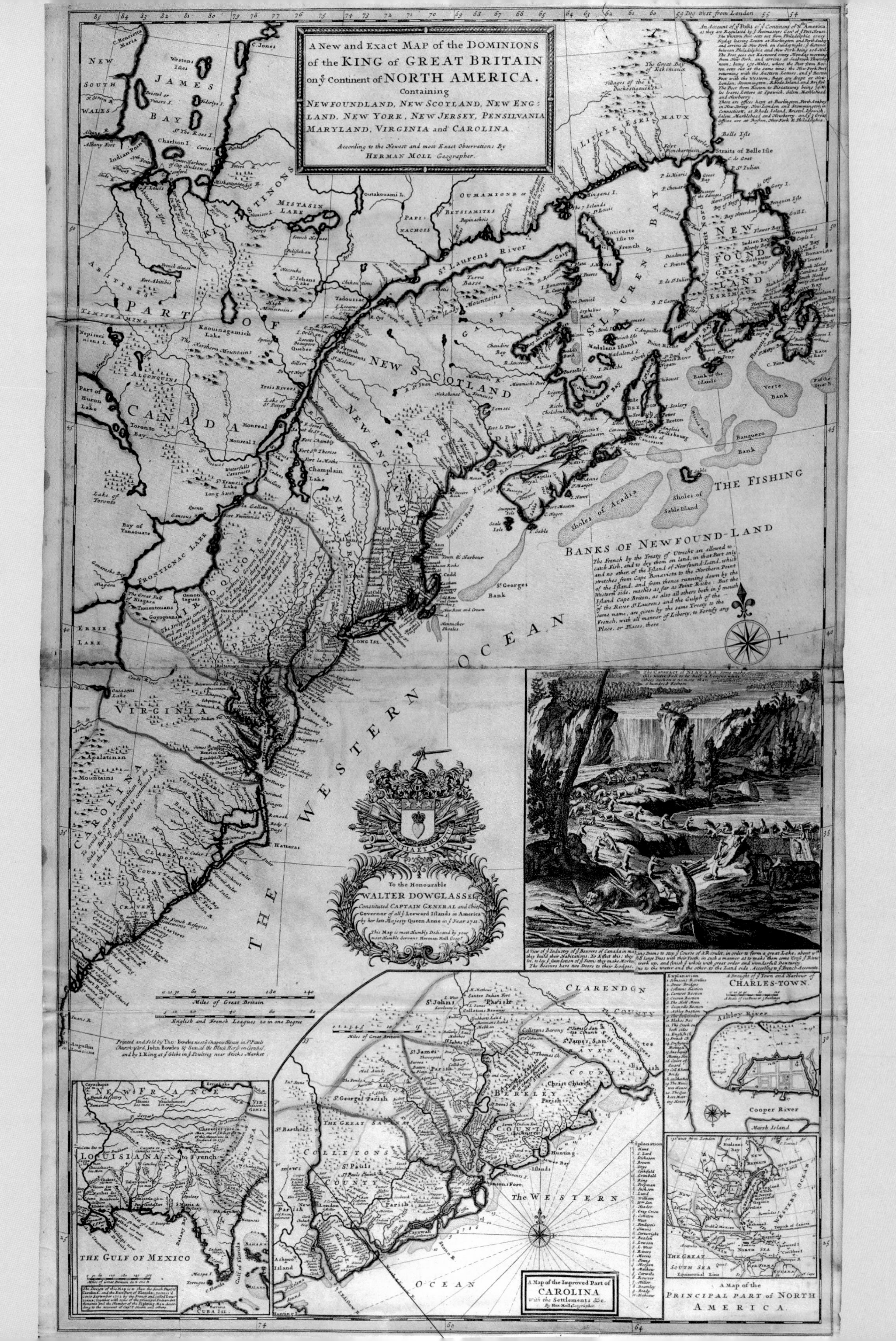

A New and Exact MAP of the DOMINIONS of the KING of GREAT BRITAIN on ye Continent of NORTH AMERICA.
Containing NEWFOUNDLAND, NEW SCOTLAND, NEW ENGLAND, NEW YORK, NEW JERSEY, PENSILVANIA MARYLAND, VIRGINIA and CAROLINA.
According to the Newest and most Exact Observations By HERMAN MOLL Geographer.
Deg. West from London
JAMES BAY
NEW SOUTH WALES
KILISTINONS
MISTASIN LAKE
St. Laurens River
PART OF CANADA
NEW SCOTLAND
NEW ENGLAND
NEW YORK
IROQUOIS
PENSILVANIA
VIRGINIA
CAROLINA
FRONTIGNAC LAKE
ERRIE LAKE
Lake of Toronto
Champlain Lake
Monreal
Quebec
LONG ISL.
NEW FOUND LAND
ST. LAURENS BAY
LITTLE ESKIMAUX
Anticoste
Banks of Newfound-Land
The Fishing
Sholes of Acadia
Sholes of Sable Island
St. Georges Bank
Ranquero Bank
Verte Bank
THE WESTERN OCEAN
C. Hatteras
Miles of Great Britain
English and French Leagues 20 in one Degree
Printed and sold by Tho. Bowles next ye Chapter House in St. Pauls Church yard, John Bowles & Son at the Black Horse in Cornhill, and by I. King at ye Globe in ye Poultry near Stocks Market
To the Honourable WALTER DOWGLASS Esq. Constituted CAPTAIN GENERAL and Chief Governor of all ye Leeward Islands in America by her late Majesty Queen Anne in ye Year 1711.
This Map is most Humbly Dedicated by your most Humble Servant Herman Moll Geogr.
The Cataract of NIAGARA
NEW FRANCE
LOUISIANA
THE GULF OF MEXICO
CUBA ISL.
CLARENDON COUNTY
BERKLEY COUNTY
COLLETONS COUNTY
Ashley River
Cooper River
A Draught of ye Town and Harbour of CHARLES-TOWN
A Map of the Improved Part of CAROLINA with the Settlements &c. By Hen. Moll Geographer.
A Map of the PRINCIPAL PART of NORTH AMERICA
THE GREAT SOUTH SEA

A modest estate map from Suffolk is swamped by an excess of mathematical information, a by-product of the Scientific Revolution.

The Cult of Science

EIGHTEENTH-CENTURY EUROPE was the setting for the Scientific Revolution when a belief in the power of the human mind came close to supplanting the belief in God as the prime mover of the universe. But for every Isaac Newton there were innumerable Isaac Caustons.

This map must be Causton's masterpiece. Drawn in the months when the enormous financial confidence trick known as the South Sea Bubble was bewitching investors in the City of London, there is evidence in the form of nail marks that the map was once hung from a bar for display. It shows the landholdings of the Bonnell family some 7 miles (11 kilometres) north-east of Sudbury and Lavenham in Suffolk. Each field is clearly differentiated and named, while a letter refers to tables giving the acreages of each field, listed under the names of the farmers John Jacob, Richard Death and John Woods who leased the lands from the Bonnell family. The 'Procession Path', separating the two parishes, is shown by parallel dotted red lines. Arable, pasture and woodland are distinguished and buildings and field gates are shown in elevation. North is at the top and the scale, of 1:3200, is a common one to find on estate maps.

All this information is, however, almost submerged beneath a dense mathematical grid drawn in red and black ink. The text panel contains a complex explanation of the mathematics underlying the grid and instructions on how to use it to locate fields and estimate acreages without the use of dividers or rulers. Causton emphasized the mathematical credentials of his map by drawing the figures of Geometry (holding a copy of Euclid) and Algebra, separated by triangulation diagrams, over the scale bar.

Mathematical skill can be an essential virtue for a cartographer but that hardly justifies such an extravaganza. The map is on a large enough scale for a grid not to be necessary. The mathematical paraphernalia may give an exaggerated idea of the accuracy of the plan which was probably constructed using centuries-old methods, such as a plane table and chains, that called for only basic mathematical skills. Causton's patron, Margaret Bonnell, however, founded a grammar school that still survives. Could Causton have been hoping for the post of mathematics teacher, the main occupation of many surveyors before 1800?

Right Isaac Causton, *A Map of Madam Margaret Bonnell's land lying in the parishes of Brent-Eligh and Preston in the County of Suffolk.* **Above** Detail showing the figures of Geometry and Algebra over the scale bar.

A MAP OF MADm MARGARET BONNELL'S LAND LYING IN THE PARISHES OF BRENT ELIGH AND PRESTON IN THE COUNTY OF SUFFOLK.

Surveyed pr Isaac Causton

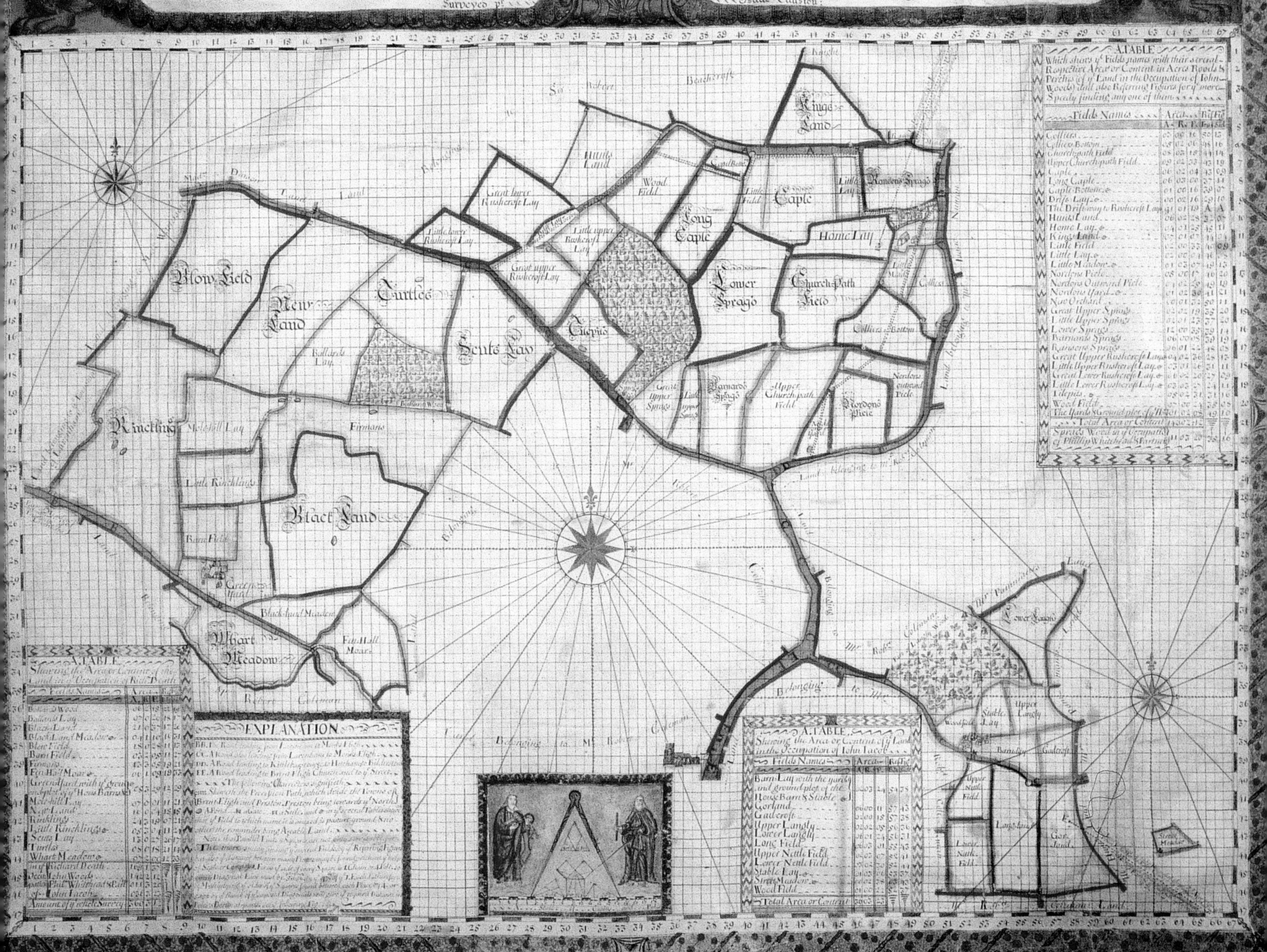

A decorative English local map made by a young schoolmaster-surveyor from a dynasty of surveyors.

Surveying and Teaching

THE EIGHTEENTH CENTURY was the golden age of the British local land surveyor, who produced a variety of large-scale manuscript maps. Many were estate maps, drawn to show a landowner's property and often beautifully decorated. Maps were thus both tools of estate administration, and symbols of status and power.

In June 1720, the Commissioners of Sewers of Ashford Levels in Kent met in Canterbury. They discovered malpractice by one of the collectors of drainage taxes and realized they needed to know the acreage of each level. Therefore they commissioned maps of the areas.

They chose a local man, Thomas Hogben, to carry out the work. He was about eighteen and was the son of a land surveyor who died in 1703. This map shows that he was already a skilled practitioner. His map is drawn on parchment and shows the local topography, including stretches of water that did not flow properly. A table lists the tenant of each field and its acreage. Hogben's map is beautifully decorated in a contemporary style with a compass rose and a scale bar surmounted by dividers. His title cartouche is distinctive, as is the note naming him as surveyor. He drew similarly decorated notes on his other maps for the commissioners, and on maps drawn for others from 1720 to 1748. Thereafter, Hogben's style became more restrained as did that of other mid-eighteenth-century surveyors, and they started drawing buildings in plan rather than in perspective.

Right Thomas Hogben, *The Topography and Mensuration of all the Flow'd Lands lying in the Valley between Ashford Bridge and Long Bridge*. **See also** 350, 1597 (p. 136), 1608, 1719.

Hogben earned £2 5*s* 4*d* for the map illustrated here (a rate of 4*d* per acre). By the 1760s he was charging 6*d* per acre and mapping a few thousand acres for the Level of Romney Marsh, but even then he would rarely have earned over £100 per annum, less than many tradesmen and farmers, and about the same as lawyers and innkeepers. He did not travel far, and probably acquired business through local recommendations: some of the marshland lords became his private clients.

Hogben also received income from his school: by 1730 he had moved from Ashford to nearby Smarden, and become the village schoolmaster to perhaps forty children. Hogben also made sundials and several can be found locally.

The youngest of Hogben's eight children, Henry, was baptized in 1741. As in many other families, Hogben passed on his skills to Henry, who became a land surveyor and regularly helped his father from the mid-1760s. Hogben died at Smarden in 1774.

Note when you see this Algebreical Sign or Character viz + it signifieth that the Piece doth not all Flow.
N
E
W
The Topography and Mensuration of all the Flow'd Lands Lying in the Valley betwee Ilford Bridge and Long Bridge Bidges Measured by Order of the Honble Commissioners of Sewers. Containing One Hundred Thirty Six Acres; Three Roods & Thirty five Perches
Actually Surveyed Delineated & Admeasured by Thomas Hogben 1720
A Table of Contents
Occupiers
A
R
P
K
L
M
N
O
P
Q
R
S
T
V
U
Y
X
Z
I
G
H
F
E
D
A

The bustling cities of London and Westminster portrayed not totally accurately by a firm of German map publishers.

Hanoverian London

IN ABOUT 1730 a plan and prospect of London was produced by the Nuremberg family firm of Johann Baptist Homann. London was the Hanoverian capital of Britain associated with the journals *Tatler* and *Spectator*, and the coffee shops frequented by Addison, Steele, Pope and Defoe.

Homann's map combines a detailed street plan with a splended panorama taken from south of the Thames. Here the dome of Christopher Wren's new St Paul's Cathedral is prominent, together with nearly forty other numbered churches, mostly built after the Great Fire of 1666. London Bridge carries a massive superstructure of houses, and the impaled heads of traitors can still be seen on the southern entrance gate – very probably a detail copied from an earlier panorama by Visscher. The fictitious northwards bend of the river beyond the Tower and the retention of Shakespeare's Globe Theatre (destroyed in 1643) show that Homann's map is not to be relied on for accuracy. Nevertheless this in no way detracts from the interesting bustle of activity on the Thames where there are boats of all kinds and capacity: galleons and rowing boats, cutters and sloops, a great hay barge and the celebrated floating banqueting hall which doubled as a bordello or brothel.

Right A map and panorama published by the heirs of J. B. Homann in Nuremberg.
See also 1574, 1748, 1793, 1800 (p. 246), 1891.

At the top of Homann's map are two vignettes. The left one shows the King's residence in Whitehall according to an earlier print by Wenceslaus Hollar dating from the 1640s. Homann makes no mention of the old Palace's destruction in 1698, leavng only Inigo Jones's central Banqueting Hall which is all that remains today. On the right, the handsome Italianate building is the new Royal Exchange, built in 1669 by Edward Jarman to replace Sir Thomas Gresham's Tudor building. The Exchange was the traditional meeting place for money dealers and merchants engaged in trading and commercial business throughout the world. Jarman's sructure, with its spacious courtyard, grand portico and tower was in its turn carried away by fire in 1838.

On the main map there are two areas of particular interest. East of London the urban spread of the cloth and silk weaving trade can be traced in the housing developments around Spitalfields and Spiers (or Spicer's) Field. To the west, new estates for the wealthier gentry were springing up around Soho and Hanover Square, the latter laid out between 1717 and 1719.

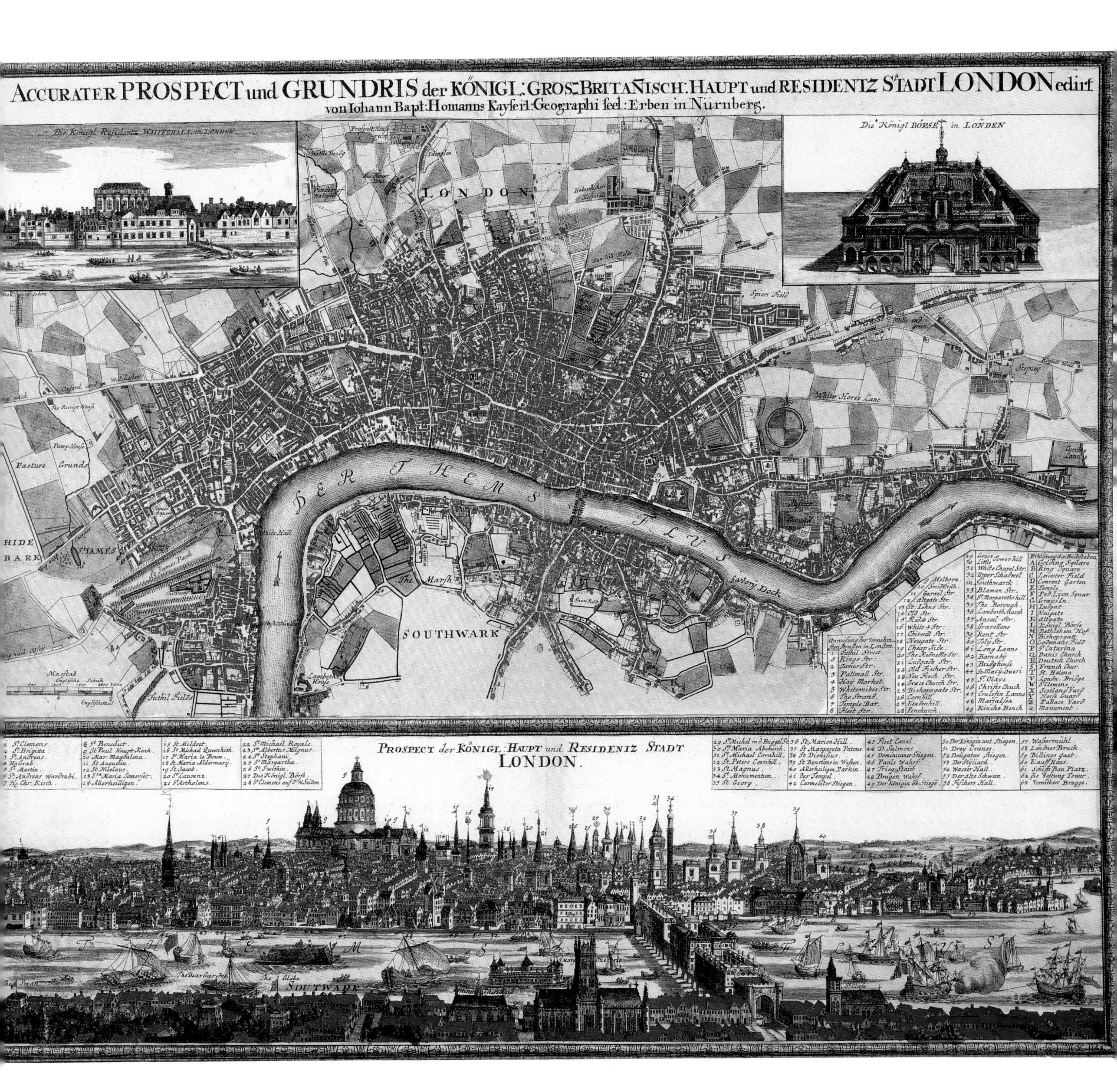
ACCURATER PROSPECT und GRUNDRIS der KÖNIGL: GROS: BRITAÑISCH: HAUPT und RESIDENTZ STADT LONDON edirt
von Iohann Bapt: Homanns Kayseil: Geographi seel: Erben in Nürnberg.
Die Königl BÖRSE in LONDEN
LONDON
Spiers Field
White Horse Lane
Pasture Grunds
HIDE BARK
DER THEMS FLUS
The Marsh
SOUTHWARK
Stepney
PROSPECT der KÖNIGL: HAUPT und RESIDENTZ STADT
LONDON.
SOUTWARK

A survey of colonial Texas by a Spanish military engineer which helped the government in Mexico City maintain a tenuous authority over its outposts.

Spanish Texas

IN THE REIGN OF PHILIP II (1556–98), the Spanish monarchy developed a large and skilful service of military engineers. At first they were largely employed in fortifying the Habsburg strongholds in Europe, particularly in the Low Countries, on the Danube valley and in Italy. But after Sir Francis Drake's piratical raid in the West Indies of 1585–6, Philip II sent some engineers as well to the Spanish islands and to the Main, where they worked for many years on such sites as San Juan (Puerto Rico), Cartagena (Colombia), Vera Cruz (Mexico) and Havana (Cuba).

In Europe, the engineers had often drawn maps and plans as well as fortifying sites, and so it was across the Atlantic. They played a part in the huge effort to reduce to paper the vastness of countries ranging from northern California to Patagonia; in this they were joined not only by sea-captains but also by many Jesuit priests. Francisco Alvarez Barreiro was a late representative of these engineers. Born in Spain, he received an appointment in New Spain in 1716, and the following year accompanied Martín de Alarcón on an expedition into the new province of Texas. Here, the Spaniards were establishing themselves at San Antonio, in response to the new French settlement at New Orleans. Remaining in Texas until 1720, Barreiro returned there between 1724 and 1728, compiling five provincial maps now at the Archive of the Indies in Seville. From these was compiled the general map shown here; one copy of this is at the British Library and another at the Hispanic Society of America in New York.

Barreiro has a good idea of the general line of the rivers in Texas, from the Rio Grande in the west to the Sabine River in the east. He also shows the two *caminos reales* (royal roads) which crossed the country from west to east, allowing the government in Mexico City to maintain a tenuous authority over its outposts towards the French settlements on the Mississippi River. Note, too, how he shows the extensive settlements in Mexico with church symbols, and the settlements of the indigenous peoples to the north and east with varying numbers of well-arranged huts.

Right Francisco Alvarez Barreiro, *Plano corographico é hydrographico de las provincias de el Nuevo Mexico...* **Above** Detail showing symbols for settlements.
See also 1579, 1780 (p. 230).

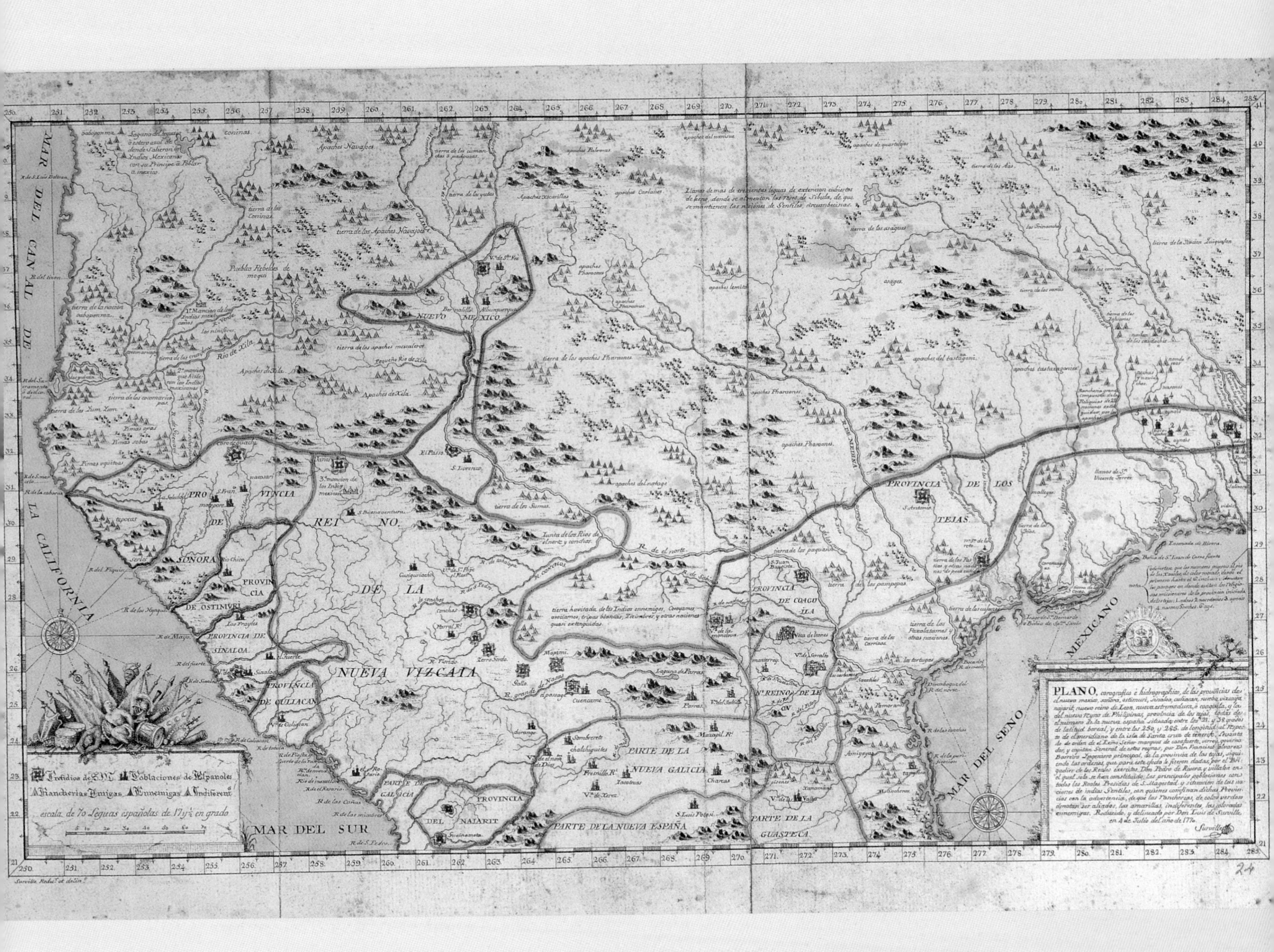
PLANO, corografico è hidrographico, de las provincias de el nuevo mexico, sonora, ostimuri, sinaloa, culiacan, nueva vizcaya, najarit, nuevo reino de Leon, nueva estremadura, ò coaguila, y la del nuevo Reyno de Philipinas, provincia de los tejas, todas de el numero de la nueva españa situadas entre los 21. y 38 grados de latitud boreal, y entre los 250, y 285. de longitud al respecto de el meridiano de la isla de Santa cruz de tenerife. Levantado de orden de el Exmo Señor marques de casafuerte, virrey, governador, y capitan General de estos reynos, por Don Francisco Alvarez Barreiro Ingeniero principal, de la provincia de los tejas, siguiendo las ordenes, que para este efecto le fueron dadas, por el Brigadier de los Reales exercitos Don Pedro de Rivera y villalon en el qual solo se han constituido, las principales poblaciones con todas los Reales Presidios de S. Magestad y situacion de las naciones de indios Gentiles, con quienes confinan dichas Provincias con la advertencia, de que las Rancherias, de color verdes denotan ser aliados, las amarillas, indiferentes, las coloradas enemigas. Reducido, y delineado por Don Luis de Surville en 4 de Julio del año de 1770.
Surville
Presidios de S.M. Poblaciones de Españoles
Rancherias Amigas Enemigas Indiferent
escala de 70 Leguas españolas de 17½ en grado
MAR DEL CANAL DE LA CALIFORNIA
MAR DEL SUR
MAR DEL SENO MEXICANO
NUEVO MEXICO
PROVINCIA DE SONORA
PROVINCIA DE OSTIMURI
PROVINCIA DE SINALOA
PROVINCIA DE CULIACAN
REINO DE LA NUEVA VIZCAYA
PARTE DE LA NUEVA GALICIA
PROVINCIA DEL NAIARIT
PARTE DE LA NUEVA ESPAÑA
PARTE DE LA GUASTECA
PROVINCIA DE COAGUILA
N. REINO DE LEON
PROVINCIA DE LOS TEJAS
tierra de los Apaches Navajoes
Apaches Navajoes
Rio de Xila
tierra de los cocomaricopas
tierra de los Zum Zum
Pimas sobas
Pimas ygitoas
tierra de los apaches Pharaones
apaches Pharaones
El Passo
R. de el norte
R. grande de Nasas
Mapimi
tierra de los Sumas
apaches del natage
tierra de los apaches del tuelagani
apaches tastaxagonies
tierra de la Nacion Luiquesan
los Grinanches
tierra de los cursiros
tierra de los vanitis
tierra de los Tejianes
nacion de los caudachos
R. de la trinidad
Rancheria grande Compuesta de las Reliquias de 22 naciones extinguidas por los apaches
llanos de Sn Vicente Ferrer
R. de S. Marcos
Lago de Sn Bernardo ò Bahia de Spto Santo
coro de guachi
La Soledad
S. Fran.
tepocas
R. del Sacramento ò del coral
R. de S. marcelo
R. de la caborca
tierra de la nacion oabopon ma

The best plan of the Battle of Culloden by a mapmaker who fought in the battle at Bonnie Prince Charlie's side also displays his Jacobite sympathies.

The Last Battle

Right John Finlayson, *Plan of the Battle of Culloden.*

CULLODEN, THE LAST BATTLE in mainland Britain, is often misleadingly portrayed as a fight between Scots and English. In fact, Scots were on both sides and the complex roots of conflict go back at least two centuries, as opposing religious, ideological and political beliefs divided the nation. Believers in the divine right of kings, the Roman Catholic Stuart kings represented tradition, hierarchy, a monarchy of absolute power. But the mood of the country was changing, and James II (or VII in Scotland) was deposed in 1689 in favour of a Protestant monarchy, whose powers were tempered by Parliament.

Jacobites supported James II and his heirs (*Jacobus* is Latin for James). Several 'Risings' or 'Rebellions' – terminology depending on allegiance – attempted to reinstate James II or his son James, the 'Old Pretender' (from the French *prétendre* meaning 'to claim'), culminating in July 1745 with the landing in north-west Scotland of Prince Charles Edward Stuart, popularly called Bonnie Prince Charlie. The 'Young Pretender', grandson of James II, accompanied by a handful of French and Irish supporters (both nationalities desiring to destabilize the British government) attracted Scottish support mainly from the disaffected Highlands and islands where the clan system, based on kinship, was under threat from increasingly centralized government.

Initial military success and a rapid advance took the Prince and his assorted army to within a few days of storming London, when a crisis of confidence, lack of support and extended lines of supply entailed a rapid retreat north from Derby. On 16 April 1746 at Culloden, near Inverness, a depleted force of perhaps six thousand weary and hungry Jacobites was routed in about an hour, then butchered, by about ten thousand government troops. The Prince fled and was hunted for several months, until escaping to France.

Artillery ordnance master in the battle, John Finlayson's map was acknowledged by contemporaries as the best plan of the encounter. A mathematical instrument maker, he also served as the Prince's engineer and commissar. In the decorative cartouche, poignant symbols of the end of the Jacobite dream display his sympathies: a snuffed out candle; a chained lion and broken unicorn horn (the Scottish lion rampant coat-of-arms has unicorns as heraldic supporters); broken thistles (symbol of Scotland, according to legend, since cries of pain from barefoot attackers standing on thistles alerted sleeping Scots). Finlayson's use of 'The Pr.' is cleverly ambiguous – Prince or Pretender?

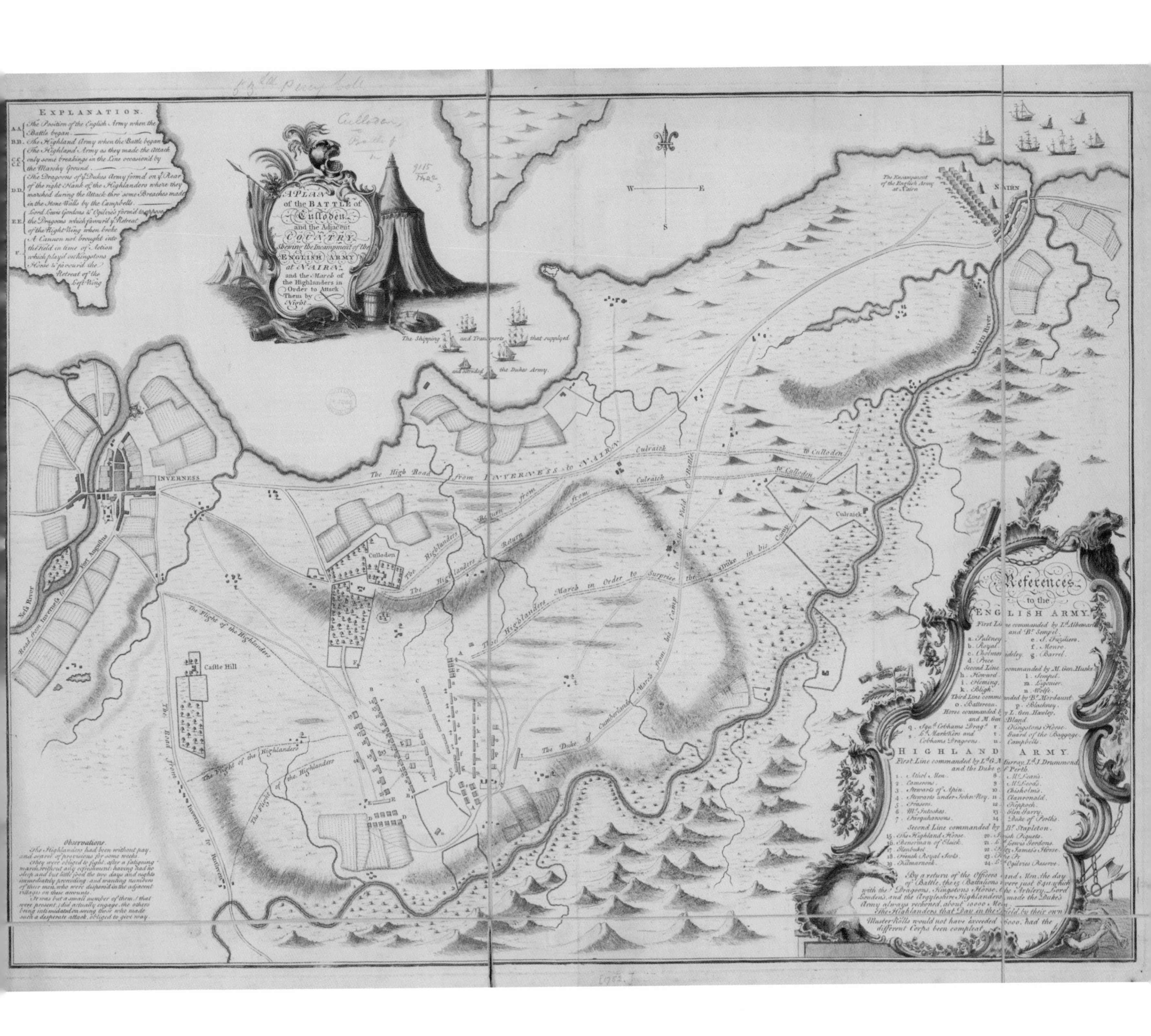
A PLAN of the BATTLE of Culloden and the Adjacent COUNTRY, Shewing the Incampment of the ENGLISH ARMY at NAIRN, and the March of the Highlanders in Order to Attack Them by Night.
EXPLANATION.
References to the ENGLISH ARMY.
HIGHLAND ARMY.
Observations.
INVERNESS
Culloden
Castle Hill
NAIRN
Nairn River
Culraick
The High Road from INVERNESS to NAIRN
The Flight of the Highlanders

A rare survival of a cartographic map screen probably created for a London merchant with commercial interests in North America and Central Europe.

A Cartographic Draught Excluder

Right A map screen assembled by John Bowles. See also 1715.

MAPS COME IN many forms. One of the most curious is the map screen. Though known in Europe as early as the sixteenth century (Henry VIII is recorded as having one), and to be found in seventeenth-century Japan, they became particularly popular in Europe in the eighteenth century. They combined utility with fashion and sometimes, as in this case, self-advertisement. Screens helped to shield people in the rooms of eighteenth-century houses, however elegant, from the draughts that whistled through sash windows, under doors and down chimneys. During the craze for all things Chinese, their backs were painted black and overlaid with flowers in imitation of oriental lacquer, even though they had been assembled only a few streets away. Finally the maps themselves advertised the owner's knowledge of the world, his civic patriotism (some screens were covered with multi-sheet images of cities like London) or his commercial interests.

This mid-eighteenth-century screen was put together by the London mapseller John Bowles who, as a member of the Joiners Company, also had links with furniture making. He seems to have viewed the map screen as a way of using up maps he had in stock. Some of these, dating back to the previous century, would have been obsolete and might well have been almost impossible to get rid of had they not been used in this way. Spreading out from a map of the world (top centre) are a map of the British Isles, a map of the American colonies, views of London landmarks and, beneath them, maps of those perennial European battlegrounds, the Netherlands, northern Germany and Catalonia, with another map of the American colonies. The purchaser could well have been a London merchant, who might perhaps have been involved in the government of London, and who had commercial links to the American colonies and central Europe.

Some maps were purpose-made for screens like a series which was published by a succession of London mapsellers from 1721. When mounted they effectively constituted a vertical world atlas with a markedly Anglocentric bias, working outwards geographically (and at an ever-decreasing scale) from maps of London, Oxford and Cambridge. The maps were repeatedly reprinted and such a screen was mentioned as late as 1777 in Richard Brindsley Sheridan's popular play *The School for Scandal*. The screens were, however, cumbersome and very fragile. Very few indeed seem to have survived into the twenty-first century.

A detailed map published within weeks of the great fire of 1748 brings the mid-eighteenth century City of London to life.

Disaster in London

NOT THE MOST POLISHED performance ever to emanate from the hand of Thomas Jefferys (1719–71), future Geographer to the King and among the most renowned mapmakers of the period – but a map produced in a hurry, and for the best of causes. The fire that swept the tangle of alleys and courtyards between Lombard Street and Cornhill in the early hours of that fateful morning of 25 March 1748 began in the powdering-room of a peruke-maker. It destroyed, with consequent loss of life, eighty or more dwellings in the City of London, just across the road from the Royal Exchange.

The purpose of the plan – compiled, engraved, printed, published and on sale within a week of the tragedy – was to launch an appeal for the survivors. But it gives us far more than a glimpse of this characteristic manifestation of eighteenth-century altruism – it opens a door on a lost world. Individual occupants and occupations are identified and, to our surprise, these are people and places we know. Here are the princely coffee houses – Garraway's, Jonathan's, The Jerusalem and The Jamaica chief among them – where anything in the world at this date might be bought, sold, traded, jobbed, bartered, exchanged, auctioned, lotteried or insured. It was at Garraway's, they say, that tea was first sold in England and the South Sea Bubblers bubbled. In among the coffee houses are the insurance offices, The London dwarfing its smaller rivals, and stitched between them in the weft of this densely occupied area are milliner, hosier, haberdasher, toyman, music-shop, fruit-stall, cabinet-maker, and all the accessories of a civilized life down to the improbable Weston's Elixir Warehouse.

And last of all the booksellers lined along Cornhill – the Scotsman Robert Willock at Sir Isaac Newton's Head (he had published Newton's assay calculations of foreign coinage early in his career); William Meadows at The Angel next door, publisher of *A Sermon against the Dangerous and Sinful Practice of Inoculation* in 1722; old John Brotherton, originally trained as a bookbinder in the 1680s, at The Bible; George Strahan at The Golden Ball – 'Honest Strahan ... always for doing the fair thing', as Dunton recalled; John Walthoe, whose catalogue survives in the British Library; and Thomas Astley, publisher of the splendid *A New General Collection of Voyages and Travels* (1743–7), more often known simply as *Astley's Voyages*. If the map can be a species of time travel, we have arrived back in 1748.

Right Thomas Jefferys, *A Plan of all the Houses Destroyed & Damaged by the Great Fire, which begun in Exchange Alley Cornhill, on Friday March 25, 1748.*
Above A view of the same area.
See also 1574, 1715, 1730 (p. 196).

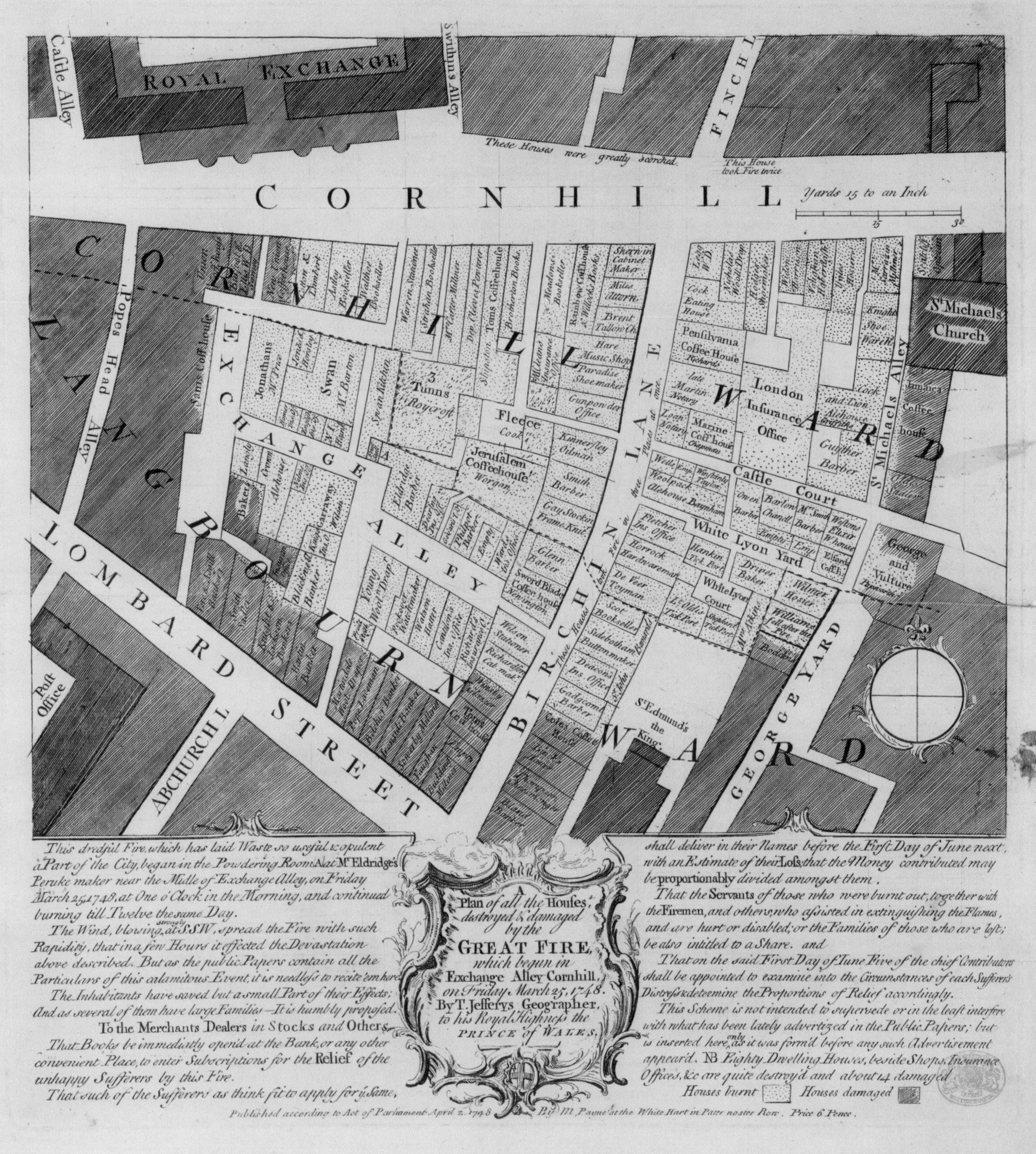

ROYAL EXCHANGE
Castle Alley
Swithin's Alley
FINCH L
These Houses were greatly scorched.
This House took Fire twice
CORNHILL
Yards 15 to an Inch
15
30
CORNHILL WARD
LANGBOURN WARD
BIRCHIN LANE
EXCHANGE ALLEY
LOMBARD STREET
ABCHURCH L.
GEORGE YARD
Popes Head Alley
St. Michaels Alley
St. Michaels Church
Castle Court
White Lyon Yard
White Lyon Court
St. Edmund's the King
Post Office
Jonathans
Swan
3 Tunns
Fleece
Jerusalem Coffeehouse
Pensilvania Coffee House
London Insurance Office
Marine Coff. house
George and Vulture
Sword Blade Coffee house
Jamaica Coffee house
Cock and Lion Alehouse
Woolpack Alehouse
Toms Coffeehouse
Rainbow Coffeehouse
Gunpowder Office
Hare Music Shop
Paradise Shoemaker

A Plan of all the Houses, destroyed & damaged by the GREAT FIRE, which begun in Exchange Alley Cornhill, on Friday March 25, 1748. By T. Jefferys Geographer, to his Royal Highness the PRINCE of WALES.

This dredful Fire, which has laid Waste so useful & opulent a Part of the City, began in the Powdering Room A at Mr. Eldridge's Peruke maker near the Midle of Exchange Alley, on Friday March 25, 1748, at One o'Clock in the Morning, and continued burning till Twelve the same Day.
The Wind, blowing strongly at S.S.W., spread the Fire with such Rapidity, that in a few Hours it effected the Devastation above described. But as the public Papers contain all the Particulars of this calamitous Event, it is needless to recite 'em here.
The Inhabitants have saved but a small Part of their Effects; And as several of them have large Families — It is humbly proposed.
To the Merchants Dealers in Stocks and Others,
That Books be immediatly open'd at the Bank, or any other convenient Place, to enter Subscriptions for the Relief of the unhappy Sufferers by this Fire.
That such of the Sufferers as think fit to apply for ye Same, shall deliver in their Names before the First Day of June next, with an Estimate of their Loss; that the Money contributed may be proportionably divided amongst them.
That the Servants of those who were burnt out, together with the Firemen, and others, who assisted in extinguishing the Flames, and are hurt or disabled; or the Families of those who are lost; be also intitled to a Share. and
That on the said First Day of June Five of the chief Contributors shall be appointed to examine into the Circumstances of each Sufferer's Distress & deteemine the Proportions of Relief accordingly.
This Scheme is not intended to supersede or in the least interfere with what has been lately advertized in the Public Papers; but is inserted here only as it was form'd before any such Advertisement appear'd. NB Eighty Dwelling Houses, beside Shops, Insurance Offices, &c are quite destroy'd and about 14 damaged.
Houses burnt
Houses damaged
Published according to Act of Parliament April 2. 1748 By M. Payne at the White Hart in Pater noster Row. Price 6 Pence.

XXI
9

London

A cosmological map in many ways similar to a complicated board game that illustrates the beliefs of the Jains.

The Habitable World According to the Jains

Right Jain cosmological map from western India. See also 1265, 1290, 1300.

THIS IS BOTH a map illustrating 'sacred geography' as well as a depiction of the world as conceived by the Jains, followers of a non-theistic religious system finalized by Mahavira, the last of the twenty-four Jinas or 'Saviours', in the sixth/fifth century BC in northern India. The Jains worried constantly about the physical appearance of the universe and successive generations of Jain monks and scholars worked out its different 'categories' along with their precise features and measurements. The universe comprised not only the visible world spread out around them (Manusya-loka, the world of men) but the countless continents spread out beyond it and the innumerable heavens and hells above and below it.

Manusya-loka consists of the inner two and half a continents (*adhai-dvipa*) of a system of annular (ring-like) continents (*dvipa*) each separated from its neighbours by oceans. In the middle of the central circular continent, Jambu-dvipa, rises the axis of the world, Mount Meru, inhabited by gods. Jambu-dvipa is divided into seven lands by parallel mountain ranges. Bharata (corresponding to India) occupies the lowest segment of Jambu-dvipa beneath the Himavat mountains. From Himavat descend the two rivers Sindhu and Ganga to west and east, which make their way through the parallel subsidiary Vaitadhya mountains to the ocean. Bharata is thus divided into six regions, but civilized men (*Arya*) dwell only in the central coastal region, the land between the Indus and the Ganges, corresponding in many details to India itself, while barbarians (*Mleccha*) occupy the other five. By an easy progression India (i.e. Bharata) came to be seen as the lowest segment of a much larger circular continent. These physical arrangements are repeated with necessary adjustments in all the other lands in Jambu-dvipa. The other annular continents are divided into two halves by mountains running north–south and each half has its own seven lands arranged similarly to Jambu-dvipa. The third continent is also divided into two concentric parts by the Manusottara ('Limits of man') mountains marking the limit of possible human existence.

Large diagrams such as this with their images of the Jinas remind the faithful that only in Manusya-loka can Jinas, the saviours of mankind, be born. Accordingly, after innumerable lives elsewhere in the universe, all must finally be reborn here as men in the 'two and a half continents' to obtain release (*moksa*) from transmigration.

The first scientifically measured British navigational chart aimed to provide safe passage for ships through the treacherous seas around the Scottish islands.

For the Benefit and Safety of Navigation

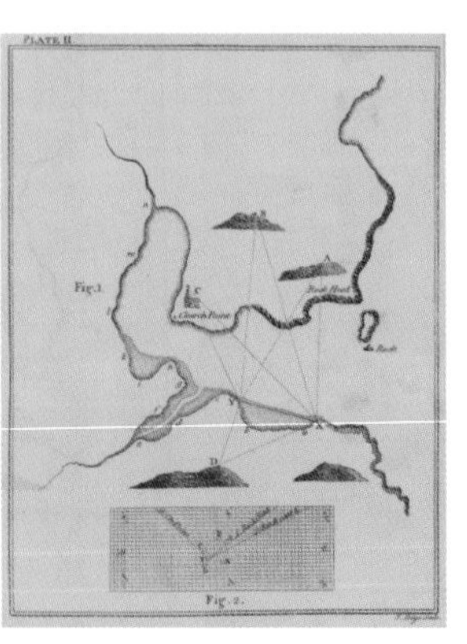

Right *The north east coast of Orkney with the rocks tides soundings &c. surveyed and navigated by Murdoch Mackenzie.* **Above** Mackenzie's diagram, from *A Treatise of Maritim Surveying*, shows triangulation from a base-line measured on a beach.

'FOR THE BENEFIT and safety of navigation the Orkney Isles are proposed to be exactly surveyed and navigate [*sic*] by subscription. All the rocks, shoals, soundings and courses of the tides will be justly marked and represented, several prospects of the land taken from the sea, and directions given for sailing through the several channels and friths [*sic*]...'

This modest advertisement in the *Caledonian Mercury* in April 1743 signalled a new era for British hydrographic surveying. Murdoch Mackenzie (1712–97) was the first to prepare sea charts of the British coast using triangulation from a land-based survey – by measuring one length and two angles of a triangle, he could calculate the other sides and angles using trigonometry, without actual measurement. This was considerably more accurate than previous methods, which involved sketching charts from boats, at the mercy of wind, tide and currents.

Mackenzie published *Orcades* in 1750, containing five charts of his homeland, Orkney, with three, less detailed, of Lewis. Map IV is illustrated, showing navigational aids, such as profile views, soundings and the nature of the seabed (e.g. sand, rock). Many of the symbols are still in use today.

Orcades was a success and Mackenzie was commissioned by the Admiralty from 1751 to 1757 to chart the west of Scotland, as the Jacobite Risings had revealed serious gaps in government information. Until retiring from active surveying in 1771, he went on to survey the west coast of England and around Ireland, publishing these charts in *A maritim* [sic] *survey of Ireland and the west of Great Britain in 1776*.

Mackenzie provided relatively few soundings as they were very time-consuming to locate and record; his aim was to show routes of safe passage rather than information about the entire seabed. This problem was solved by the invention of the station-pointer, a three-pronged instrument which could measure two angles simultaneously, thus locating points off-shore more easily. Attributed to a young relative, Graeme Spence, assistant hydrographer, the station-pointer's introduction came too late for Mackenzie to use himself, but is first described in what is perhaps his greatest legacy, *A treatise of maritim surveying* (1774). For over fifty years this practical handbook trained British navigators in the art and science of hydrographic surveying, during the period when Britannia came to rule the waves.

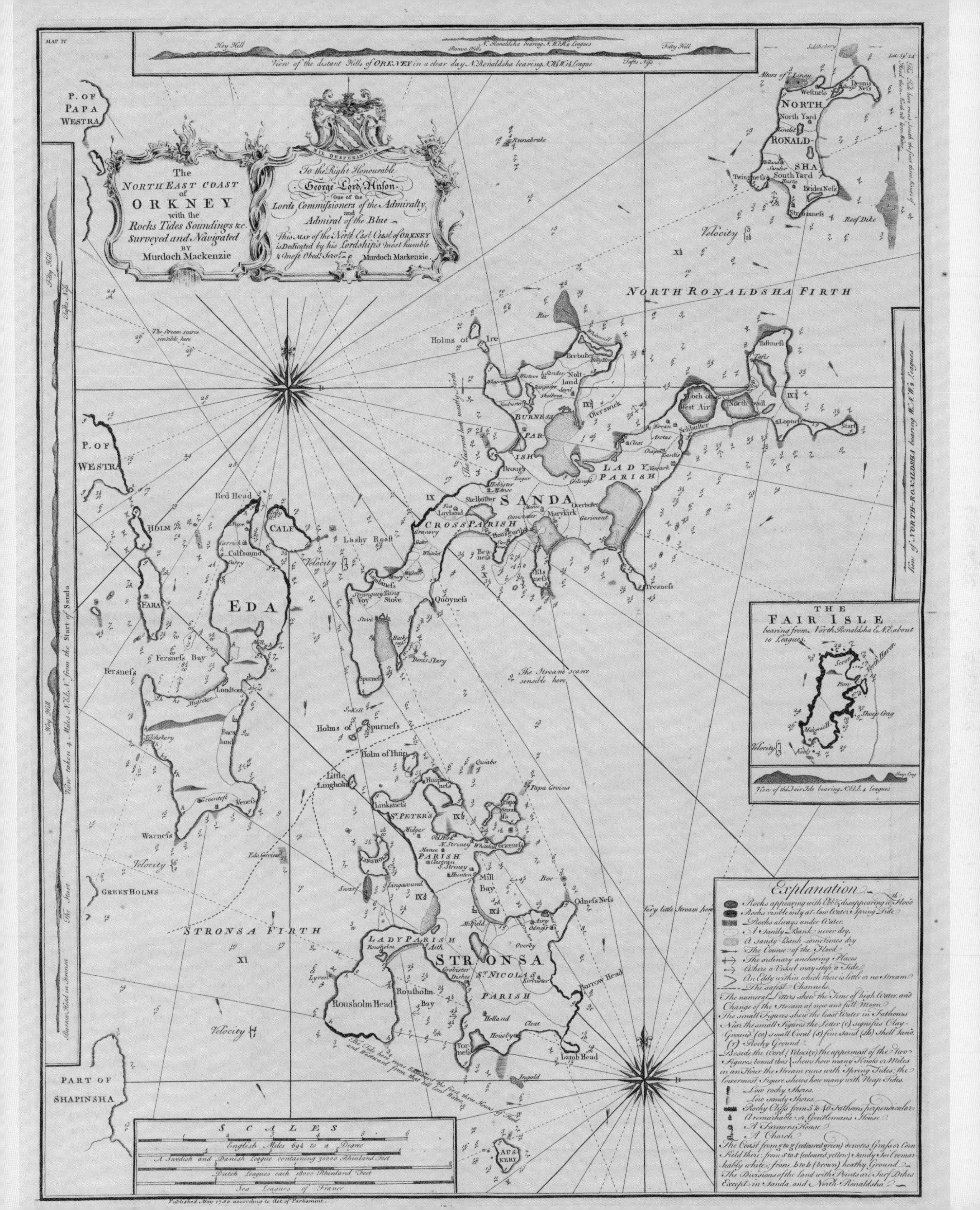
MAP IV
View of the distant Hills of ORKNEY in a clear day N. Ronaldsha bearing N.b.W. 1/4 League
Hoy Hill
Fitty Hill
The NORTH EAST COAST of ORKNEY with the Rocks Tides Soundings &c. Surveyed and Navigated BY Murdoch Mackenzie
NIL DESPERANDUM
To the Right Honourable George Lord Anson One of the Lords Commissioners of the Admiralty and Admiral of the Blue
This MAP of the North East Coast of ORKNEY is Dedicated by his Lordship's most humble & most Obed.t Serv.t Murdoch Mackenzie.
P. OF PAPA WESTRA
P. OF WESTRA
NORTH RONALDSHA
NORTH RONALDSHA FIRTH
SANDA
CROSS PARISH
LADY PARISH
BURNESS PARISH
EDA
CALF
HOLM
FARA
STRONSA
ST. NICOLAS PARISH
LADY PARISH
ST. PETER'S PARISH
STRONSA FIRTH
PART OF SHAPINSHA
GREEN HOLMS
AUSKERY
Velocity
The Stream scarce sensible here
THE FAIR ISLE bearing from North Ronaldsha N.E. about 10 Leagues.
View of the Fair Isle bearing N.E.b.E. 4 Leagues
View taken 4 Miles N.E.b.N. from the Start of Sanda
View of North Ronaldsha bearing W.N.W. 2 Leagues
Explanation.
Rocks appearing with Ebb & disappearing w.th Flood
Rocks visible only at low Water Spring Tide
Rocks always under Water.
A Sandy Bank never dry.
A sandy Bank sometimes dry
The Course of the Flood
The ordinary anchoring Places
Where a Vessel may stop a Tide.
An Eddy within which there is little or no Stream
The safest Channels.
The numeral Letters shew the Time of high Water, and Change of the Stream at new and full Moon
The small Figures shew the least Water in Fathoms
Near the small Figures the Letter (c) signifies Clay-Ground (co) small Coral (s) fine Sand (sh) Shell Sand. (r) Rocky Ground
Beside the Word (Velocity) the uppermost of the two Figures, bound thus { shews how many Knots or Miles in an Hour the Stream runs with Spring Tides; the lowermost Figure shews how many with Neap Tides.
Low rocky Shores.
Low sandy Shores.
Rocky Cliffs from 5 to 40 Fathoms perpendicular
A remarkable or Gentleman's House
A Farmers House
A Church
The Coast from g to g (coloured green) denotes Grass or Corn Field there; from s to s (coloured yellow) Sandy Soil remarkably white; from b to b (brown) heathy Ground
The Divisions of the land with Points are Turf Dikes Except in Sanda, and North-Ronaldsha
SCALES
English Miles 69 1/2 to a Degree
A Swedish and Danish League containing 30000 Rhinland Feet
Dutch Leagues each 18000 Rhinland Feet
Sea Leagues of France
Published May 1750 according to Act of Parliament.

Maps painted on silk to delight and inform a Chinese Emperor planning to move between his travelling palaces.

An Imperial Progress

Right Map of Hangzhou and the imperial 'travelling palaces' prepared for the Qianlong emperor with the hardwood travelling case and descriptive text.

INSIDE THE WOODEN case are five maps, bound in yellow brocade and wrapped in yellow satin. The maps, painted on silk and folded concertina-style, depict routes around the West Lake at Hangzhou, in Zhejiang province, from one imperial 'travelling palace' to another. Each map is accompanied by a manuscript on paper, similarly folded and bound, describing the major landmarks as well as distances between buildings and sights on the route in *li* (about one-third of an English mile).

The West Lake has been visited and admired by painters and poets for thousands of years and the Qing emperors, Manchus from the north-east who ruled China from 1644–1911, paid many visits to Hangzhou, staying in lodges specially constructed for them around the lake. A 'xing gong' or travelling palace, a rectangular enclosure with red walls and grey roofs, can be seen on the extreme right of the map, inside the battlemented walls of Hangzhou. The imperial route (in yellow) leads leftwards, up towards the military drill ground, and out through the Qiantang gate (its lower walls painted blue), towards the willow-fringed lake. It then leads over a long dyke, the 'Bai' causeway, named after Bai Juyi (772–846) one of China's greatest poets who, when Governor of Hangzhou, ordered the strengthening of the dyke (a fact noted in the accompanying text). The route ends on Gu shan (Lonely mountain) with another more spacious 'travelling palace' (extreme left).

Pavilions were built on Gu shan when the Kangxi Emperor visited in 1699 but the Qianlong Emperor, for whose first visit in 1751 these maps were probably prepared, greatly expanded the imperial lodge and in 1790, as a mark of special favour, deposited in a specially constructed pavilion one of the sets of the imperial collectanea, *Siku quanshu*, in some 36,000 volumes.

Though a mathematical tradition of mapping, with an understanding of scale, was established in China before the Han dynasty (206 BC–AD 220), pictorial representation was not considered to detract from the utility of a map as in this example where willows and mountains are depicted in a painterly manner. It was also common to find a considerable mass of explanatory text associated with maps, providing accurate detail of relative distance and scale which was not always a primary consideration in Chinese mapmaking. Here, as well as in the explanatory booklet, the distances between places are noted on narrow slips of gold paper pasted on the map itself.

Maps, Mammon & Monarchs

Just as there is a traditional view of an artist or composer heroically creating masterpieces in face of adversity and lack of appreciation, so a similar view has built up of the mapmaker. He stands on the prow of a ship, in the midst of a storm, charting outlines of the dimly perceived coastline of unknown lands. Alternatively, unrecognized and without financial backing, he doggedly braves steaming jungles, Antarctic storms, pouring rain and savage dogs to create trailblazing and amazingly accurate maps.

This book contains examples demonstrating that such maps and mapmakers – and such hazards – have existed (1500, 1564, 1750 p. 208, 1873, 1914 p. 310), but mostly the reality has been different, though no less interesting. In the eighteenth century there were a surprisingly high number of female mapmakers (1764) and the primary concern of them and their male counterparts was to earn a living through the sale of their maps. In late medieval Europe, following the decline of the monastic scriptoria, and afterwards as engravers, mapmakers hired out their skills and created maps designed and commissioned by others, very occasionally leaving their own traces through their doodles (1415, 1669). After the introduction of printing most map publishers were all too aware of the need to keep down their costs. Copperplates were expensive and this frequently meant the re-use of old plates, long after the information on them had ceased to be current, sometimes refreshed with new decoration and new information (1707), but often left untouched in the hope that the purchasers would not notice (1785, 1890). Another ploy was to raise income through flattery, dedicating the map to powerful individuals (1669), indicating wealthy merchants' houses (1790) or showing their coats of arms. Advertising was another alternative (1890, 1930, 1941 p. 332). Nevertheless sponsorship sometimes enabled them to undertake surveys for which there would otherwise not have been the money. Thus only a loan from the Phoenix Fire Office, anxious to know the whereabouts of the properties it had insured, enabled Richard Horwood to complete the house-by-house survey of London for which he is famous (1793).

Most important of all publishers needed to include features likely to appeal to the potential buyer. The Strasbourg publisher Johannes Grüninger outraged the Nuremberg humanist, Willibald Pirkheimer, in 1524, by inserting depictions of monsters and fabulous creatures on the up-to-date maps that Pirkheimer had sent him for engraving: Grüninger insisted that the public would not buy the maps unless they saw them! More constructively, in the early 1600s, John Speed encouraged the sale of his county maps, which were largely copied from Saxton's (1576), by appealing to local patriotism by adding plans of the relevant county towns, the arms of major figures from the past, and views of local antiquities or of stirring events from the county's past.

It was rare indeed for mapmakers to be left free to indulge their intellectual interests unconcerned by worldly considerations as Ringmann and Waldseemüller were enable to at St Dié through the financial support of the Duke of Lorraine (1507). Maps of outstanding importance, like the Ebstorf and Hereford world maps (1290, 1300) and the pioneering map-view of Augsburg of 1521 might nonetheless be created, but their content was influenced by the political and personal need of the patrons to advertise their status through them. It was the essentially the same story with many of the most famous maps. The now-lost world and the partially lost town maps created in Imperial Rome (200, 300, 1025), the lost world maps in Henry III's palace at Westminster (see 1265), in the town hall of the republic of Siena, the ducal palace of Renaissance Venice, and the world maps in the town hall (now royal palace) of Amsterdam were created to underline the importance of Imperial Rome, Angevin England, medieval Siena, Renaissance Venice and the seventeenth-century Dutch Republic. Part of the purpose of the maps in the Gallery of Maps in the Vatican (1580) was to impress and possibly intimidate visitors with the Pope's political as well as spiritual power. Vincenzo Coronelli (1688) created the largest surviving set of world globes for Louis XIV and the celestial globe records the position of the stars at the time of the King's birth in 1638.

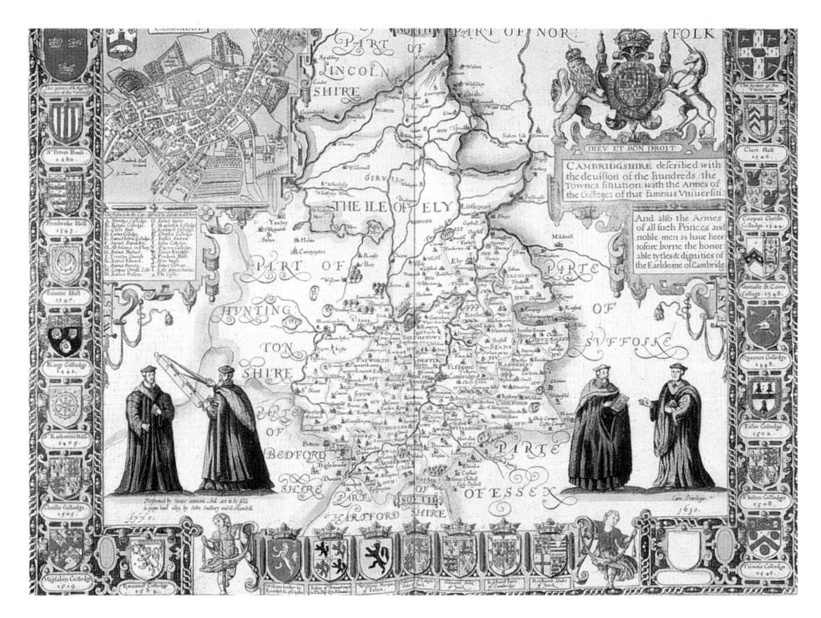

John Speed's decorative map of Cambridgeshire.

Sometimes maps and atlases have served as commercial and political tools. In the fourteenth century we know that handsome, Catalan-style portolan maps (1465) were commissioned as gifts to flatter the kings of France and Italy into granting Italian merchants commercial privileges – and the largest known atlas was assembled for a consortium of Dutch merchants as a present for Charles II of England in 1660 (see p. 164). In the mid-sixteenth century Charles V used enormous manuscript maps of the world by the most outstanding mapmakers of the time to win himself allies in his struggle with Portugal for mastery of the Spice Islands (now in Indonesia) (1525), while in the next century the Dutch Republic used magnificent atlases as diplomatic gifts. The beauty of the title-page and a world map from a Mercator-Hondius atlas (1623) suggest the reason, though the Dutch government made use of the multi-volume *Atlas Maior* published by the rival firm of Blaeu, who were official mapmakers to the Republic and to the all-powerful Dutch East India Company (see 1599 and 1658 p. 160).

It was war (1684, 1707, 1746, 1815, 1900, 1941 pp. 328, 330) – or the prospect of it (1544 p. 100, 1940, 1973) – that has, however, been the single greatest inspiration for mapping. It accounted for most of the now largely lost mapping created under the Roman Empire, but also for the first modern mapping of England and France under Henry VIII, Elizabeth I and Henri IV respectively (1544 p. 100, 1576, 1600), the mapping of British colonies (1658 p. 138, 1782) and for much of the early mapping of South America (1524, 1558, 1579, 1730 p. 198, 1780 p. 230).

St Petersburg portrayed on the fiftieth anniversary of its foundation confirming its status as a European centre for the arts and sciences.

1753

An Imperial Newcomer

Right I. Truscott, Plan of St Petersburg.

THIS BEAUTIFUL AND decorative map shows St Petersburg as a large and important city a mere fifty years after its foundation. The map accompanied an album with views of important buildings issued to mark this anniversary, and copies were distributed to many important European libraries and courts.

Peter the Great founded St Petersburg in 1703, and made it the country's capital in 1712, intending to focus Russia more towards western Europe. Many things about the map confirm St Petersburg's status as a European capital city. Important maps had recently been published of Paris (by Turgot, in 1739) and London (Rocque, 1746), at a similar scale and in comparable style; this map could be regarded as an equivalent representation of St Petersburg. The skills and equipment used in surveying and mapmaking are represented in both the cartouche at bottom left, and the decorative group at the top right. The latter displays a diverse collections of objects, from the severed head of an enemy, through weapons and banners, to maps, scientific instruments and an artist's palette; collectively they symbolize the city's role as a centre for the arts and sciences and as a national bastion against invasion. By the time this map was published Peter's daughter Elisabeth was Empress; she is represented standing on a pedestal to the bottom left of the map. Unusually for such a late date, the map is oriented with south at the top.

The map was drawn by Ivan Fomich Truscott, also known as John Truscott, a member of the St Petersburg Academy of Sciences (proposed by Peter the Great but not founded until after his death) from 1742. Born in St Petersburg, the son of a British merchant who had settled there, Truscott was responsible for many of the Academy's maps over the next few decades. Many professionals or skilled craftsmen in the city at the time originated from western Europe; their specialist knowledge was appreciated by the Russians, and many made a good living as a result. Although it was some years before French would become the official language of the court, the map and its accompanying text are in both Russian and French, and west European languages were widely used in the city.

It was engraved on copper by the engravers of the Academy of Sciences, notably Ivan Solokhov.

МОСКОВСКАЯ СТОРОНА
ЛИТЕЙНАЯ СТОРОНА
БОЛЬШАЯ
НЕВА РѢКА
БОЛЬШАЯ НЕВА РѢКА
КРОНШТАТСКОЙ ЗАЛИВЪ
МАЛАЯ НЕВА

The first sheet of the first modern national topographical survey, for nearly fifty years the most celebrated map in France, shows Paris and its surroundings.

A National Survey

Right Detail from the first sheet of the Cassini map of France showing the area between Paris and Versailles. **See also** 1930.

THE CASSINI MAP of France at the scale of 1:86,400 was the first detailed printed map to cover a whole European country. A typical product of the European Enlightenment, it was a logical extension of the French Royal Academy of Science's ambition to measure the world accurately. This had begun in 1668 with the work of abbé Jean Picard (1620–82) and was continued in 1683 by Louis XIV's famous astronomer Giovanni-Domenico Cassini. The main aim of the designer and director of the map of France, Cassini's grandson, César-François Cassini de Thury (1714–84), was to give the exact positions of about three hundred points on each of its 175 sheets. These were created between 1749 and 1790 by surveyors who had also to collect information about place names from local guides.

Louis XV commissioned Cassini de Thury in 1747, after checking a battle map where both the army and landscape were precisely represented. But in 1756, when Cassini de Thury presented one of the first printed sheets to the King, Louis XV announced that royal financing would cease. However he helped the scientist to found a society to support the survey composed of high-ranking officials, military dignitaries, scientists and upper court magistrates, who were all interested in science.

The Cassini map of France stressed the unity of the kingdom under royal authority. But Cassini de Thury could only balance his books with the help of contributions from provincial administrations keen to use the maps as complements to local histories which drew attention to their particularities. In some cases, he had to produce special local maps. At the same time national mapping was needed when replacing the old administrative provinces with new *départments* in 1789, at the beginning of the French Revolution. Dumez and Chanlaire used the Cassini map, now considered as the 'National Property', to produce the *Atlas national* which put an end to provincial claims and projects.

Military engineers criticized the Cassini map because it revealed frontier details on freely accessible sheets. In 1793 the transfer of the whole map (copperplates and printed sheets) to the War Archives ('Dépôt de la Guerre') marked the militarization of French mapping. During the nineteenth century the *Carte de l'État-Major* was to replace the Cassini map of France.

PARIS
Montesson
Carrieres
Chatou
Croisy
Nanterre
Ruel
Putaux
Mont Valerien
Surelnes
Courbevoye
Garenne de Colombes
Asnieres
St. Ouen
Clichy
Villiers
Neuilly
le Roule
Chaillot
Montmartre
la Chapelle
Clignancourt
Boulogne
Passy
Auteuil
Gros Caillou
Invalides
Ecole Militaire
Grenelle
Vaugirard
Billancourt
Issy
Vanves
Montrouge
Gentilly
Arcueil
Chatillon
Clamart
Meudon
Sevre
Bellevue
St. Cloud
Garches
Bougival
Louvecienne
Villedavre
Marne
St. Antoine
Montreuil
VERSAILLES
Viroflay
Chaville
Velisy
Villacoublay
Cour Roland
Jouy
les Loges
Bievres
Verrieres
Chatenay
Sceaux
Plessis Piquet
Fontenay aux Roses
Bagneux
Bourg la Reine
Antony
Fresnes
Rungis
Chevilly
Villejuif
l'Hay
Massy
Vauhallan
Igny
Saclé
Villiers le Bacle
Wissous
Parey
Ivry
Route d'Orleans
Route de Fontainebleau

A map compiled for the French Navy shows the detailed plans for an invasion of Britain that envisaged landings near Hastings and in Scotland.

The Battle of Hastings that Never Was

THE FRENCH HAD had a distinctly bad war since the outbreak of hostilities with Britain in 1756, and had lost their strongholds in Canada and India and several of their islands in the Caribbean. To make matters worse, they could no longer rely on a fifth column of Jacobites, the supporters of the exiled Stuart dynasty, inside Great Britain to make trouble for the Hanoverian British government. Hope springs eternal, however.

This plan comes from a large, elegantly produced volume compiled for the French Navy in about 1762, and presumably intended for presentation at court, showing how Great Britain might be invaded. It is extremely rare for such an openly aggressive document to survive, even if the basic strategy in it, echoing that of William the Conqueror in 1066, is somewhat lacking in originality. Most of the volume is taken up with detailed lists of the precise numbers of ships, munitions and sailors available in each of the French ports. The three plans at the end, of which only two now survive, explained how the invasion was to take place.

This plan is the first and shows how squadrons from ports in northern France, Normandy and Britanny were to cross the Channel. The size of the red circle by the name of each port indicates the size of squadron it was expected to furnish. The targets on the English coast are highlighted in yellow. The main objectives were Hastings and Pevensey. Meanwhile the Rochfort and Brest fleets were to blockade Plymouth and Portsmouth respectively and a Spanish fleet was to head for Scotland presumably in the hope of warming the dying embers of Jacobitism.

The second plan, which is clearly copied from secret British manuscript maps which a spy must have smuggled out of the Board of Ordnance Drawing Room in the Tower of London, describes the proposed landing and the surrounding terrain in considerable detail, showing how a beach head was to be established. The third plan, which has not survived, showed the advance on London.

The planning was to no effect. At about the time when the maps were being drawn, the peace negotiations that were to be concluded in Paris in 1763 were already under way. In 1793, however, the cartoonist James Gillray mapped his vision of an invasion in reverse, with an English fleet heading for Normandy from Portsmouth.

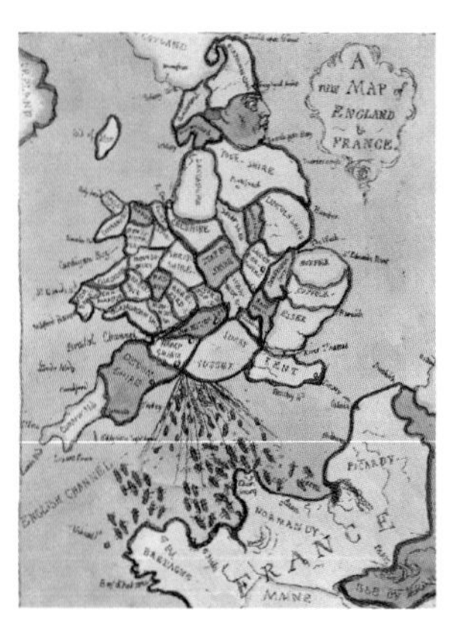

Right A map for a proposed French invasion of England.
Above James Gillray, *A New Map of England: the French Invasion.*

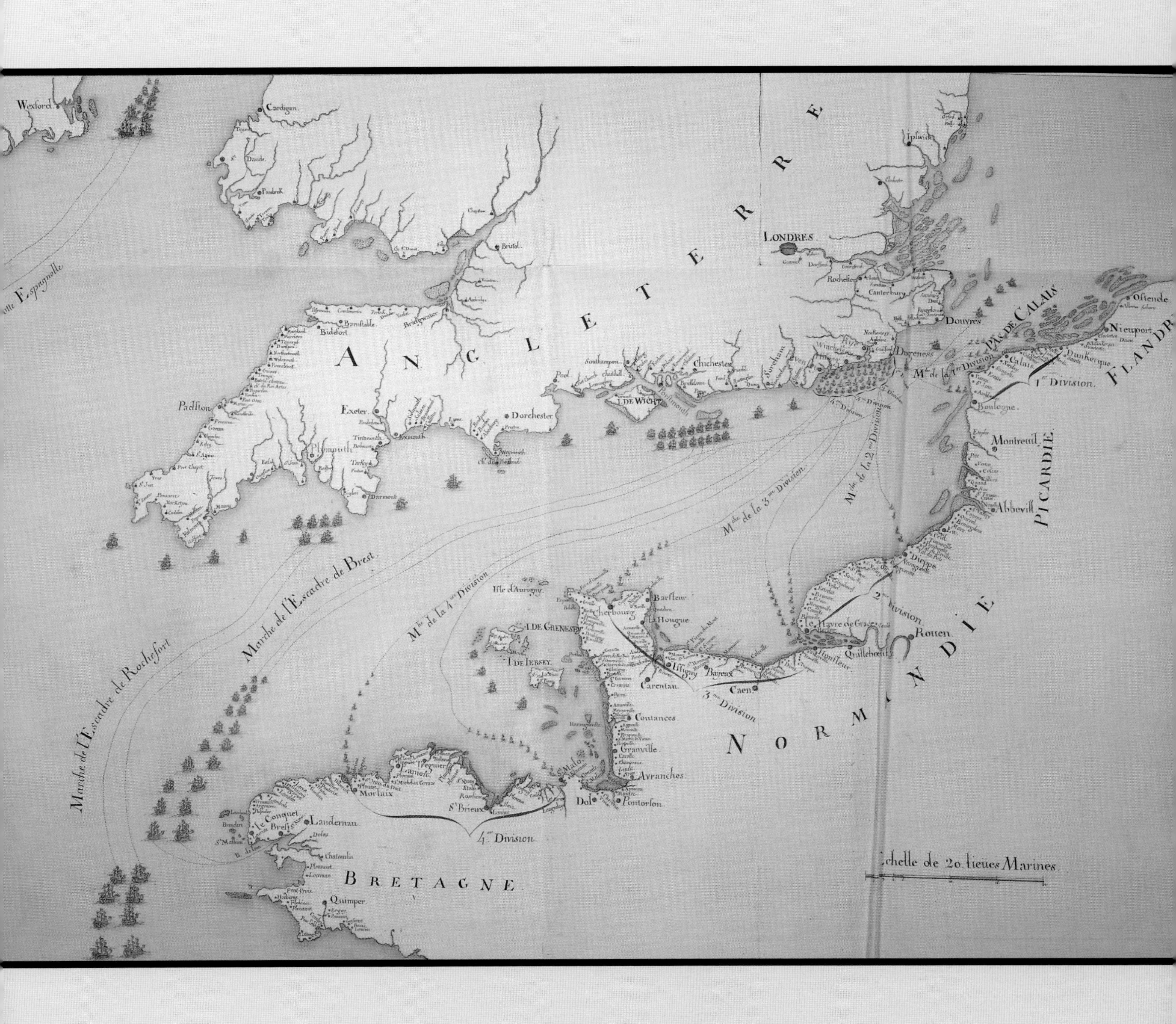
ANGLETERRE
LONDRES
Ipswich
Rochester
Canterbury
Douvres
Dogeness
Rye
Winchelsea
Hastings
Chichester
Southampton
I. DE WICHT
Portsmouth
Pool
Dorchester
Weymouth
Exeter
Exmouth
Tinmouth
Plymouth
Darmont
Padston
Bidefort
Barnstable
Bridgewater
Bristol
Cardigan
Pembrok
Wexford
PAS DE CALAIS
Calais
Dunkerque
Nieuport
Ostende
FLANDRE
Boulogne
Montreuil
Abbeville
PICARDIE
Dieppe
Le Havre de Grace
Honfleur
Quilleboeuf
Rouen
NORMANDIE
Caen
Bayeux
Isigny
Carentan
Barfleur
La Hougue
Cherbourg
Isle d'Aurigny
I. DE GRENESEY
I. DE JERSEY
Coutances
Granville
Avranches
Pontorson
Dol
S. Malo
S. Brieux
Morlaix
Landernau
Le Conquet
Brest
Quimper
BRETAGNE
1re Division
2me Division
3me Division
4me Division
Mche de la 1re Division
Mche de la 2me Division
Mche de la 3me Division
Mche de la 4me Division
Marche de l'Escadre de Brest
Marche de l'Escadre de Rochefort
otte Espagnolle
Echelle de 20. lieües Marines

A map nicely balancing the artistic with the scientific and produced by a female astronomer, a female engraver and a female publisher predicts an eclipse.

A Female Enterprise

Right Madame Le Pauté Dagelet, *Passage de l'Ombre de la lune au travers de l'Europe.*

DURING THE EIGHTEENTH century the French were world leaders in the science of cartography. Although Edmond Halley pioneered the scientific eclipse map in England from 1715 onwards, the French were the first to achieve colour printing on eclipse maps by using two separate, successively printed, copperplates. In keeping with eighteenth-century France's enlightened attitude towards the position of women, this map predicting the eclipse of 1764 was produced by three women: Madame le Pauté Dagelet, Madame Lattré and Elisabeth Claire Tardieu.

Little is known about Madame Le Pauté Dagelet, an astronomer and member of the Académie Royale des Sciences (Béziers). Two of her maps have been recorded: this one predicting the 1 April 1764 eclipse, which was total but 'annular' (i.e. like a narrow blazing ring around the moon) over parts of south-east England and north-west France, and the other, predicting the 24 June 1778 eclipse which was total over parts of Africa and the United States but partial across France and Great Britain. The map predicting the 1764 eclipse was based on Halley's design and engraved by Madame Lattré and Elisabeth Claire Tardieu. Madame Lattré was the wife of the distinguished Jean Lattré (d. 1782), 'Graveur Ordinaire du Roi', who specialized in engraving town plans and maps for leading mapmakers Robert de Vaugondy and Bonne. The Tardieus were a well-established dynasty of map publishers.

This map, which was copied, likewise in black and sepia, by map publisher Louis-Charles Desnos, shows the track of the annular eclipse in dark grey. The black lines drawn parallel to the path of centrality (centre of eclipse track) indicate the decreasing amount by which the sun will be obscured by the moon. As Paris is on the edge of the track the eclipse should have been partial. The track of centrality on the map was about 50 miles (80 kilometres) too far west, however, and Paris actually experienced a full annular eclipse. The map has a more scientific appearance than earlier maps but the title cartouches are very decorative and impart a good balance of the artistic and the scientific.

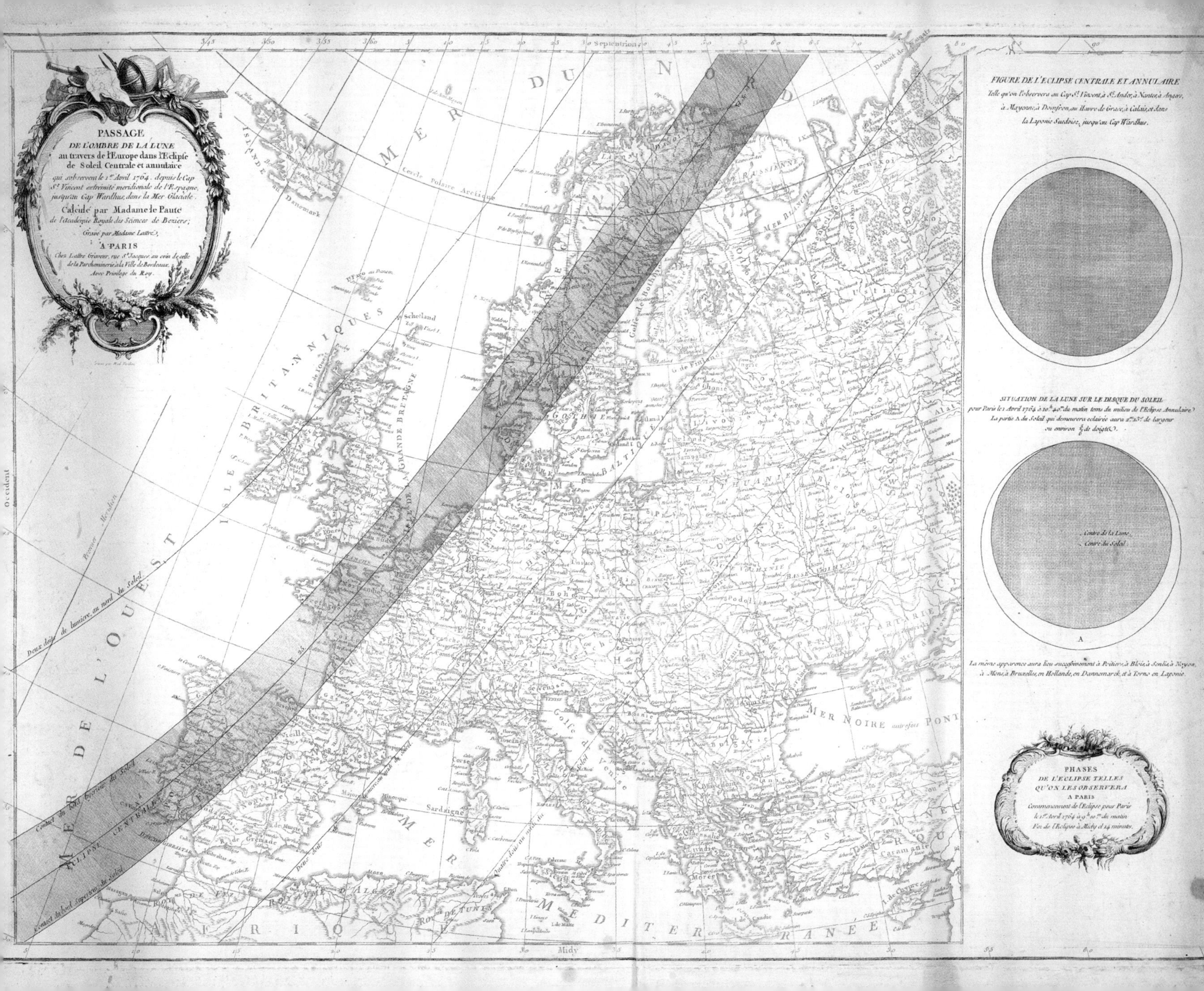
PASSAGE
DE L'OMBRE DE LA LUNE
au travers de l'Europe dans l'Eclipse
de Soleil Centrale et annulaire
qui s'observent le 1er Avril 1764 depuis le Cap
St Vincent extrémité meridionale de l'Espagne,
jusqu'au Cap Wardhus, dans la Mer Glaciale.
Calculé par Madame le Paute
de l'Académie Royale des Sciences de Beziers;
Gravé par Madame Lattré,
A PARIS
Chez Lattré Graveur, rue St Jacques, au coin de celle
de la Parcheminerie, à la Ville de Bordeaux.
Avec Privilege du Roy.
FIGURE DE L'ECLIPSE CENTRALE ET ANNULAIRE
Telle qu'on l'observera au Cap St Vincent, à St Ander, à Nantes, à Angers,
à Mayenne, à Domfron, au Havre de Grace, à Calais, et dans
la Laponie Suedoise, jusqu'au Cap Wardhus.
SITUATION DE LA LUNE SUR LE DISQUE DU SOLEIL
pour Paris le 1 Avril 1764 à 10h 40m du matin tems du milieu de l'Eclipse Annulaire.
La partie A du Soleil qui demeurera eclairée aura 2m 13s de largeur
ou environ 4/6 de doigts.
Centre de la Lune
Centre du Soleil
A
La même apparence aura lieu successivement à Poitiers, à Blois, à Senlis, à Noyon,
à Mons, à Bruxelles, en Hollande, en Dannemarck, et à Torno en Laponie.
PHASES
DE L'ECLIPSE TELLES
QU'ON LES OBSERVERA
A PARIS
Commencement de l'Eclipse pour Paris
le 1er Avril 1764 à 9h 10m du matin
Fin de l'Eclipse à Midy et 14 minutes.
MER DU NORD
MER DE L'OUEST
ISLES BRITANNIQUES
GRANDE BRETAGNE
MER BALTIQUE
MER BLANCHE
MER NOIRE autrefois PONT
MER MEDITERRANÉE
AFRIQUE
Cercle Polaire Arctique
Septentrion
Midy
Occident
Orient
Premier Meridien
Deux doigts de lumière au nord du Soleil
ECLIPSE CENTRALE
Contact du bord Superieur du Soleil
Contact du bord Inferieur du Soleil
Schetland
Sardaigne
Corse
Minorque
Majorque
Golfe de Bothnie
G. de Finlande
LITHUANIE
Royaume d'Alger
Royaume de Tunis
I. de Malte

An up-to-date pair of terrestrial and celestial globes 'contrived to solve the various phenomena of the earth and heavens'.

A Scientific World and Universe

IN 1766 GEORGE ADAMS, mathematical instrument maker to King George III, published a treatise on the use of globes to accompany two new pairs of globes which he had made.

This handsome pair of globes are examples of the larger size, with a diameter of 18 inches (46 centimetres). The other pair had a diameter of 12 inches (30 centimetres). They were the first new globes to be published in England since the early part of the eighteenth century and though their appearance is similar to other globes of the time, Adams pronounced in his treatise that these globes were 'of a construction new and peculiar, being contrived to solve the various phenomena of the earth and heavens, in a more easy and natural manner than any hitherto published.'

The globes now have the brown appearance of an old and yellowed varnish but it is still possible to see the information they contain. On the terrestrial globe, Captain Cook's voyages to the southern seas are not yet depicted and New Zealand is shown as nothing more than a few vague coastlines. Very little is shown of New Holland (Australia), and there is no passage separating Van Diemen's Land (Tasmania). These deficiencies were soon corrected in subsequent editions of the globe after Cook's return to England in 1771.

But whereas the terrestrial globe soon became out of date, the celestial globe showed the latest astronomical discoveries of the time. In 1763, three years before the appearance of Adams's globes, the work of the French astronomer Nicholas de Lacaille was posthumously published. Lacaille, on a four-year expedition to the Cape of Good Hope organized by the French Academy of Sciences, had charted the stars in the southern hemisphere and he had devised several new constellations which he based on contemporary instruments and tools of the arts and sciences. Adams depicts these and therefore, amongst the older constellations of animals and mythical figures, one can see the Sculptor's Workshop, the Painter's Easel, the Clock, the Chemical Furnace, the Air Pump, Hadley's Octant, the Sextant, the Telescope, the Microscope, the Mariner's Compass, Euclid's Square and the Rhomboidal Net. The drawings seen on the globe of the microscope and air pump are in fact based on the instruments Adams made and sold in his shop, and which he advertised at the end of his treatise on the globes.

Right George Adams, Terrestrial and Celestial Globes. c.1766.
See also 1756.

OCEANUS PACIFICUS

An rare early Indian world map created probably by Hindus from a very wide variety of Asian and European sources for a Muslim patron.

Fusion in India

SURVIVING PRE-MODERN world maps of Indian provenance are rare. While those that are known are so variously executed that one cannot categorize them neatly, most that were executed for Muslim patrons, including this one, have roots in the Ptolemaic tradition. The map is oriented with south at the top and presumably dates from the latter half of the eighteenth century. Painted in tempera on cloth, it incorporates a wealth of geographic and mythological detail depicted in the miniature style then characteristic of Rajasthan and the Deccan. But its use of the place name Amer (Amber), rather than Jaipur, which succeeded it as a Rajput capital in 1728, argues more for a Deccani source.

The map is unique in that its text is in Persian, Arabic and Hindi (written in the Devanagari script). The painters were probably Hindus, many of whom were versed in the languages of their Muslim patrons. But its eclectic aspect stems also from its many sources. The map text cites a no longer extant work, *Secrets of the Sea*, by the fifteenth-century pilot, Ibn Majid, and also includes knowledge derived from European contacts. France, Germany and Austria are named and the word Portugal is inscribed near the red caravel anchored in the Indian Ocean, while several Portuguese islands with misplaced names appear in the Atlantic Ocean. Approximately fifty place names appear in South Asia; these alone are written in Devanagari. Middle Eastern place names are also abundant. Constantinople, shown by a large rectangle joined to the mainland by a narrow tongue of land, is especially prominent. Mecca is indicated by a depiction of the Ka'ba. In Africa, we see the Nile, flowing out of the Mountains of the Moon. Ceylon appears twice, recalling the dual depiction of Taprobana on many Ptolemaic maps. Japan is represented by a vertical island wherein a group of dog-faced creatures are seated. China forms an arc along the left margin of the earth disc.

Right Eclectic world map from India. **Above** Africa. **See also** 150, 1025.

The map's mythological content derives largely from the *Iskandarnamah*, the Alexandrian Romance, many versions of which were composed in Asia. Among the elements depicted are the 'Spring of Life', allegedly discovered by Moses, shown by the black rectangle in the bottom centre, the wall built by Alexander to protect people from the giants, Gog and Magog; the aforementioned dog-faced men; and Alexander's palace atop the Mountains of the Moon.

A political cartoon utilizes a map as a way of satirizing and condemning the attitudes of the main actors in the partition of Poland.

Paper Partition

THE 'RATIONAL' EIGHTEENTH century was the age above all others when lands were transferred from one ruling dynasty to another and states were negotiated into or out of existence, regardless of the wishes and cultures of the peoples involved, for the avowed purpose of maintaining or restoring international peace. The most grotesque and shameful example was Poland, formerly one of the largest – but weakest – states in Europe, which was partitioned in 1772, 1792 and again and finally in 1795, when it ceased to exist until 1918.

This finely engraved cartoon uses a map, with the Baltic Sea (north) at the right, to satirize the first partition. Maps are emotive objects and lend themselves to uses of this kind: a century earlier a map of England and Wales replaced the King on the English Great Seal after the execution of Charles I. Here the attitudes of the main participants are finely caught. Russia, Prussia and Austria had decided on the partition in order to avoid war between themselves over Poland, which had become a power vacuum because of its internal instability. Thousands of lives and vast amounts of money were thereby saved. Yet none of the male participants look happy even though the winged figure of Fame trumpets their humanitarian triumph abroad. Frederick II of Prussia (right) soberly indicates the Baltic area he is interested in with his sword. The Holy Roman Emperor, Joseph II, looks away, shame-faced, as he points to his prize, the province of Galicia. Stanislas Poniatowski, the elected King of Poland, almost knocks his crown off his head as he looks away distraught at what his former lover, Catherine the Great of Russia, is claiming. But she, unconcerned at her vandalism, actively tears the provinces that she wants from the map – as though she were only tearing paper. The attitudes satirized are not far removed from those held by the actual participants.

This version of the cartoon was published in London by the print and mapseller Robert Sayer. It was entitled *The Royal Cake* and was probably known to the outstanding British satirist James Gillray who was later to draw a plum pudding, representing the world, being partitioned between the French ruler, Napoleon Bonaparte and the British Prime Minister William Pitt.

Right *The Royal Cake/Le Gateau des Roys.* **Above** James Gillray, *The Plumb-pudding in Danger*, 1805.

Leopol
Luck
Rava
Lecyca
Varsovie
Cracovie
Sieradz
Guesne
Bug R.
Thorn
Netze R.
Bielsk
Novogrodek
Minsk
Vitebsk
Polock
Dyna R.
Vilia R.
Vilna
Brandebour
Culm
Marienbourg
Dantzik
Konigsberg
POMERANIE
RUS
SIE
Riga
Mittau
POLOGNE
en 1772

A French commercial map depicts the latest news from America and encourages the French government to support the American rebels.

An Appeal for Foreign Intervention

THIS FRENCH IMAGE of the harbour of Boston, Massachusetts, demonstrates a map's multiple meanings. On one level, it portrays the current news of the early phases of the revolution of American colonists against the British mother country. On another, its decorative title cartouche encourages French intervention in the war on the side of the Americans.

The map's author used British sources to portray the early skirmishes around Boston that precipitated the War of American Independence, in particular, the battle of Bunker Hill (letter H), the burning of Charlestown (letter I) and the siege of Boston by the American rebels, shown by bold red American forces encircling blue British forces around the perimeter of the harbour. This forecast of a potential British defeat, while premature at the moment of the map's publication, would appeal to the King of France, to whom the map is dedicated.

The map's cartouche takes the cartographic argument a step further by urging the King to enter the nascent war on the side of the Americans. The cartouche depicts civil war. A colonial soldier clings to the banner of Massachusetts, with its native white pine, topped by the pilleus, the Liberty Cap. He grapples with a British soldier who grasps the banner of the British lion rampant. At their feet lie the fasces, symbolizing government authority, in danger of being trampled. A horrified Native American watches the struggle, his thought expressed in a Latin tag running cartoon-like from his bow: '*Quo scelesti ruitis?*' This is the first line of Epode VII by the ancient Roman poet Horace, written after the assassination of Julius Caesar, spelling out the dangers of civil strife. In his poem, Horace warns that the only one to benefit from civil war is a country's enemy. The learned Indian reminds the viewer that France can profit from the internecine British struggle and thus avenge France's humiliating loss of Canada to the British in 1763, at the close of the Seven Years' War.

Beaurain's map, in fact, confirmed publicly what already was happening privately. The French government had been secretly mobilizing aid for the colonists. By the time of the map's publication, emissaries of the American Continental Congress had arrived in Paris to seek French intervention, formally announced in the Treaty of Amity in February 1778.

Right Le Chevalier Jean de Beaurain, *Carte du Port et Havre de Boston.*

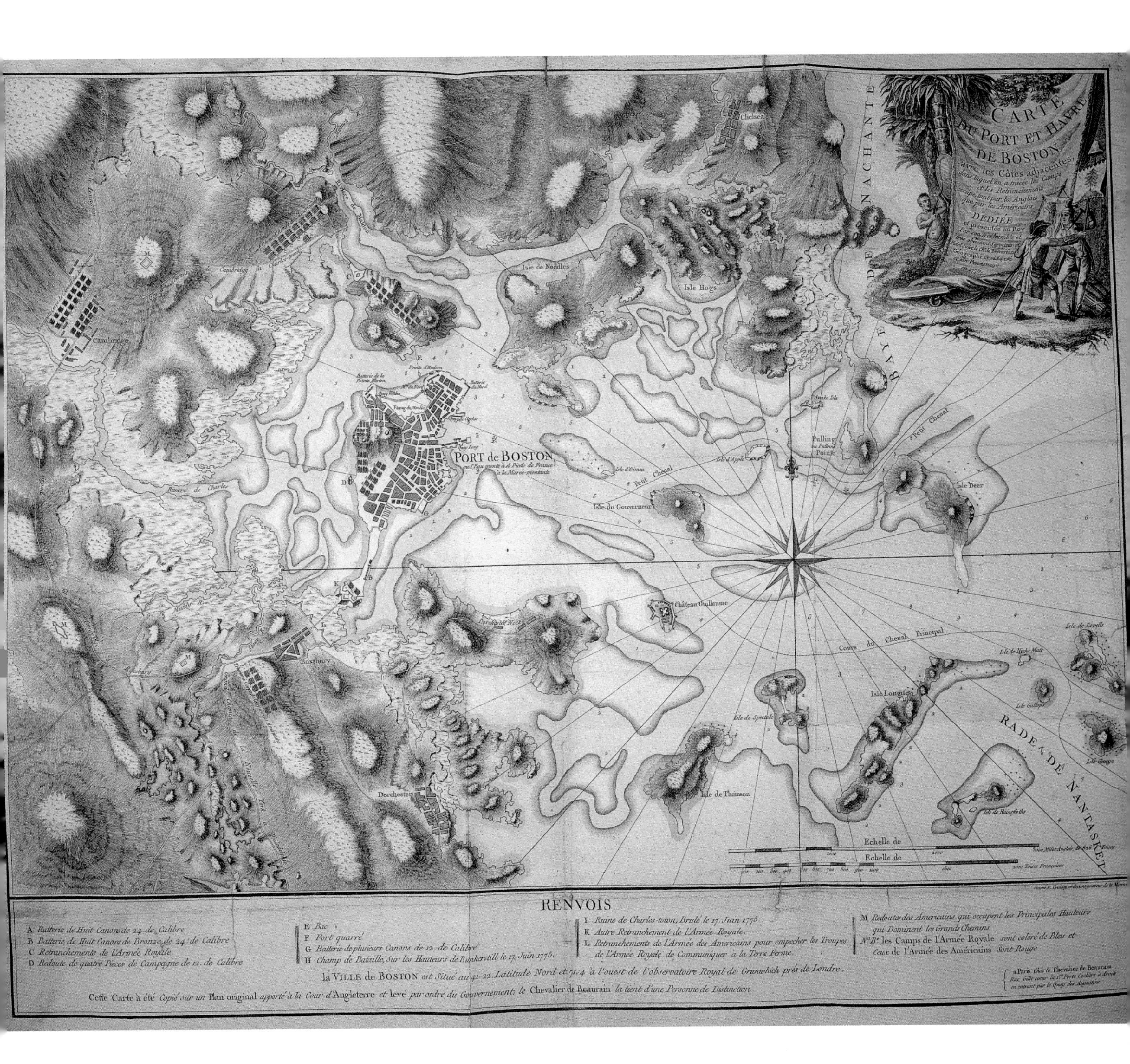

CARTE
DU PORT ET HAVRE
DE BOSTON
avec les Côtes adjacentes,
dans laquel on a tracée les Camps
et les Retranchemens
occupé tant par les Anglois
que par les Américains
DEDIÉE
et presentée au Roy
Par son tres humble et
tres obéissant Serviteur et
Fidel sujet le Ch. de Beaurain
Géographe de sa Majesté
et son Pensionnaire
en 1776.
BAYE DE NACHANTE
RADE DE NANTASKET
Chelsea
Cambridge
Isle de Noddles
Isle Hogs
PORT de BOSTON
Isle du Gouverneur
Isle Deer
Petit Chenal
Pulling Pointe
Château Guillaume
Cours du Chenal Principal
Isle Longue
Isle de Spectacle
Isle de Thomson
Isle de Levells
Isle de Nicks Mate
Rivière de Charles
Roxbury
Dorchester
Echelle de
Echelle de
2000 Toises Françoises
RENVOIS
A Batterie de Huit Canons de 24 de Calibre
B Batterie de Huit Canons de Bronze de 24 de Calibre
C Retranchements de l'Armée Royale
D Redoute de quatre Pieces de Campagne de 12. de Calibre
E Bac
F Fort quarré
G Batterie de plusieurs Canons de 12. de Calibre
H Champ de Bataille, Sur les Hauteurs de Bunkerstill le 17. Juin 1775.
I Ruine de Charles-town, Brulé le 17. Juin 1775.
K Autre Retranchement de l'Armée Royale.
L Retranchements de l'Armée des Americains pour empecher les Troupes de l'Armée Royale de Communiquer à la Terre Ferme.
M Redoutes des Americains qui occupent les Principales Hauteurs qui Dominent les Grands Chemins
N.B. les Camps de l'Armée Royale sont coloré de Bleu et Ceux de l'Armée des Américains Sont Rouge
la VILLE de BOSTON est Situé au 42. 22. Latitude Nord et 71. 4 à l'ouest de l'observatoire Royal de Grunwhich prés de Londre.
Cette Carte à été Copié Sur un Plan original apporté à la Cour d'Angleterre et levé par ordre du Gouvernement; le Chevalier de Beaurain la tient d'une Personne de Distinction
a Paris Chés le Chevalier de Beaurain

A colonial map portrays South America, for commercial purposes, as an equal pillar, with Madrid, of the Spanish Empire.

Mapping the Material Wealth of Spain's American Empire

1780

THIS ANONYMOUS chorography of El Collao – a region in the Andean altiplano north of Lake Titicaca near the present-day border between Bolivia and Peru – exemplifies the way that geography and resource allocation merged in the imperial ideology of the Spanish Empire. The district of Collao had been a zone of agricultural prosperity from the time of the Incas. For Spaniards, however, the region's ample mineral wealth was even more important than exotic fruits and plants, represented in this map by textual descriptions and graphic depictions of gold nuggets, silver bars and other metallic resources. Also shown are cattle and sheep that provided clothing and textiles for the native, European and mestizo populations. In the eighteenth century, reforms instituted by the Bourbon King Charles III (1759–88) attempted to squeeze greater profit from Spanish dominions in the Americas as a counterbalance to augmented fiscal anxiety and significant Spanish military commitments in the Caribbean and Atlantic theatres due to rising British maritime power. This attention to the productive power of Spain's American possessions is exemplified by this map's iconographic interest in natural resources, native labour and the interconnections between individual regions within the Spanish Empire.

Rather than portraying Madrid as the undisputed centre of the empire, the symbolic structure held aloft by the indigenous figure in the map's lower left-hand corner portrays El Collao and Madrid as two equal and interconnected components of the imperial system. El Collao is the focal point of a radiating American network, the key conduit of a global commercial alliance without which Cuzco, La Paz, Buenos Aires, Lima and Huancavelica would be disconnected and peripheral imperial possessions. The visual rhetoric of the map, echoing an earlier Incan world-view itself, thus argues for the centrality of this highland Andean region within a global economic system: the Andes were connected to the rest of the world through the successful management of their agricultural and mineral resources, just as they had been during the time of Inca domination. In the eighteenth century, these connections were controlled by ecclesiastical leaders, viceregal governors, Bourbon reformers and other administrative agents of the Spanish crown, individuals whose power would soon be challenged by insurrections and revolts throughout the former provinces of the Inca Empire.

Right An anonymous map of El Collao. **See also** 1579, 1730 (p. 198).

EL COLLAO
PARTE D LAS MICIONES D APOLOBAMBA
PARTE D LOS CHUNCHOS
PARTE DE LARE CAJA
PARTE DE QUIS PI CAN CHE
PROUINCIA DE CARABAIA
SAN DIA
PROUINCIA DE AZANGARO
PROUINCIA D LAMPA
PARTE D TINTA
PARTE D CAILLOMA Juridicion de Arequipa
PARTE DE MOQUEGUA
PARTE DE PUNO
LA PROUINCIAS DEL COLLAO
Aporoma
Azillo
Santiago Pupuja
Caminaca
Aiaviri
Pucara
Nicasio
Guaca
Caracoto
Juliaca
Umachiri
Vilavila
LAMPA
Umpuco
Ucubiri
Macari
Cupi
Llalli
Pomasi
Cavanilla
Cabana
Vilque
Mañaso
Pocomoro
Nuñoa
Orurillo
Sania Rosa
Muñani
Potoni
Ananea
Chupa
Putina
Arapa
Samán
Crucero
Coaza
Limbani
Para
Ituata
Ollachea
Macusani
Aiapata
Ajoiani
El Collao
CURATOS
LAMPA 13
Lampa Cap.l con Calapuja
Cavanilla
Cavana
Mañaso con Vilque
Atuncolla
Juliaca
Caracoto con Guaca y Llalli
Pucara con Vilavila
Aiaviri
Orurillo
Nuñoa con Sta Rosa
Umachiri con Ucubiri
Macari con Cupi y Llalli
AZANGARO 9
Azangaro con Muñani Poto y Ananea
Asillo
Pupuja
Chupa con Putina
Arapa con la villa de Betanzos
CARABAIA 6
Sn Juan del Oro
Aporoma
Aiapata con 6

A plan of an English encampment just north of London which met the requirements of commanders, officers and men.

Soldiers of the King

Right I. W. Green, *Plan of the Encampment on Finchley Common under the command of Major Genl Faucit from 10th Augt to 20th Octr. 1780.*

IN THE EIGHTEENTH CENTURY English military encampments gave commanders the chance to display their prowess at manoeuvres, soldiers and local girls the opportunity to meet at the accompanying reviews and balls – and military mapmakers the occasion to create the sort of plans that they would be expected to produce on campaign.

This map, drawn for presentation to George III, shows a small encampment on a common north of London, lining the main road to Scotland, at a time when the major powers on the European mainland were joining forces against Great Britain in support of the rebellious American colonists. Like estate maps it depicts fields and single buildings, but in most respects it could hardly be more different. Field divisions are vague, where those on estate maps are precise. The physical relief is accurately depicted, where it is ignored on most estate maps. The plan concentrates on what commanders needed to know, presented in a manner calculated to please the eye. Relief was essential information for the General, as were the sources of water – here in the form of streams and ponds – for horses and men. The plan picks out the well used by one of the regiments and the location of forage for the horses. The landscape shows the balance between enclosed and unenclosed land. There is the scrub of Finchley Common, over which infantry and cavalry could roam freely, green pastureland where there was some freedom of manoeuvre and where horses could graze, and yellow arable land (yellow), ultimately a source of bread at the time when this encampment took place, and best left untouched by the military. Buildings show where the officers could be billeted, with the grandest houses such as 'Fryant' (Friern) House being reserved for commanders or visiting royalty. The Swan and The Horse show where the common soldiers could relax from the rigours of military life.

The surveyor of this map was almost certainly taught by Paul Sandby, often regarded as the founder of the English school of watercolourists. Its colouring certainly reflects Sandby's palette. The features depicted and the cartographic style in many ways anticipate the work of the Ordnance Surveyors who were to begin mapping the British Isles just a few years later.

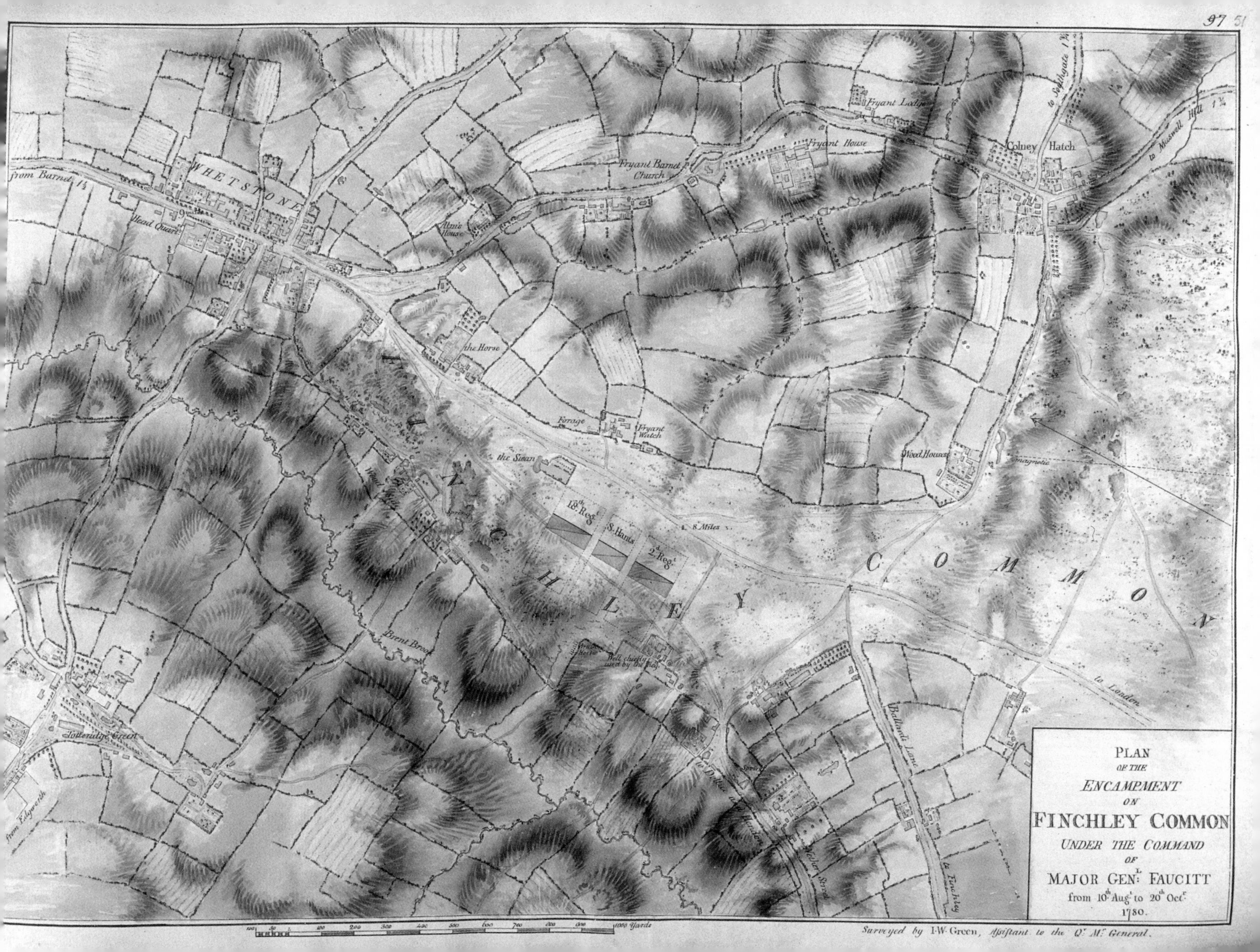

97
51
from Barnet 1¼
WHETSTONE
Head Quart.
Alm's House
Fryant Barnet Church
Fryant Lodge
Fryant House
to Southgate 1¾
Colney Hatch
to Muswell Hill 1¼
the Horse
Forrage
Fryant Watch
the Swan
Wood House
magnetic
18th Regt
S. Hants
2d Regt
8 Miles
F I N C H L E Y
C O M M O N
Brent Brook
to London
Ballards Lane
to Finchley
Totteridge Green
from Edgworth
PLAN OF THE ENCAMPMENT ON FINCHLEY COMMON UNDER THE COMMAND OF MAJOR GENL. FAUCITT from 10th Augt. to 20th Octr. 1780.
100 50 100 200 300 400 500 600 700 800 900 1000 Yards
Surveyed by I.W. Green, Assistant to the Qr. Mr. General.

James Rennell's map of 'Hindoostan' was the first step towards the detailed mapping of British India required to administer a vast subcontinent extending from Afghanistan to Burma.

British India

1782

JAMES RENNELL WAS appointed Surveyor General of Bengal in 1767, having come to India with the British Navy three years earlier. Under his leadership, the territories acquired by the East India Company were surveyed and accurately mapped for the first time. Not happy with the way his information had been assembled by cartographers in England, on his return home in 1778 Rennell set about organizing the many sketches made in the field into a single large map of the subcontinent, 'Hindoostan'.

Jean Baptiste Bourguignon d'Anville, cartographer to the French King, had already published a large four-sheet map of India in 1752, with a 'Memoir' giving his sources, and this had been copied in England soon after. Many areas were left blank on d'Anville's map as his scientific approach did not permit him to include any speculative information, but now Rennell was in a position to fill them in and provide the first large, detailed map of the whole country.

The British and the French had been competing for the rich trade with India for many years, ostensibly aiding local rulers in their territorial quarrels, but in fact capturing large areas for themselves. Robert Clive had defeated the French in south India, and then moved to Calcutta where he defeated the Mughal governor at Plassey in 1757. Now the British were tax collectors and administrators, and they required accurate maps of their newly acquired territory.

Rennell's map covered a far wider area than that of d'Anville, extending from Afghanistan to Burma, and north to the border with Tibet. His own surveys had provided information for the north-east, and British Army surveyors were now travelling across the south. With the map came a Memoir, again providing the sources, and also a list of all places named on the map and the square where they might be found. The map was redrawn in a larger size six years later, and there were several updated editions of the Memoir.

The large cartouche shows the Brahmins of India handing over their sacred books to the embodiment of Britain in the form of a Greek goddess backed by an imperial lion, watched by British gentlemen and with a British ship at anchor behind.

Right James Rennell, *Hindoostan*.
See also 1619.

HINDOOSTAN
By J. Rennell F.R.S. 1782.
EXPLANATION.
THIBET
Country of the Grand Lama
CABUL
LAHORE
MOULTAN
DELHI
AGIMERE
SINDY
GUZERAT
OUDE
NAPAUL
BOOTAN
ASSAM
BAHAR
BENGAL
MECKLEY
BURMAH
ARACAN or RECCAN
PEGU
CANDEISH
BERAR
ORISSA
AMEDNAGUR
VISIAPOUR
GOLCONDA
MYSORE
CEYLON
BAY of BENGAL
LACCADIVE ISLANDS
MALDIVE ISLANDS
NICOBAR ISLANDS
Province of YUNAN
CHINA
SIAM
Part of SUMATRA ISLAND
STRAIT of MALACCA
Gulf of SINDE

An out-of-date map still limping on after seventy years and used largely for decorative rather than scientific purposes.

Neither New Nor Correct...

THIS MAP IS A GOOD example of the large two-sheet world and continental maps, first created in France, which were produced and sold in large numbers in the eighteenth century. They were multi-purpose and were sold singly, as part of sets, or in atlases. They were also used as decorative wall maps in the home, in schools and as a focal point on screens often surrounded by maps of the various countries of the world. However, seen by their creators primarily as a method of making money from the less well educated, they were often of poor quality.

The maps were often slavishly and shamelessly copied by others, complete with errors, despite the first British Copyright Act of 1734. Sometimes the copying was to replace worn out plates but more often competition with other publishers was the motive, whatever the quality of the maps copied. Émigré mapmaker Hermann Moll working in London condemned this as an 'evil practice' – despite being guilty of it himself.

The plate from which this impression is pulled was originally drawn by William Godson in about 1715 and sold by a well-known and commercially minded publisher and bookseller, George Willdey. The two large hemispheres were copied from a map by John Senex, who had a reputation for scientific precision, with the smaller astronomical and terrestrial hemispheres in the margins being added to give a more scientific appearance. This example shows that by the 1780s the plate had been used so much that a crack due to wear and tear had to be repaired with rivets. Marshall made attempts at updating the map, for example with the tracks of Cook's voyages of 1768–79, to give the impression of modernity, but the gross errors of showing California as an island, more than eighty years after being proved to be a peninsula, and the fictitious and archaic Strait of Anian to the north, betray a cynical carelessness about the accuracy of geographical information. The map's life was extended still further in 1794 when it was almost exactly copied onto a new plate by Richard's brother John Marshall, who took over the family business.

Significantly, these maps were described by the publisher John Bowles as 'being cheap and proper ornaments for rooms, halls and stair-cases' generally being sold for the almost derisory sum of 1*s* 6*d* each.

Right Richard Marshall, *A New and Correct Map of the World*. **Above** The marks left by the rivets holding the copperplate together. **See also** 1890.

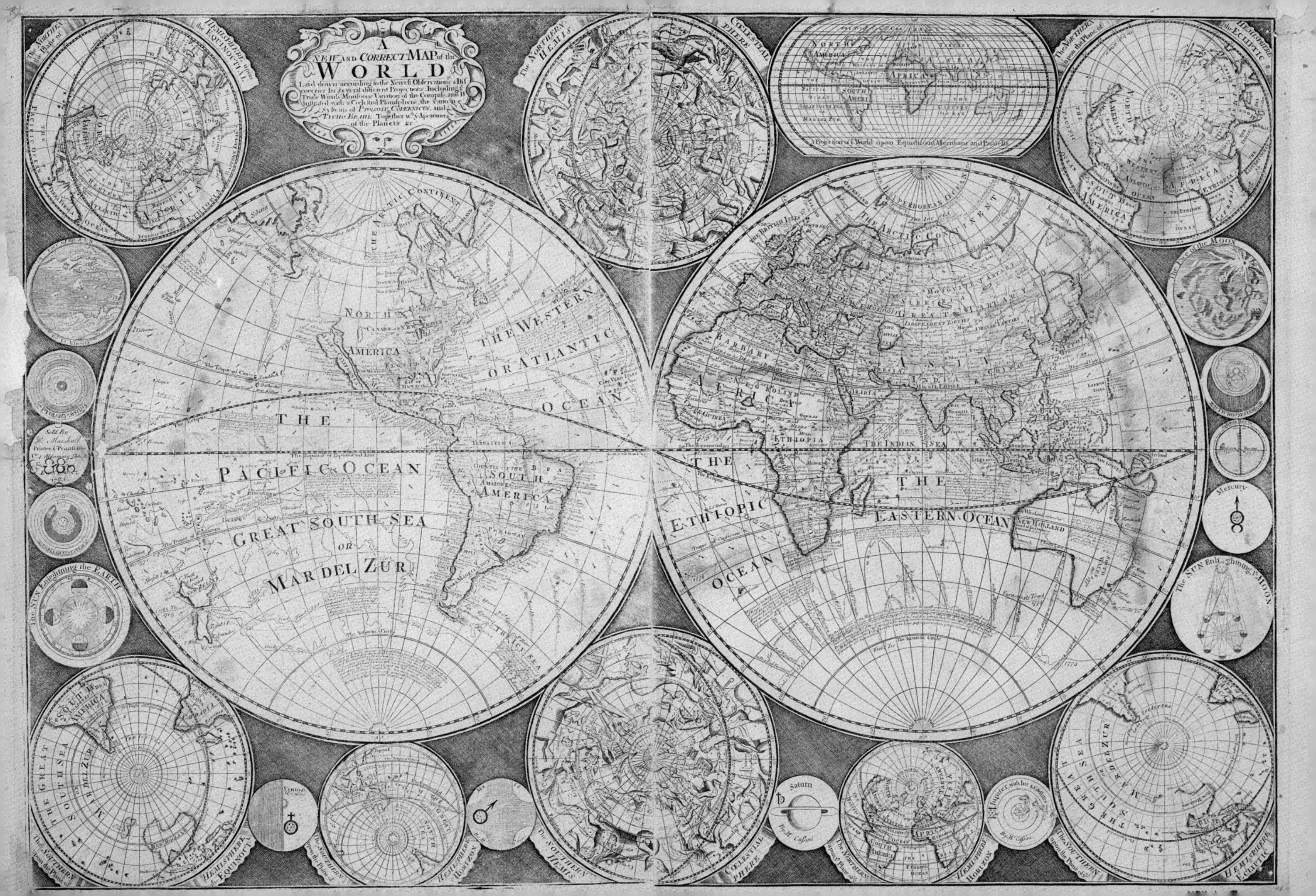
A
NEW AND CORRECT MAP of the
WORLD
Laid down according to the Newest Observations & Discoveries In several different Projections Including ye Trade Winds Monsoons Variation of the Compass and Illustrated with a Celestial Planisphere the various Systems of PTOLOMY, COPERNICUS, and TYCHO BRAHE Together wth ye Apearance of the Planets &c.
Sold By
Printer & Printseller
THE WESTERN OR ATLANTIC OCEAN
THE PACIFIC OCEAN
GREAT SOUTH SEA OR MAR DEL ZUR
NORTH AMERICA
SOUTH AMERICA
THE ARCTIC CONTINENT
AFRICA
ASIA
ETHIOPIA
THE INDIAN SEA
THE ETHIOPIC OCEAN
THE EASTERN OCEAN
NEW HOLLAND
The Face of the MOON
Mercury
The SUN Enlightning the EARTH
The SUN Enlightning the MOON
Venus
Saturn
Jupiter with his Satellites
The NORTHERN HEMISPHERE upon the Plane of the EQUINOCTIAL
The NORTHERN HEMISPHERE upon the Plane of the ECLIPTIC
The SOUTHERN HEMISPHERE upon the Plane of the EQUINOCTIAL
The SOUTHERN HEMISPHERE upon the Plane of the ECLIPTIC
The NORTHERN HEMISPHERE upon the Plane of the HORIZON
The NORTHERN CŒLESTIAL HEMISPHERE
The SOUTHERN CŒLESTIAL HEMISPHERE
Projection of ye World upon Equidistant Meridians and Parallels

John Cary tried to recoup the cost of his route maps by indicating inns and the homes of the wealthy, shown here in a 'neat Pocket Volume', later much reissued.

Sightseeing en Route

Right The roads to Highgate and Hampstead from *Cary's Survey of the High Roads from London to Hampton Court...* **See also** 1675, 1890, 1930, 1941 (p. 332), 1998.

ALTHOUGH JOHN CARY'S *Survey of the High Roads from London to Hampton Court...* and twenty-five other specified destinations around the capital is one map at a scale of one inch to one mile, it is divided into forty small strips to form a delightful 'neat Pocket Volume'. It first appeared in 1790 and was obviously popular, being reissued in 1799, 1801 and 1810.

Cary's *Survey* reveals the main concerns of contemporary road mapmakers and their customers.The map focuses on turnpikes, with notes and distances which were particularly useful where turnpikes overlapped and differing rates were charged. This allowed travellers to work out how to achieve the maximum distance from any turnpike ticket. Cary advertised that this information would prevent the frequently complained of 'unpleasant altercation' which all too often resulted 'from the incivility as well as imposition of the Toll-gatherers' when calculating charges.

'Every Gentleman's Seat, situate on, or seen from the Road, (however distant)' was 'laid down, with the Name of the Possessor'. It was Cary's stated contention that the traveller was fascinated by 'the numberless villas which so often attract his attention' and that 'his enquiry' was 'naturally directed to whom do they belong'. Consequently, a 'principal object' of Cary's road publications was to specify house ownership. In truth, the main reason was to attract as many affluent customers for the map as possible through flattery, as many mapmakers had done before. Country houses are represented pictorially with lines of sight showing their viewpoints from the road, even indicating in the map's frame houses that were visible but beyond the limits of the map's coverage.

Public inns are noted, not only because of their importance as stops for coaching services but also 'with a view to utility, as it enables a party to form a meeting with certainty, and gives them a choice of pursuing their pleasures to a greater extent than they otherwise would do'.

However, in some ways the *Survey* was a disappointment. It concentrated only on high ground, heath and rivers, ignoring other landscape features. More importantly for the wheeled traveller, it gave no indication of the quality of the road surface – a vital consideration at a time of very variable road maintenance. With road charging by the mile once again in sight, it may well be that modern versions of Cary's *Survey* may yet appear.

LONDON to HIGHGATE & HAMPSTEAD continued to HENDON

Published by J. Cary, July 1st 1790.

LONDON to St. ALBANS

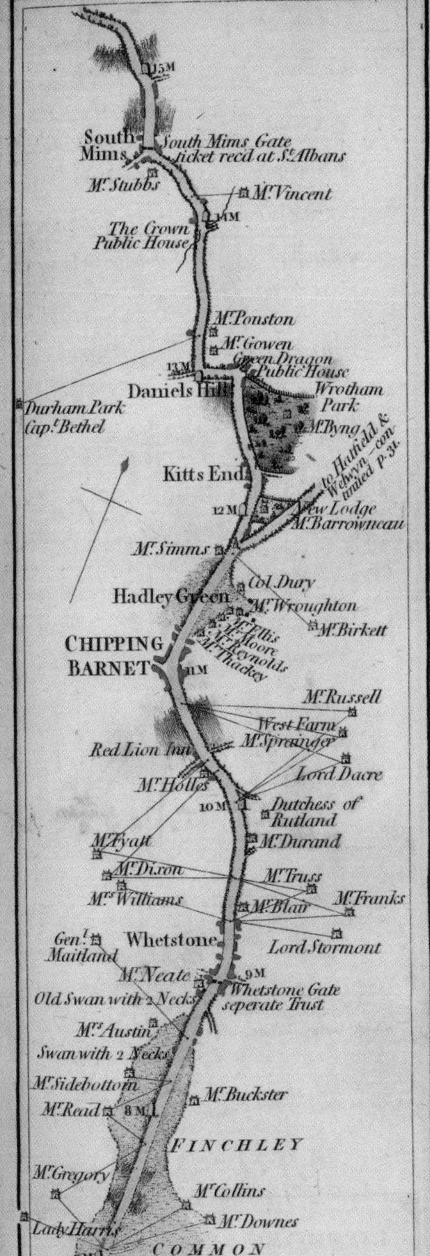

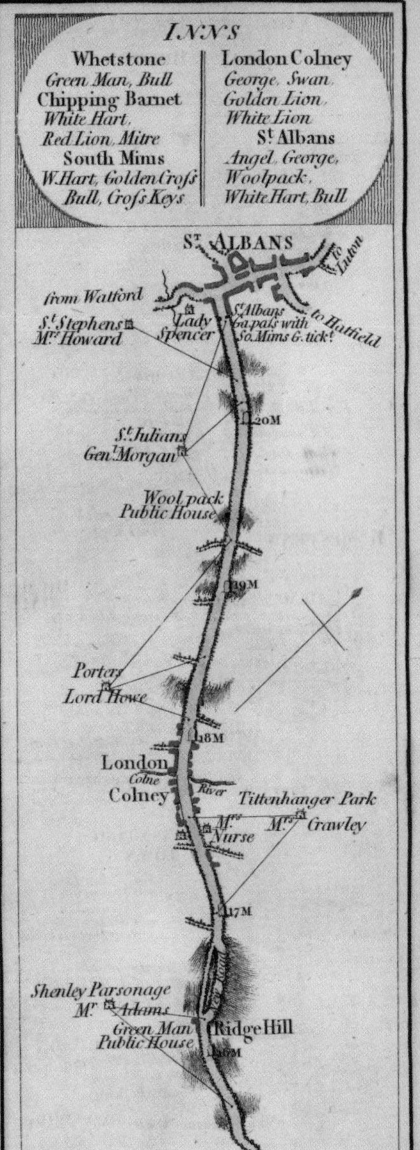

Published by J. Cary, July 1st 1790.

Horwood's exhaustive, detailed map of late Georgian London shows every home – 'by name or street number' – in a rapidly expanding city.

Every House in London

SHOWING EVERY HOUSE – and not just showing, but identifying each one by name or street number; such was Richard Horwood's extraordinary claim in his prospectus for this compelling map of what was then the largest city in the world. Fittingly enough, when the thirty-two sheets of completed map were fitted together they became the largest map ever produced in the British Isles.

Begun in 1790, with the production of a specimen sheet to encourage subscribers, the individual sheets emerged in fits and starts in a saga of tribulation, vicissitude and heroic perseverance. Horwood (1758–1803) eventually needed a loan of £500 from the Phoenix Fire Office to carry the work to completion in 1799. As he later recorded, 'It is an Undertaking which cost me 9 years of the most valuable part of my Life – I took every Angle [and] meas.^{d} almost every line'. How valuable that part of his life was he was not yet to know: he lived only long enough to complete a similar map of Liverpool, where he died in 1803.

The portion of the map shown here (published in 1793), rendered in exquisite detail on a scale of 26 inches (66 centimetres) to the mile (1.6 kilometres), depicts the area around Fitzroy Square, in the angle between Tottenham Court Road and what is now the Euston Road – then simply the New Road. Fitzroy Square itself is an exact contemporary of the map, the first phase of building (the south and east sides) being completed to the designs of Robert Adam between 1790 and 1800. The image is a haunting one – a part of London at once instantly recognizable and familiar – the streets and many of the buildings (certainly those away from the main thoroughfares) not much changed since Horwood's day. But it is a London wholly different, because beyond the New Road and Kendal's Farm it is a London that simply stops. We are in open country. Large as London was, it was still only a fraction of the size it was to become as the nineteenth-century suburbs rolled ever outwards.

Although there were some to quibble, then and now, as to how well and how accurately this obscure and previously unknown surveyor from Aylesbury was successful in his monumental task, there is no doubt among historians of the London landscape. This is the map we best like to use to plot our urban past.

Right Fitzroy Square and surroundings from Richard Horwood, *Plan of the Cities of London and Westminster.*
Above View of Fitzroy Square by Thomas Malton.

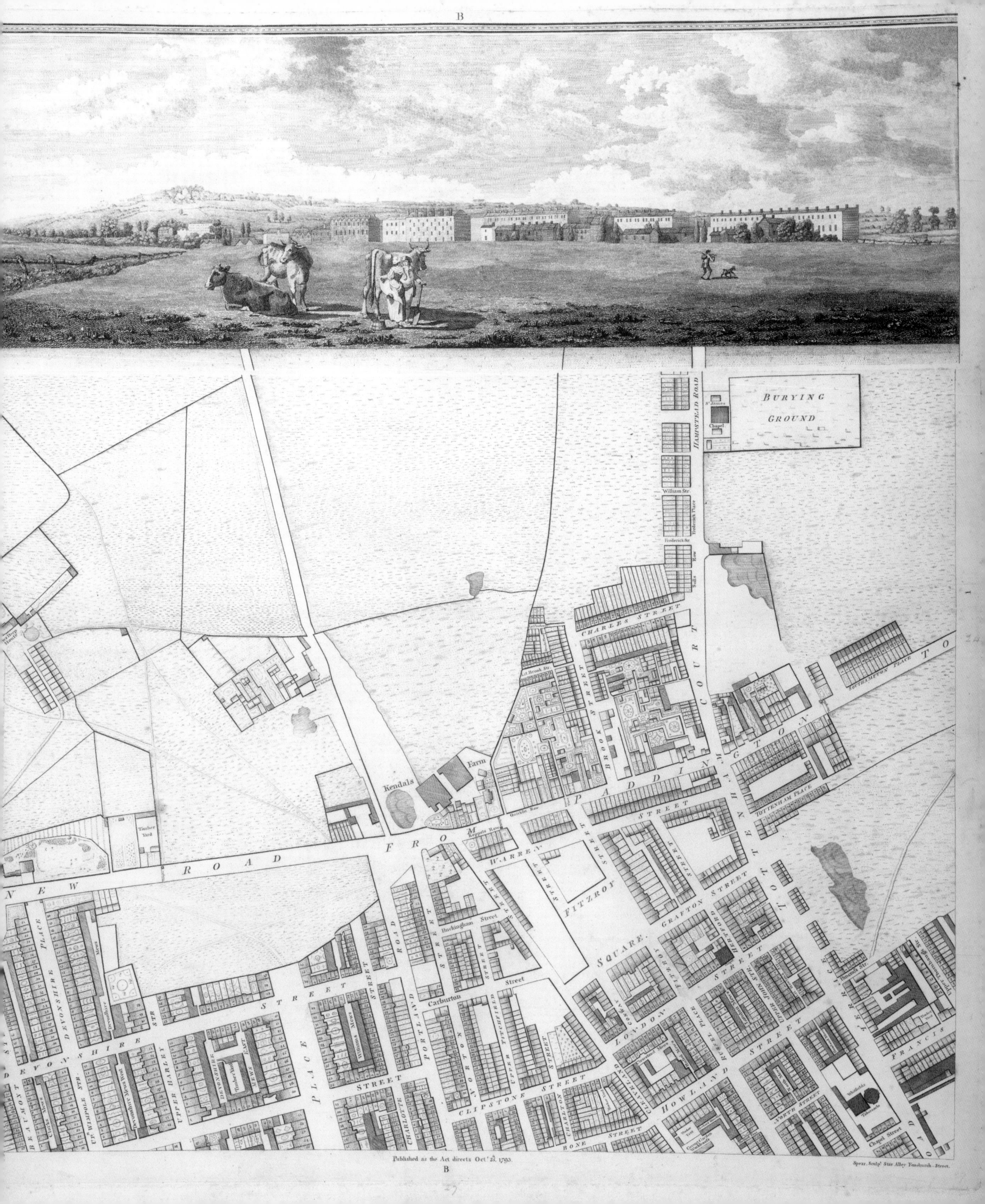
B
BURYING GROUND
St. James Chapel
HAMPSTEAD ROAD
William Str
Frederick Str
Frederick Place
CHARLES STREET
BROOK STREET
TOTTENHAM COURT
PADDINGTON STREET
SOUTHAMPTON PLACE
TOTTENHAM PLACE
Farm
Kendals
Quickset Row
Keppels Row
NEW ROAD FROM PADDINGTON TO
Timber Yard
WARREN STREET
FITZROY SQUARE
FITZROY STREET
GRAFTON STREET
Buckingham Street
Carburton Street
PORTLAND ROAD
NORTON STREET
UPPER TITCHFIELD STREET
CLIPSTONE STREET
LONDON STREET
DEVONSHIRE PLACE
DEVONSHIRE STR
UPPER HARLEY STR
UP. WIMPOLE STR
BEAUMONT STREET
Devonshire Mews Street
Williams Mews
CHARLOTTE STREET
CHARLTON STREET
HOWLAND STREET
NORTH STREET
BONE STREET
Tabernacle
Chapel Street
FRANCIS
Published as the Act directs Oct.r 24. 1793.
B
Spear, Sculp.t Star Alley Fenchurch Street.

A Gujarati chart depicting the Gulf of Aden, probably used to enable pilgrims making the Hajj to reach Mecca in safety.

Indian Navigation

REFERENCES IN INDIGENOUS texts and European commentaries on Indian manuscripts suggest that nautical charts may have been used by Indian navigators as far back as the fifth century. The earliest surviving chart, however, dates only from 1664. Of the small corpus of known extant works, all are of Gujarati provenance, reflecting Gujarat's formerly dominant position in the commerce of the Indian Ocean Basin. Even after the advent of the Portuguese, Gujarati seafarers long continued to trade over much of the region, relying on sailing manuals (*pothi*) with both cartographic and textual components, as well as separate maritime charts, to supplement their memorized navigational lore. Their navigational aids appear to have been based largely on Arabic prototypes; but Chinese and other sources may also have been utilized in their construction.

The long but narrow chart illustrated here was presented to Sir Alexander Burnes in 1835 by a pilot from Kutch (in northern Gujarat). It mainly depicts the Gulf of Aden, the Strait of Bab el Mandeb, leading to the Red Sea, being represented by the constriction towards the left edge. Jiddah, the port for Mecca, appears at the northern limit of the chart, suggesting that it may have been used in transporting pilgrims making the Hajj. As on many European charts, ships are prominently portrayed. The land features depicted are primarily those of interest to sailors: the nature of shorelines, profiles of prominent inland topographic features visible from the sea, flags of local potentates and the occasional portside mosque or other important edifice. Conventional symbols also locate reefs, rocky shoals and other maritime hazards. Of particular interest is the system of directional lines showing the horizontal angles to be followed on successive legs of a voyage. Such lines are marked at both ends by conventional symbols, utilizing signs for clusters of stars on compass cards with thirty-two equal sectors. The apparent direction of these lines on the map had no close relation to those that would appear on modern nautical charts of the same area. Thus, it mattered little that the chart suggests that the Gulf of Aden and the Red Sea are aligned in a continuous direction, rather than at more or less right angles to one another as they actually are.

Right A Gujarati chart of the Red Sea and Gulf of Aden. **Above** Detail of a ship.

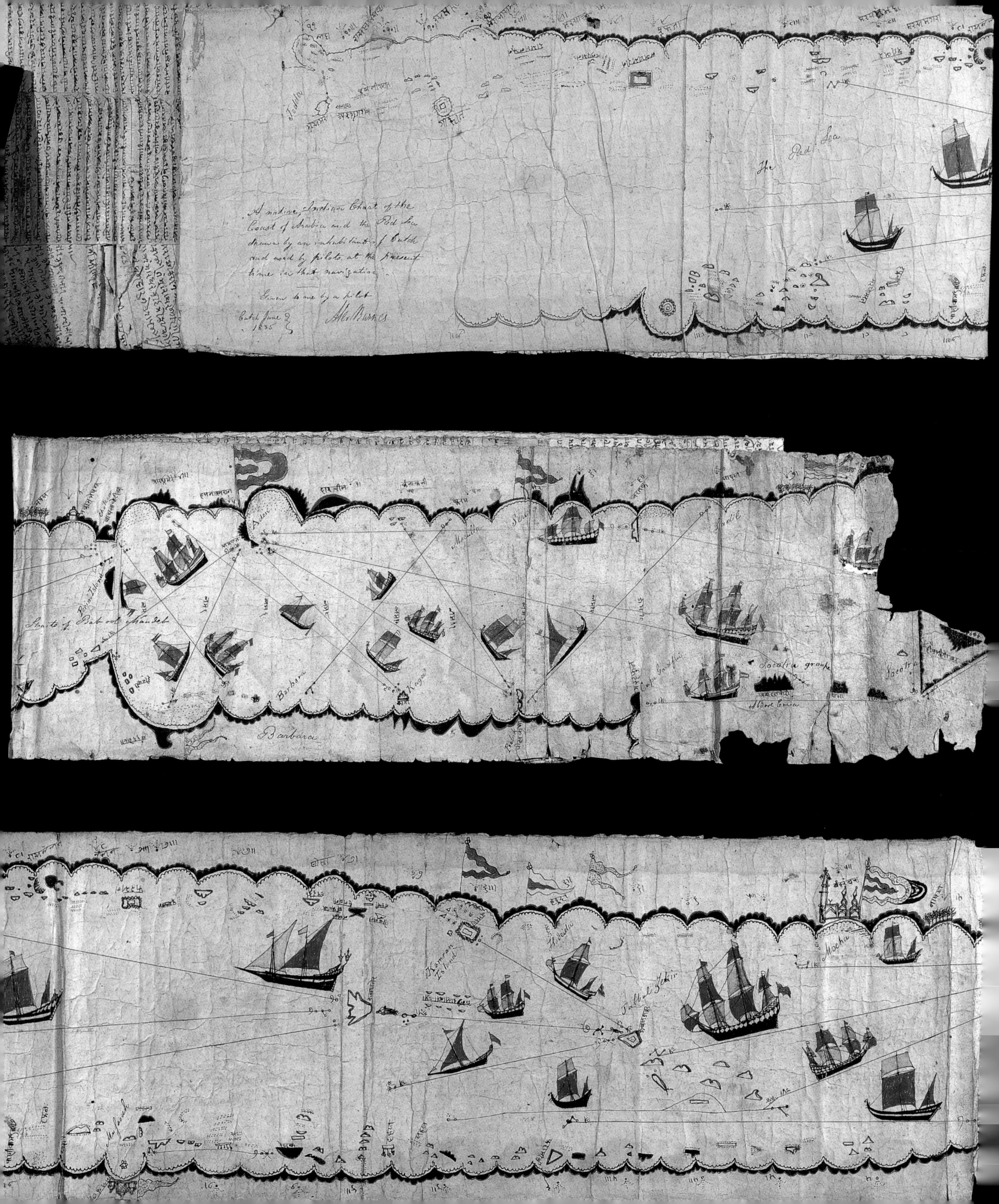
A native, Indian Chart of the
Coast of Arabia and the Red Sea
drawn by an inhabitant of Cutch
and used by pilots at the present
time in that navigation.
Given to me by a pilot
Cutch June 2
1835
Alex Burnes
The Red Sea
Straits of Bab-ool-Mandeb
Aden
Barbara
Socotra groupe
Socotra
Kamran Island
Hodeida
Mocha

A Korean cosmographic map mixing the real and the imaginary showing an outsized China as the centre of the known universe.

Ch'onhado: 'All Under Heaven'

THIS NINETEENTH-CENTURY Korean map entitled *Ch'onhado* ('Map of All Under Heaven') provides a revealing glimpse into an Asian world-view that persisted through until modern times. While maps based on Western science and cartographic principles had been circulating in Asia from as early as the sixteenth century, traditional forms of Asian mapping continued, in particular in works relating to popular culture, and in those maps deriving from Buddhism, indigenous or local religions, and cosmology.

This map from the year 1800 maintains China at the centre of the known universe. It reconfirms the importance that the Chinese have had in the political, social and cultural development of Asia as a whole. China is identified here by the calligraphic characters of 中原 meaning the 'Central field'. Notable details include the Great Wall and the Yangtze and Yellow rivers. Throughout the entire map, topography is amorphous and stylized. This reflects the dominating role that calligraphy has had in pictorial art and written language in Asia.

This map is unique because of its mixture of the real and the imaginary. China, the 'Land of Korea' (国鮮朝), the 'Land of Japan' (国本日), and 'Land of the Ryūkyū (国球琉) are correct in their relative position. Surrounding them, however, are mythical islands and mountains that were originally mentioned in an ancient Chinese geographical text known as the *Shanhai jing*. Among the wondrous places depicted include the 'Land of Giants' (国人) and the 'Land of Little People' (国人小) to the south-east. The 'Land of Hairy Peoples' (国民毛) is situated in the north-east, while the 'Land of Immortals' (国死不) lies immediately to the south of China. In the north-west, we find the 'Land of One-eyed People' (国目一) beside the 'Land of White Coloured People' (国民白). Further towards the periphery, past the source of 'Cold and Hot Water' (水暑寒), is the 'Land of Refined Elegant Ladies' (国女淑) located on a mysterious peninsula. This intriguing place should not be confused, however, with the 'Land of Women' (国人女) that is found on the opposite side of this universe.

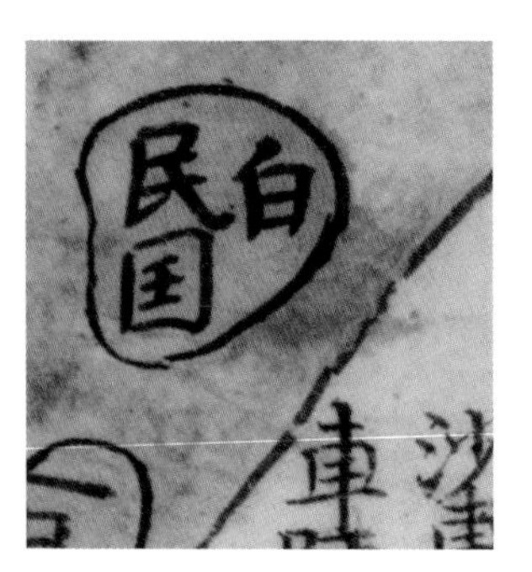

Right Korean world map.
Above The 'Land of White Coloured People'.

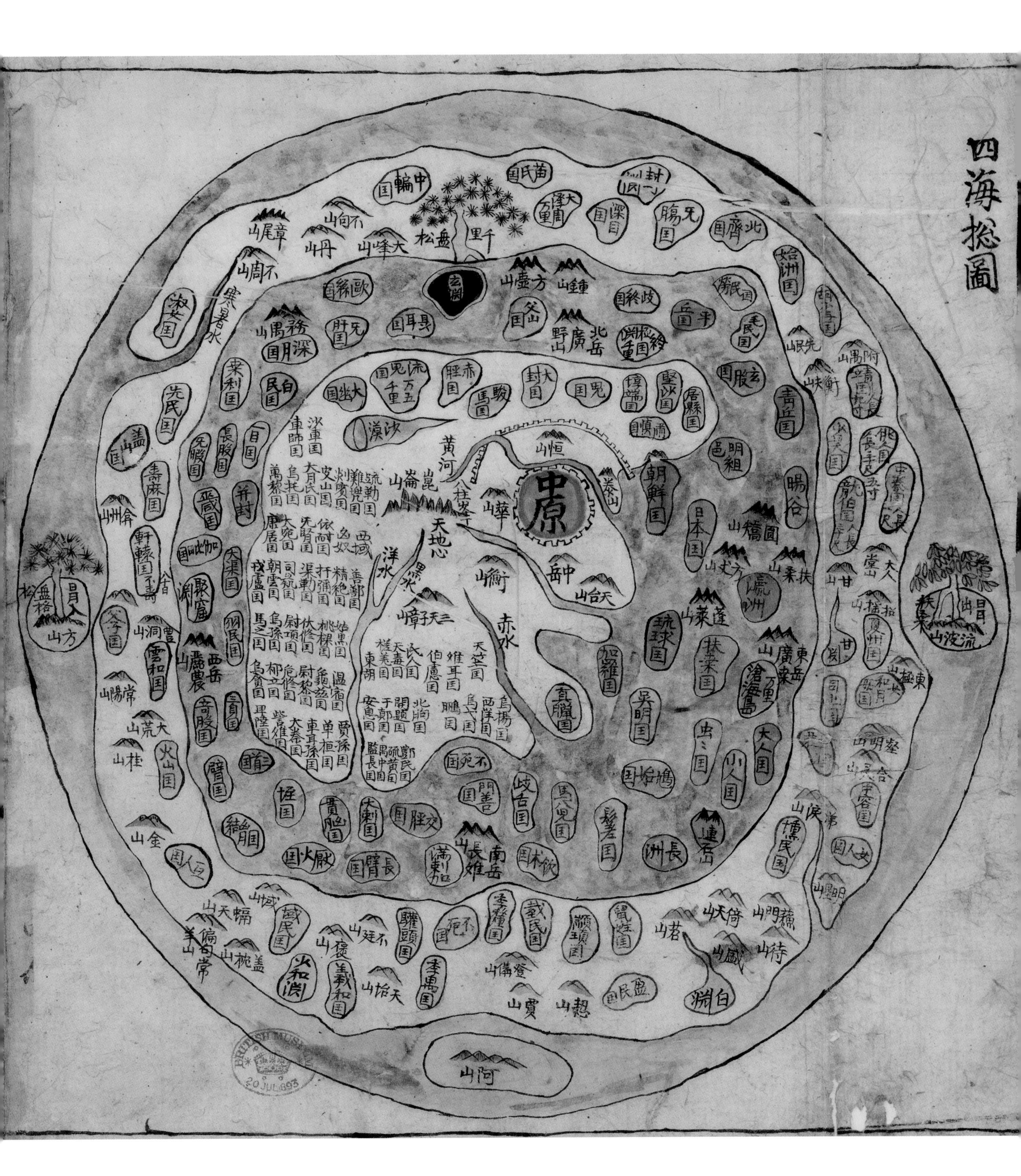
四海捴圖
中原
朝鮮国
日本国
黃河
赤水
天地心
崑崙山
恒山
泰山
華山
衡山
中岳
北岳
東岳
南岳
西岳
琉球国
扶桑国
蓬萊山
瀛洲
長洲
大人国
小人国
流波山
方山
阿山
白淵

A detailed map, never published, precociously identifies the wide variety of land uses from the packed inner city to the open fields in the vicinity of London.

Analyzing Town and Country

FROM ABOUT 1780 Britain began to emerge as a leader in the field of mapping, commercially and as far as techniques, concepts and scientific accuracy were concerned. This advance was exemplified in the work of Thomas Milne. He began his career as a traditional estate surveyor. By the 1780s, working with William Faden, the principal mapseller of the day, Milne was busy with the creation of accurately measured, relatively large-scale (generally 1 inch/2.5 centimetres to the mile/1.6 kilometres) county maps, in many cases the first fresh surveys in two hundred years.

In the 1790s he was closely involved with Richard Horwood's survey of London and with the Ordnance Surveyors who were beginning work in the south of England. The exact nature of the relationship is unclear, but fragments of manuscript surveys of the streets of London and the only known complete version of his amazing land-use survey of London and its environs survive in the British Library. The land-use map portrays the capital and its surroundings in hitherto unparalleled detail and for the first time shows field divisions accurately – undoubtedly on the basis of the work of the Ordnance Surveyors.

Right Extract from Thomas Milne's *Land-Use Plan of London*, showing west and south-west London. **Above** A view over south London, 1790. **See also** 1793, 1943 (p.334).

The map does much more. Through the sophisticated use of four types of hachuring it distinguishes different types of built-up area, ranging from the packed inner city to the speculative developments and ribbon development on the fringes. Even more striking is the depiction of the suburban and rural areas. Using lettering conventions underpinned by hand-colouring that anticipated those used from the 1940s, he showed precisely how, moving out from London, clay and gravel pits, providing bricks for new houses and improved road surfaces, were succeeded by nurseries and market gardens, producing food and plants for Londoners, pastures and meadows ultimately supplying milk for Londoners and hay for their horses, the parks and paddocks of wealthy Londoners with suburban homes, woods, orchards, hopfields and arable land, distinguishing as he went between enclosed fields, commons (uncoloured) and the open fields that despite the enclosure movement, still survived in surprising numbers.

Milne's map was never published, very possibly because Ordnance Survey objected to his attempt to utilize their data before it had been published.

a arable enclosed (yellow)
caf common arable fields (brown)
g enclosed market gardens (light blue)
h hopfield (dark green)
m enclosed meadow and pasture (light green)
ma drained marshland (yellow green)
n nurseries (orange)
o osier beds or orchards (green)
p paddocks and little parks (pink)
w woodland (green, usually over dots)
commons are left uncoloured.
Abbots Kensington
Knightsbridge
Pimlico
Tothill Fields
North End
Earls Court
Walham Green
Sandford
Parsons Green
Putney Bridge
Battersea Bridge
Nine Elms
South Lambeth
Stockwell
Manor of Wick
Vauxhall
Clapham Common
Battersea Rise
Wandsworth Common
Allfarthing
Battersea Common
Brixton Causeway
Balham
Dunsford
Garrett
Wimbledon Park
Tooting Com.
Streatham Common
Tooting Graveney
Tooting Bec
Norwood Common
Merton Street
Abbey
Pigs

A plantation map of Jamaica identifies the plantation owners whose names are now borne by the descendants of their slaves.

Slavery by Any Other Name

Right Detail from James Robertson's *Map of the Island of Jamaica.*

PRINTED IN 1804, still three years before the abolition of the Atlantic slave trade, this twelve-sheet map of Jamaica illustrates one of Britain's West Indian colonies.

As well as recording the basic geographical features of this island, it also reflects the importance of the slave trade for the British Empire. The map's key identifies the relative size of each plantation; from a single oxen mill, two oxen or windmill – the mills being necessary to process the sugar cane. So by implication it shows which sugar plantations held the most slaves.

Despite its tropical location, the map also provides a familiar echo of early nineteenth-century Britain. It identifies Jamaica's three counties of Cornwall, Middlesex and Surrey. It also records the names of the plantation owners – names which would not look out of place in any nineteenth-century address book – which includes the likes of the Hon J. Cambell, Sir D. Kinlock, Lady Taylor and Dr Miller.

However, today these quintessentially British names – so proudly engraved – give this map a new use. Once the African slaves had been bought at the slave market, they were given new names that indicated whose plantation they belonged to. Many of these 'given names' were carried through generations of slaves and, following emancipation, freed families. Today many of Britain's African-Caribbean families, whose ancestors migrated to Britain from the Caribbean, still carry such 'given' names. Many of these families are now trying to trace their family's history, a history of which there are few written records, and the names of the British plantation owners can be cross-referenced to their own family's name.

So despite this map's depiction of an imperial past it can be a valuable tool for rediscovering what were previously hidden family histories.

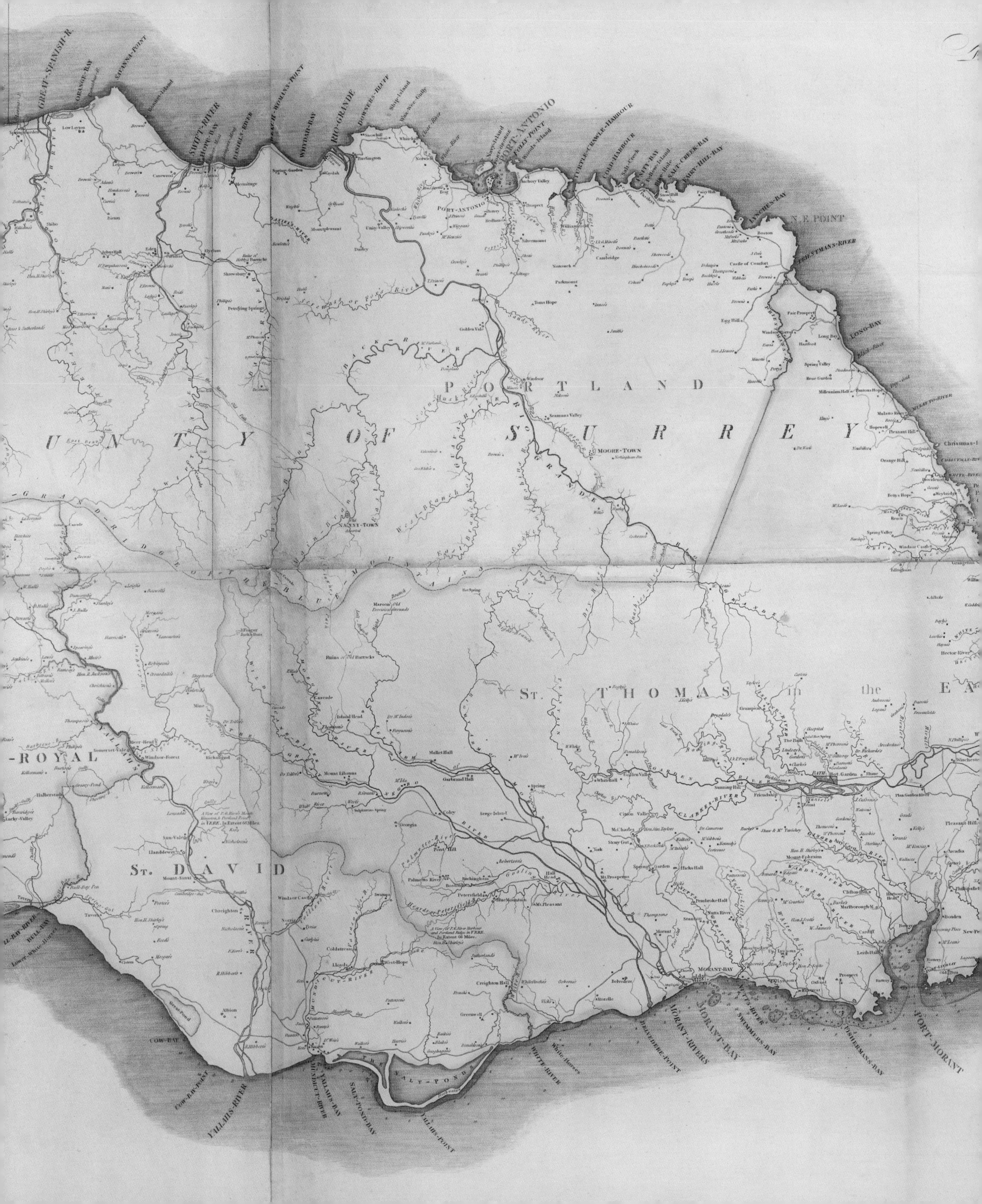

GREAT-SPANISH-R.
ORANGE-BAY
SAVANNA-POINT
SWIFT-RIVER
HOPE-BAY
RIO-GRANDE
PORT-ANTONIO
FOLLY-POINT
SALT-CREEK-BAY
FAIRY-HILL-BAY
N.E. POINT
LONG-BAY
PORTLAND
UNTY OF SURREY
MOORE-TOWN
NANNY-TOWN
BLUE MOUNTAINS
ST. THOMAS in the EA
-ROYAL
ST. DAVID
BATH
MORANT-BAY
PORT-MORANT
MORANT-RIVERS
SWIMMERS-BAY
FISHERMANS-BAY
WHITE-RIVER
THE SALT-PONDS
YALLAHS-RIVER
YALLAHS-BAY
YALLAHS-POINT
SALT-POND-BAY
COW-BAY

The map that the Duke of Wellington used to plan the Battle of Waterloo found on the body of one of his officers after the battle.

The Waterloo Map

1815

THIS TATTERED SURVIVOR from the field of the Battle of Waterloo bears pencilled marks, allegedly made by the Duke of Wellington, indicating his intentions for the disposition of his troops before the battle.

A note, written in 1846, records that it consists of reconnoitring sketches made in 1814 and 1815 by Royal Engineer officers. The work of at least three different hands can be seen on thirteen separate pieces of paper joined together. Roads and rivers mark the boundaries of each separate sketch.

The irony of this map, used by the Iron Duke to plan his tactics at Waterloo, is that the technique of combined reconaissance sketching, which it exemplifies, was first taught to the British Army by a Frenchman, General François Jarry. Jarry was an émigré, who became topographical instructor at the Royal Military College (High Wycombe) in 1799. He taught the skills of reconnaissance mapping to a generation of officers who went on to practise during the Peninsula War. Jarry influenced Robert Dawson, a young part-time instructor at the Royal Military College. Dawson became Chief Surveying Draughtsman on the Ordnance Survey, during which time (from 1803) he also taught reconnaissance, following Jarry's system, to Gentlemen Cadets of the Royal Engineers.

Right The Waterloo map.

The influence of Dawson's style of draughtsmanship can be seen in the top two sketches of this map. The summits of the hills around Mont St Jean and the field of Waterloo bear numbers. These indicate the relative heights of the ground, thus a hill numbered 4 would be higher than one numbered 2. This technique, known as 'Relative Command', was taught by Jarry and, through his student Robert Dawson, was adopted by the Royal Engineers and used on the Ordnance Survey.

A fair copy of this map was still unfinished on 16 June 1815, and the Duke requested these rough sketches to be sent to him instead so that he could plan his manoeuvres. The map has an eventful history. It was lost and recovered in the 'melée' at Quatre Bras – the action which preceded the Battle of Waterloo. Then, having on 17 June been marked up by Wellington, it was given by him to his Quarter Master General, Lieutenant-Colonel Sir William de Lancy, with orders to place the troops in position. The note on the map records that 'this plan was on the person of Sir William de Lancy when that officer was killed'.

NIVELLES
GENAPPE

A revolutionary new medium, as so often prompted by military requirements, used to map one of English mapmaking's most historic sites.

Set in stone

Right Quarter Master General's Office, *The South End of Hounslow Heath.*

HOUNSLOW HEATH, 3 miles (5 kilometres) south-east of Heathrow Airport, has a long history of association with matters military: this map opens a small window into that history and also tells a cartographic story.

The area was used for encampments from the Roman period. The cavalry barracks seen on the map date from *c.*1790 and continue in occupation by the Household Division to the present day. Two other features of military significance are the powder mills on the River Crane, and the manoeuvre ground to the south of the Roman Road. The manufacture of gunpowder is thought to have been established here during the reign of Henry VIII. The last mills were closed in 1927.

Hounslow Heath has a very specific resonance with historians of cartography. Thirty-four years before this map was printed it was the location for an event which was arguably the most significant milestone in the progress of cartographic science. The establishment and scientifically accurate measurement of the Hounslow Heath baseline by General William Roy of the Royal Engineers paved the way for the primary triangulation of Britain. This single event laid the foundation stone for the development of systematic and accurate topographic survey over the following hundred years.

This map also reflects a revolution in the technology of cartographic production. When William Roy and his associates were working on Hounslow Heath in 1784 the technology for producing multiple copies of maps employed engraved copperplates and a printing press. Excellent results could be achieved but the printing equipment was heavy and difficult to move and corrections and revisions were difficult to apply. The map described here was not printed in a press from engraved copper, but drawn onto stone and printed using a lithographic process, relatively new in 1818, but commonplace today. This new technology had been developed during the Peninsula War ten years earlier, when the military was receptive to innovative solutions to operational needs for military intelligence that could be quickly updated.

At the time this map was produced in 1818, the focus of military policy was moving rapidly from the external challenges of France and war in Europe to internal security and the maintenance of social order at home as a result of a crisis in the agricultural industry on which the majority of the population continued to depend for its livelihood.

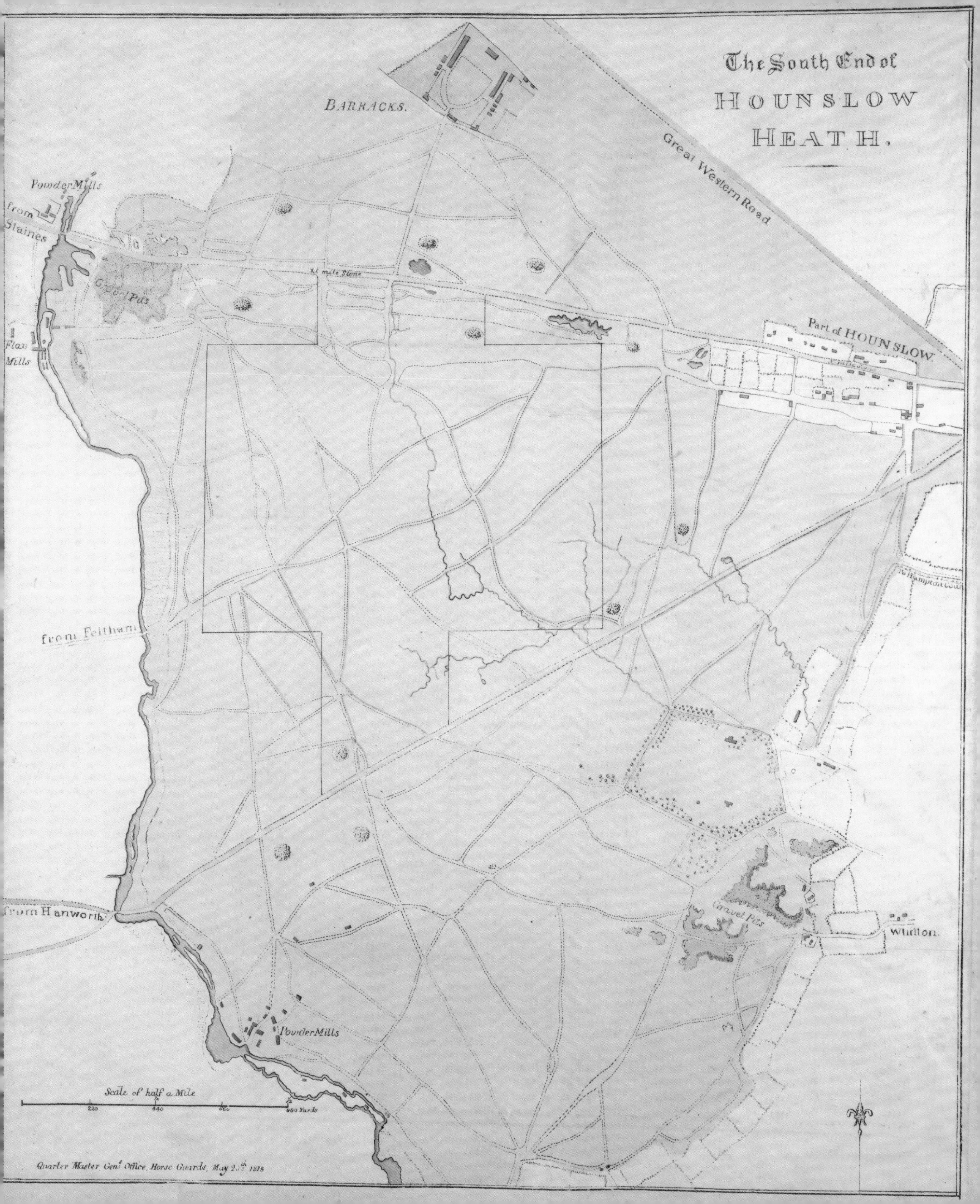

The South End of
HOUNSLOW
HEATH.
BARRACKS.
Great Western Road
Powder Mills
from Staines
Gravel Pits
XI mile Stone
Flax Mills
Part of HOUNSLOW
To Hampton Court
from Feltham
from Hanworth
Gravel Pits
Whitton.
Powder Mills
Scale of half a Mile
220
440
660
880 Yards
Quarter Master Genl. Office, Horse Guards, May 20th. 1818

An early geological map of an area around the city of Bath also represents an early step toward the multi-coloured printing of maps.

Geology Mapped

1823

THE YOUNG SCIENCE of geology developed rapidly in early ninteenth-century Britain. The pastime of collecting rocks and fossils led some amateurs to more serious study. Among them was William Daniel Conybeare (1787–1857), a clergyman who also lectured on geology. In 1822 Conybeare wrote, with William Phillips, *Outlines of the Geology of England and Wales*, a book that discussed the use of fossils to date sedimentary formations. Conybeare became the geological mentor of Henry Thomas De la Beche (1797–1855), collaborating with him on the earliest scientific descriptions of fossil plesiosaurs (published 1821 and 1824). Drawing lessons at Military College in Marlow had developed De la Beche's artistic talents (he drew the plesiosaur illustrations), while expulsion (for insubordination) in 1815 had given him leisure for travel and geology.

In 1823 Conybeare and De la Beche mapped the geology around Bath onto a map from a guidebook. Mapping local geology was a learning process, as it had been for William Smith, who had progressed from drawing a geological map of Bath around 1800 (never published) to creating the first geological map of England and Wales (published 1815). While Smith's earlier map may have inspired the 1823 Bath map, the latter bears stronger resemblance to Georges Cuvier and Alexandre Brongniart's map of Paris (published 1811). Conybeare had visited Paris in 1816 and could have seen it then. Related geology from opposite sides of the English Channel is shown similarly on both maps by distinctive, coarse red patterns atop hand-washed area colours. At that time most maps were printed from engraved copperplates and watercoloured by hand. The red patterns on the Paris map were indeed hand-drawn, but those on the 1823 Bath map were printed from a second engraved plate along with red dotted lines outlining the geological units. Overprinting geology in red made economical re-use of an existing map and represented an early step toward multi-colour printing of geological maps.

In November 1823 De la Beche went to inspect the Jamaica estate whose dwindling income would soon force him to become a professional geologist. In the late 1820s he went to work for the Ordnance Survey, mapping the geology of southern England (including Bath) onto its new one-inch (2.5-centimetre) topographic maps. The 1823 map of Bath was an early milestone in the career path that made him first director of the new Ordnance Geological Survey in 1835.

Right *Map of 24 miles round the city of Bath... by... C. Harcourt Masters. Coloured geologically by the Revd. W. D. Conybeare F.R.S. &c. and H. T. De la Beche Esq. F.R.S.*

MAP
OF 24 MILES ROUND THE
CITY of BATH.
Scale of Miles
Plastic Clay
Chalk
Portland Stone
Kimmeridge Clay
Coral Rag & Calcareous Grit
Oxford Clay
Cornbrash & Forest Marble
Great Oolite
Inferior Oolite
Lias
New Red Sandstone
Magnesian Limestone
Coal Measures
Millstone Grit
Old Red Sandstone
Trap Rocks
PART OF MONMOUTH SHIRE
PART OF GLOCESTER SHIRE
PART OF WILTSHIRE
PART OF SOMERSET SHIRE
RIVER SEVERN
THE BRISTOL CHANNEL
WELSH GROUNDS
CHEPSTOW
BRISTOL
BATH
KEYNSHAM
CHIPPENHAM
MALMSBURY
TETBURY
BROADFIELD DOWN
NAILSEA MOOR
KING'S SEDGEMOOR
EAST SEDGEMOOR
HEATH MOOR
GODNEY MOOR
MARK MOOR
SALISBURY PLAIN
GLASTONBURY
WARMINSTER

A set of cartographic playing cards reveals the English view of the peoples and countries of the world and assumes a battle for world domination between Great Britain and the USA.

The World Classified

1827

MAPS HAVE APPEARED on playing cards from at least the late sixteenth century. From the first they seem to have been intended to make education more pleasurable. This is the case with this card game, first published in the 1820s. The little book of rules accompanying the cards includes an introduction to geography with brief descriptions of the continents and of the countries depicted on the cards. These cover the geographical position, natural resources, religion, climate and constitution of each country, with an account of its population.

The Court Game of Geography, however, adds a political dimension which is very revealing of early nineteenth-century British attitudes towards the world. Each of the four continents (Australasia being considered part of Asia) has a suit to itself. Significantly Europe, 'the most important part of the globe [where] the arts and sciences have been carried to the greatest perfection' is hearts, Asia diamonds, America spades and Africa clubs. Within each suit the relative importance of the countries corresponds to the numerical value of the card, though the ace is always a map of a whole continent. Thus the British Isles, whose people are 'essentially ... commercial, entreprizing, industrious and successful, possessing the most extensive empire in the universe' are the 10 of hearts, the map being plastered with hearts, while the Netherlands, whose people are merely 'industrious, economical, and fond of the fine arts' are the 2 of hearts. In Asia, it is China which dominates, its peoples being described as 'intelligent, but deceitful, and extremely fond of litigation'. Egypt is the major African state, whose Coptic population 'the only descendants from the original Egyptians ... are of a grave deportment, but fond of sensual pleasures'. The leading American state is the USA where 'there is no religious establishment supported by the government, as it was thought more advisable to leave religion to its own operation on the minds of men'. Historical figures represent the kings, queens and knaves, with a motley crew of George IV, Catherine the Great of Russia and the French revolutionary leader Robespierre doing service for Europe. The instructions for playing show that the game climaxes with a battle for world domination between Great Britain and the USA – George Washington being King of America.

Right Selected cards from *The Court Game of Geography*.

BRITISH ISLES
ATLANTIC
OCEAN
NORT
SEA
English Channel
LONDON
TURKEY IN ASIA
BLACK
SEA
TATARIA
PERSIA
Trebisond
Bagdad
Bassora

A map drawn at a time of violent agitation for Catholic emancipation links absentee landlords to civil unrest and military oppression in Ireland.

'Where the Military Attacked the Insurgents'

Right An extract from Nicholas Philpot Leader's *Map of Munster prepared for the purpose of presenting ... the state of this disturbed and distressed province.*
See also 1658 (p. 158).

THIS MAP WAS probably drawn in the winter of 1827/8 at a time of violent agitation for Catholic emancipation in southern Ireland. The map juxtaposes different types of information in a way that the strongly Protestant King George IV, to whom it is dedicated, would have thought outrageous.

The symbols emphasize the extent of the military repression and the brutal means adopted for dealing with Irish unrest. The red lines show the newly built and the proposed military roads. Blank red squares show the locations of military stations. Triangles indicate 'places where Special Sessions are now holding under the Insurrection Act'. Blank black squares locate places of capital punishment. Most provocatively of all, the tent-like symbols show the 'places where the military attacked the insurgents'. The map further suggests that there was a link between absentee landlords (whose lands are shaded in yellow-ochre), the distress of the native Irish and subsequent Irish nationalist unrest. Among the named absentee landlords are several English and Scottish peers who held posts at court and in the Conservative government at Westminster.

The map's political message is skilfully conveyed through the selective presentation of the evidence. By altering the selection – for instance by showing the places where the insurgents had attacked the military and the homes of the landlords – quite another impression would have been fostered.

The map was almost certainly never seen by the King. It may well, however, have influenced its eventual owner the Marquess of Anglesey, better known as the Duke of Wellington's right-hand man at the Battle of Waterloo. He resigned a year after being appointed Lord Lieutenant of Ireland and the government's representative in Dublin in 1828 because the government would not heed his calls for reform. These had perhaps been partly encouraged by this map. The map remained in his descendants' hands until bought by the British Library in the 1980s.

It had been commissioned by Nicholas Philpot Leader, the Liberal MP for Kilkenny, a philanthropic squire and, with over 50,000 acres, one of the largest landowners in County Cork. He was, however, a resident landlord. His property at Dromagh Castle and Colliery is shown. Resident or not, his family was not to remain immune from the continued unrest: the castle was burnt down during the Irish Civil War of 1920–1.

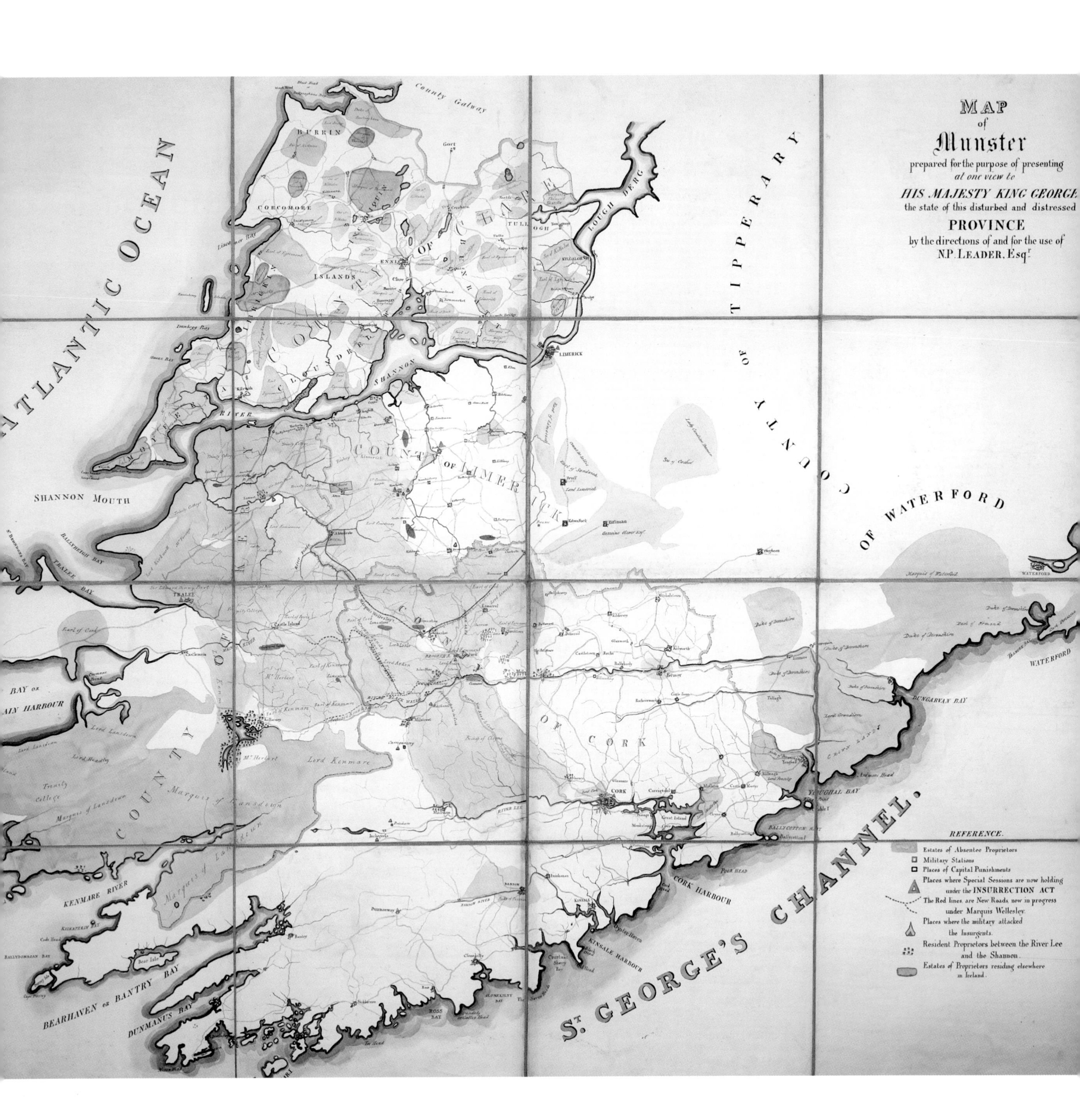
MAP
of
Munster
prepared for the purpose of presenting
at one view to
HIS MAJESTY KING GEORGE
the state of this disturbed and distressed
PROVINCE
by the directions of and for the use of
N.P. LEADER, Esqr.
ATLANTIC OCEAN
COUNTY OF CLARE
COUNTY OF TIPPERARY
COUNTY OF LIMERICK
COUNTY OF WATERFORD
COUNTY OF CORK
COUNTY OF KERRY
ST. GEORGE'S CHANNEL.
County Galway
BURRIN
CORCOMORE
TULLOGH
ISLANDS
LOUGH DERG
SHANNON
RIVER
SHANNON MOUTH
BALLYHEIGH BAY
TRALEE BAY
TRALEE
LIMERICK
CORK
WATERFORD
DUNGARVAN BAY
YOUGHAL BAY
BALLYCOTTON BAY
CORK HARBOUR
KINSALE HARBOUR
KENMARE RIVER
BEARHAVEN OR BANTRY BAY
DUNMANUS BAY
Bear Isle
Lord Kenmare
Marquis of Lansdown
REFERENCE.
Estates of Absentee Proprietors
Military Stations
Places of Capital Punishments
Places where Special Sessions are now holding under the INSURRECTION ACT
The Red lines are New Roads now in progress under Marquis Wellesley.
Places where the military attacked the Insurgents.
Resident Proprietors between the River Lee and the Shannon.
Estates of Proprietors residing elsewhere in Ireland.

A map showing natural resources and the canal network of Great Britain at the moment when the latter was about to be eclipsed by the fledgeling railway system.

Mapping Britain's Industrial Revolution

FROM THE 1760s onwards the face of Britain was being changed by the onset of industrialization, led by far-reaching innovations in the production of pottery, textiles, coal and other minerals. Meanwhile, to transport the greatly increased output a nationwide network of canals had been laboriously dug to link navigable rivers across the realm. The next revolution was in a new form of communications – the railway system.

All these three elements – canals, railways and sources of minerals – were brought together in a large folding map by a little-known compiler from Wakefield, John Walker. Dated 1830, the map contains considerable detail at the relatively large scale of just over 6 miles (10 kilometres) to the inch, with northern Scotland on a smaller scale. The overall size of the map is approximately six feet by five feet (183 centimetres by 152 centimetres). The main towns and roads and estates of the gentry are distinctively marked, and the hills are meticulously hachured. The engraving is a fine example of steelplate work, carefully wash-coloured by hand.

Walker's map traces 4,000 miles (6,400 kilometres) of canals, showing the network almost at its peak. However, railways were already taking over canal traffic and all the routes known up to 1830 are marked, including the Stockton and Darlington and the Liverpool and Manchester lines. Many railways serve collieries and quarries in industrial areas and only two lines are marked in the south-east: the Surrey line from Wandsworth to Reigate (opened 1803) and the recently completed Canterbury and Whitstable railway. The oldest railway marked is the short line from Middleton Colliery to Leeds, incorporated as long ago as 1758. Walker's map also shows mineral deposits and workings. Some are named – clay, gravel, chalk, coal, copper, ironstone, etc. – or alternatively the common chemical signs are used. Coal workings are, as expected, in the main industrial areas today. In Cornwall tin, lead and copper mines are prominently marked. More surprising is the recording of less common minerals in remote areas: strontium in Mull, silver around Ben Nevis, asbestos in Lanark, cobalt in Cheshire and septaria near Colchester. Within 30 miles (48 kilometres) of London can be found print, silk and lime works, mills for the manufacture of gunpowder, and factories for making iron, bricks, snuff, glass, oils and paper. In addition widespread deposits of chalk, gravel, clay, stone, fuller's earth, fire clay, flint and sand are located around the capital.

Right An extract from John Walker's *Map of the Inland Navigation, Canals and Rail Roads with the situations of the various Mineral Productions throughout Great Britain.* **Above** A contemporary five-coach steam train.

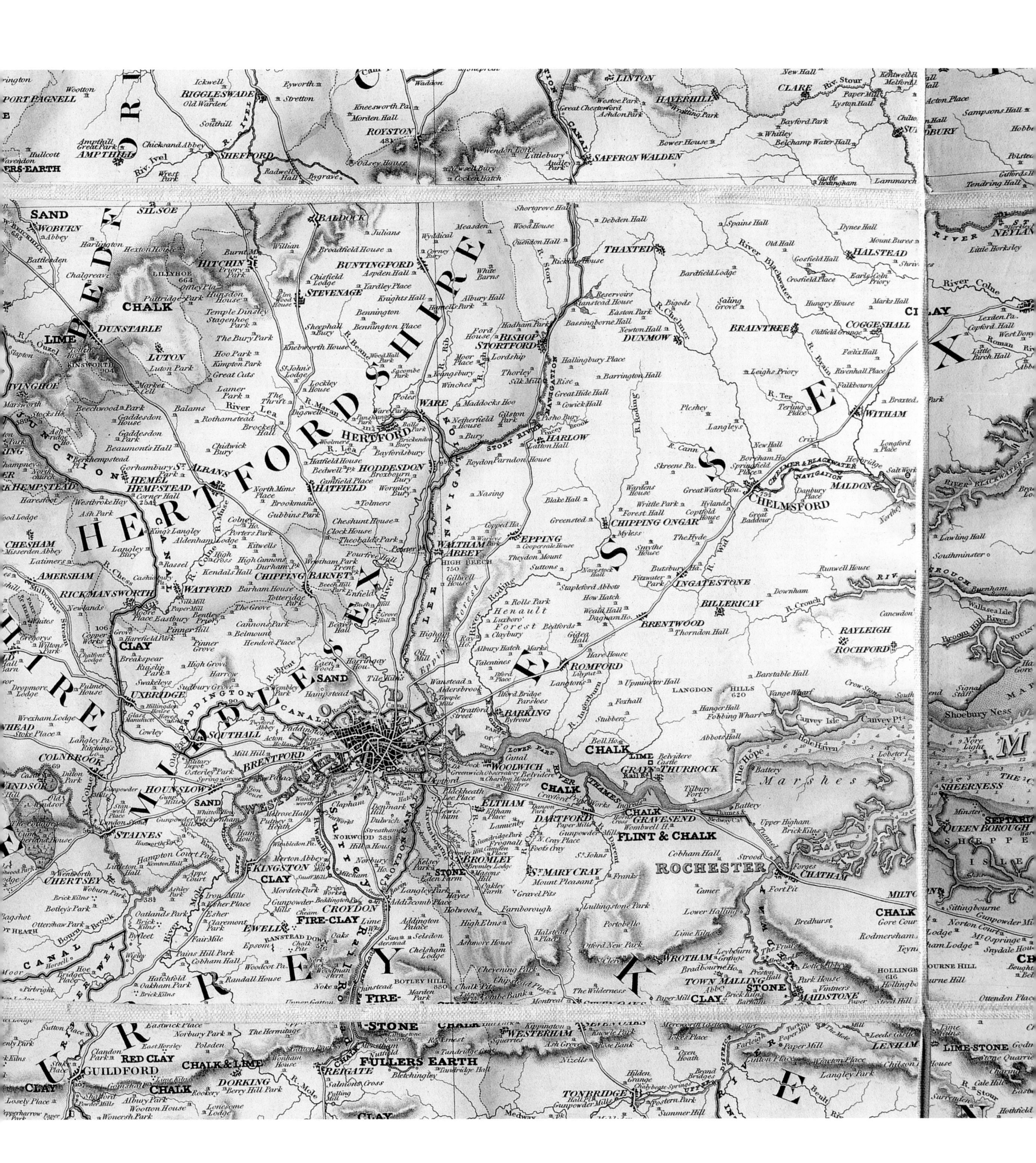
HERTFORDSHIRE
ESSEX
MIDDLESEX
SURREY
KENT
LONDON
BIGGLESWADE
ROYSTON
SAFFRON WALDEN
HAVERHILL
CLARE
LINTON
SHEFFORD
AMPTHILL
BALDOCK
HITCHIN
BUNTINGFORD
STEVENAGE
THAXTED
HALSTEAD
BISHOP STORTFORD
DUNMOW
BRAINTREE
COGGESHALL
DUNSTABLE
LUTON
WARE
HERTFORD
HODDESDON
HATFIELD
HARLOW
WITHAM
MALDON
CHELMSFORD
HEMEL HEMPSTEAD
ST. ALBANS
EPPING
CHIPPING ONGAR
WALTHAM ABBEY
CHIPPING BARNET
INGATESTONE
BILLERICAY
CHESHAM
AMERSHAM
RICKMANSWORTH
WATFORD
BRENTWOOD
RAYLEIGH
ROCHFORD
UXBRIDGE
ROMFORD
BARKING
SOUTHALL
BRENTFORD
WOOLWICH
HOUNSLOW
GRAYS THURROCK
GRAVESEND
SHEERNESS
QUEEN BOROUGH
STAINES
KINGSTON
ELTHAM
DARTFORD
BROMLEY
CROYDON
ROCHESTER
CHATHAM
EWELL
WROTHAM
TOWN MALLING
MAIDSTONE
GUILDFORD
DORKING
REIGATE
TONBRIDGE
CHALK
CLAY
SAND
FIRE-CLAY
FLINT & CHALK
FULLERS EARTH
RED CLAY
CHALK & LIME
LIME-STONE

A map issued by the Polish government in exile depicts previously held territories and the many battles for independence and betrays patriotic fervour behind a factual façade.

The Cartography of Nostalgia and Heroism

AT ONE LEVEL, this is a sober map showing the principal roads and postal routes through Poland. As well as the supposedly autonomous 'Kingdom of Poland' around Warsaw, ruled by the Czar, and the minute independent republic of Cracow, both created in 1815, it shows the areas of the former Commonwealth of Poland-Lithuania that were directly ruled by the Russians, Prussians and Austrians. Administrative divisions, provincial capitals, large towns, villages and fortresses are indicated. There is even a table, compiled by Stanislas Plater giving a detailed breakdown of the population by language and religion, revealing that historic Poland contained substantial Jewish, German and Russian minorities.

At another level the map is an exile's evocation of what had once been and a call to arms against the foreign oppressor, issued within weeks of the failure of the latest revolt. The map is dedicated by one of his former pupils, Leonard Chodzko, to one of the rebel leaders, Joachim Lelewel, a professor of geography (and one of the earliest historians of mapmaking) who had just taken refuge in France and become a member of the Polish government in exile. Inscriptions beyond the 1772 boundaries of Poland mention lands lost since the tenth century such as those in Pomerania occupied by the dukes of Stettin in the thirteenth century, those ceded to the kingdom of Bohemia in 1339 or those invaded by the Turks in 1672. Most poignantly the map lists and shows the locations of battles for Polish independence since 1768 with the names of the Polish commanders. Victories are indicated with sabres pointing upwards, defeats with sabres pointing downwards. Victories predominate by a large majority – but these were transient. There are few defeats – but these tended to be final. In the top right corner there is a miniature map of Warsaw. It is bathed in sunrays, surmounted with the names of great historic patriots such as Kosciuzko, and flanked by banners bearing the arms of the Polish-Lithuanian Commonwealth and embossed with the words Liberty, [Territorial] Integrity and Independence. There were to be fresh revolts in every generation until 1917 when an independent Poland again came into being – for a generation.

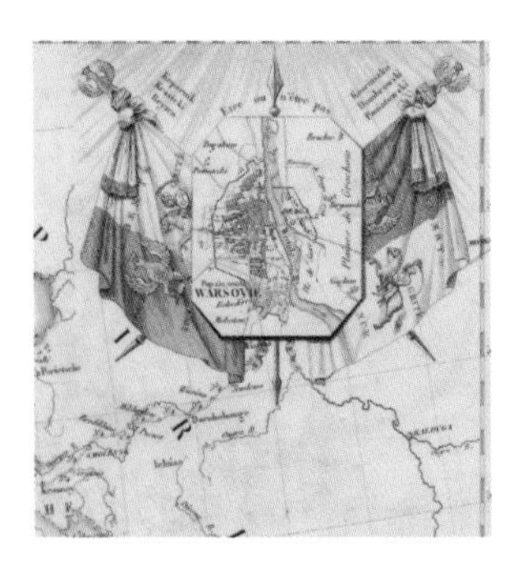

Right Extract from Auguste-Henri Dufour and Leonard Chodzko, *Carte Routiere, Historique et Statistique des Etats de l'ancienne Pologne.* **Above** Miniature map of Warsaw from the title panel.

CARTE ROUTIÈRE, HISTORIQUE ET STATISTIQUE
DES ÉTATS DE L'ANCIENNE POLOGNE
Indiquant ses Limites avant son premier démembrement en 1772
ET SON ÉTAT ACTUEL
depuis son dernier partage en 1815
Dédiée à l'Historien Géographe
JOACHIM LELEWEL
Président de la Société Patriotique Polonaise,
& Membre du Gouvernement Suprême National,
Par A.H. DUFOUR, Géographe,
Membre de la Commission centrale de la Société de Géographie
et Léonard CHODZKO, Polonais,
Elève de Joachim Lelewel.
1831.
Gravée par Flahaut – Ecrite par Hacq.
MER BALTIQUE
GOLFE DE RIGA
GOLFE DE DANTZIG
GOUVt. DE KOURLANDE
SEMIGALLE
GOUVt. DE SAMOGITIE
PRUSSE ORIENTALE
POMÉRANIE
POMÉRANIE POLONAISE
Pays envahis sur la Pologne par les Ducs de Stettin aux 13e et 14e siècles
G^d. DUCHÉ DE POSEN
PAL. DE PLOCK
PAL. DE KALISZ
PAL. DE SANDOMIR
PAL. DE KRAKOVIE
PALt. DE PODLAQUIE
GOUVt. DE GRODNO
SILÉSIE
Cédée par la Pologne à la Bohême en 1339
WARSOVIE
KRAKOVIE
LUBLIN
KÖNIGSBERG, ou Krolewiec
RIGA
…ats livrés pour l'Indépendance Polonaise
depuis 1768 jusqu'en 1831.
Dates. | Endroits. | Généraux Commandants | Ennemis
1ère Epoque, Confédération de Bar.
Berdyczew | Pulaski | Russes
21 Mai | Brzezany | J. Potocki | id.
28 Mai | Bar | id. | id.
Brzesc-Litewski | Pulaski | id.
20 Juin | Minsk (Lith.) | Bierzynski et Sapieha | id.
Czenstochowa | Pulaski | id.
26 Avril | Szrensk | Sawa-Calinski | id.
22 Juin | Lanckorona | Dumouriez | id.
6 Sept. | Radzica | M.K. Oginski | id.
14 Sept. | Stolowicze | id. | id.
22 Avril | Krakowie | Choisy | id.
25 Août | Czenstochowa | Pulaski | id.
2ème Epoque, Campagne de 1792.
10 Juin | Stolbce | Bielak | id.
11 Juin | Mir | Judycki | id.
15 Juin | Boruszkowce | M. Wielhorski | id.
18 Juin | Zielence | J. Poniatowski | id.
4 Juillet | Zelwa | J. Zabiello | id.
17 Juillet | Dubienka | Kosciuszko | id.
24 Juillet | Granne | J. Zabiello | id.
Epoque, Guerre de l'Indépendance en 1794.
4 Avril | Raclawice | Kosciuszko | id.
7 Mai | Polany | Jasinski | id.
6 Juin | Szczekociny | Kosciuszko | Rus. Pr.
8 Juin | Chelm | Zaionczek | Russes

The first detailed survey of the River Douro, combining both beauty and functionality, utilizes steel engraving to create delicate maps and views.

Careful with the Port!

THIS REMARKABLE MAP of the River Douro was produced in 1848 by Joseph Forrester (1809–61), an Englishman living in Portugal who throughout his life there demonstrated a great love for the country. A producer of port wine, he used the Douro to transport his goods, and many cargoes were lost on the treacherous and largely uncharted rapids. So he decided to map the river himself. It was the first detailed survey to be made of the river, at a time when many European countries had not been surveyed in anything like the degree of detail we take for granted now. He had to proceed almost from scratch, measuring the locations of individual landmarks by triangulation. It must have been an immense task, of the sort more often carried out by government or military agencies than by private individuals however accomplished or dedicated, and it took him up to twelve years to create. The finished result is an enormous, decorative, finely engraved and well-illustrated map, that combines beauty and functionality in a way that must be the objective of every cartographer: a small section only is reproduced here. The attention to detail shown in the marking and colouring of the rocks and shoals within the river, the precise mapping of towns so that the layout of houses and church within a village can be distinguished even at this comparatively small scale, are all remarkable.

By the time Forrester's map was published, lithography was rivalling engraving as a cheaper and faster method of map reproduction. However, Forrester chose to have the map engraved, to produce results of a degree of detail and clarity of line that lithography could not. In this case the map was engraved on steel rather than copper. This was a relatively new method of engraving, and could produce results of even greater fineness. The map was sent to England for engraving by William Hughes. Forrester was prepared to go to great efforts not just in the creation of this work but in the technicalities of its production, and seeing the finished result, it certainly seems to have been worthwhile. Ironically, a few years later Forrester was himself drowned on the River Douro; he was travelling through the narrow gorge Cachao da Valeira when his *barco* was swamped.

Right Extract from Joseph Forrester's map of the Portuguese Douro and the adjacent county. **Above** Detail showing a typical Douro *barco*.

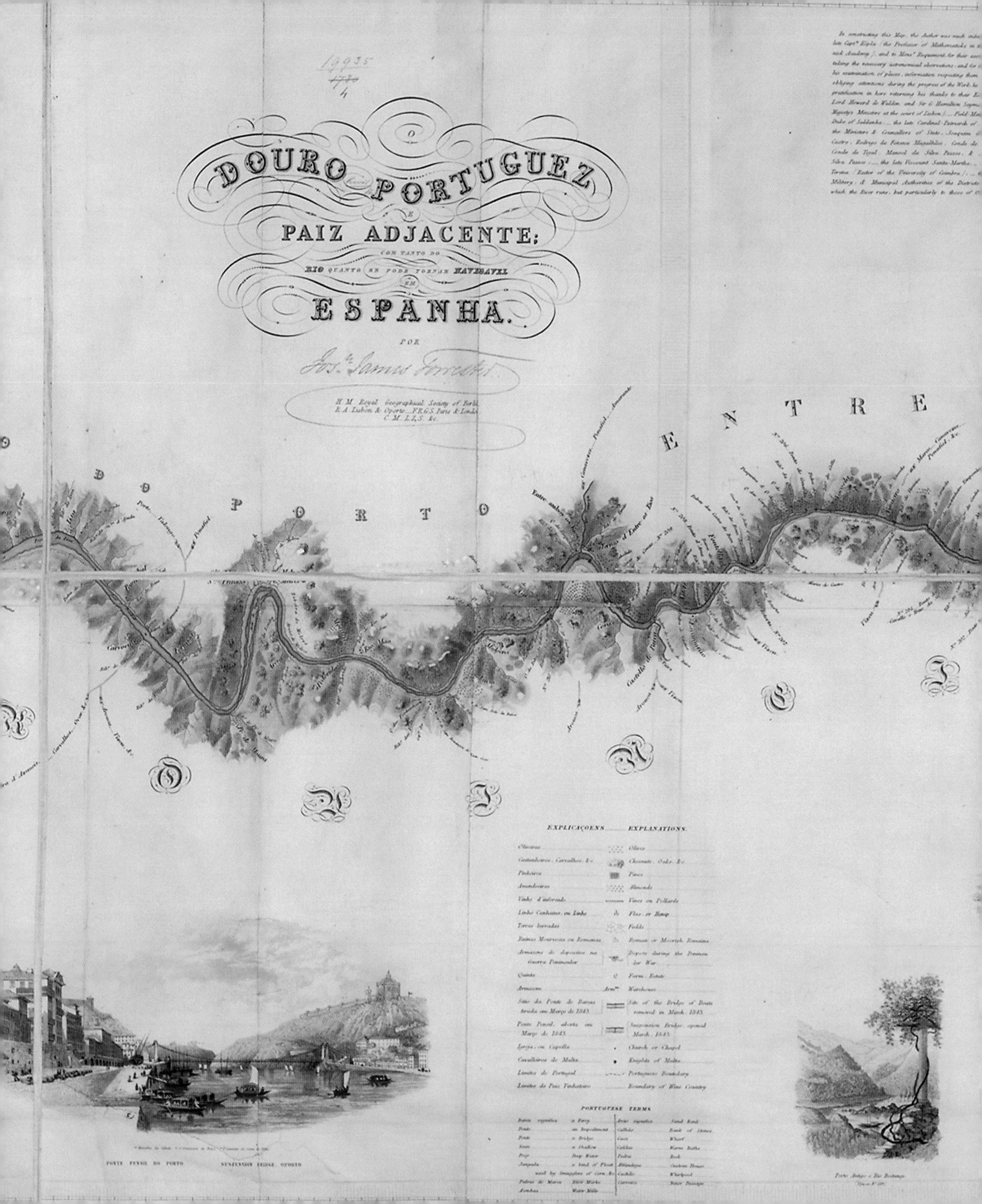

O
DOURO PORTUGUEZ
E
PAIZ ADJACENTE;
COM TANTO DO
RIO QUANTO SE PODE TORNAR NAVEGAVEL
EM
ESPANHA.
POR
Jos. James Forrester
H.M. Royal Geographical Society of Berli
R.A. Lisbon & Oporto...F.R.G.S. Paris & Lond.
C.M.L.Z.S. &c.
PORTO
ENTRE
EXPLICAÇOENS
EXPLANATIONS.
Oliveiras
Olives
Chesnuts, Oaks, &c.
Pines
Almonds
Vines on Pollards
Flax, or Hemp
Fields
Roman or Moorish Remains
Depots during the Peninsular War
Quinta
Farm, Estate
Armazem
Warehouse
Site of the Bridge of Boats removed in March, 1843
Suspension Bridge, opened March, 1843
Church or Chapel
Cavalheiros de Malta
Knights of Malta
Limites de Portugal
Portuguese Boundary
Boundary of Wine Country
PORTUGUESE TERMS
PONTE PENSIL DO PORTO
SUSPENSION BRIDGE, OPORTO

A detailed plan for a Utopian settlement, subsequently a failure, created by the Chartist movement as a practical response to an unyielding establishment.

A Rural Idyll

Right Anon., *O'Connorville, the first Estate purchased by the Chartist Co-operative Land Company, Situate in the Parish of Rickmansworth, Hertfordshire. Founded by Feargus O'Connor Esq[ui]r[e] 1846.* **See also** 1596, 1998.

THIS IS AN OPTIMISTIC map; a picture of a dream that was never fully realized. It shows O'Connorville, the first of the five settlements established by Chartists in the mid-nineteenth century. The map is surrounded by pictures of the houses, and a larger view showing people at work on their land. It's a beautiful image of happy, hard-working people in peaceful surroundings.

The story behind it is complex. The Chartist movement was a huge national organization which campaigned from the late 1830s for universal male suffrage and other electoral reforms. They presented three national petitions to Parliament, containing millions of signatures, but their quite reasonable demands were ignored by those in power. Political campaigning was the Chartists' main activity, but following the failures of their petitions and of a subsequent armed rising, the more moderate supporters tried a different approach. They bought five estates on which their members could settle. The houses were worth enough money to entitle their owners to a vote. The land was intended to give a few of the urban poor a chance for self-sufficiency, away from the grime, illness and moral corruption of the city. The houses were allocated by a lottery system; hundreds of people bought tickets and a lucky few were allocated a house and land. It was intended that the scheme would grow to accommodate far more people in due course. This settlement was named after Feargus O'Connor, an inspirational leader of the Chartist movement. This map, produced as propaganda at the time, shows what it was intended to be.

Sadly, it was a failure; the small plots of land were inadequate to support a family, especially when farmed by people with no knowledge of agriculture, and the Land Company was dissolved a few years later. A few individuals undoubtedly gained, but the wider benefits that had been anticipated were never achieved. Although it's sad that the scheme didn't succeed, the beliefs and dedication of the people behind it are still inspiring. It's also encouraging that almost all the Chartists' aims were eventually realized, although not until after the organization had ceased to exist.

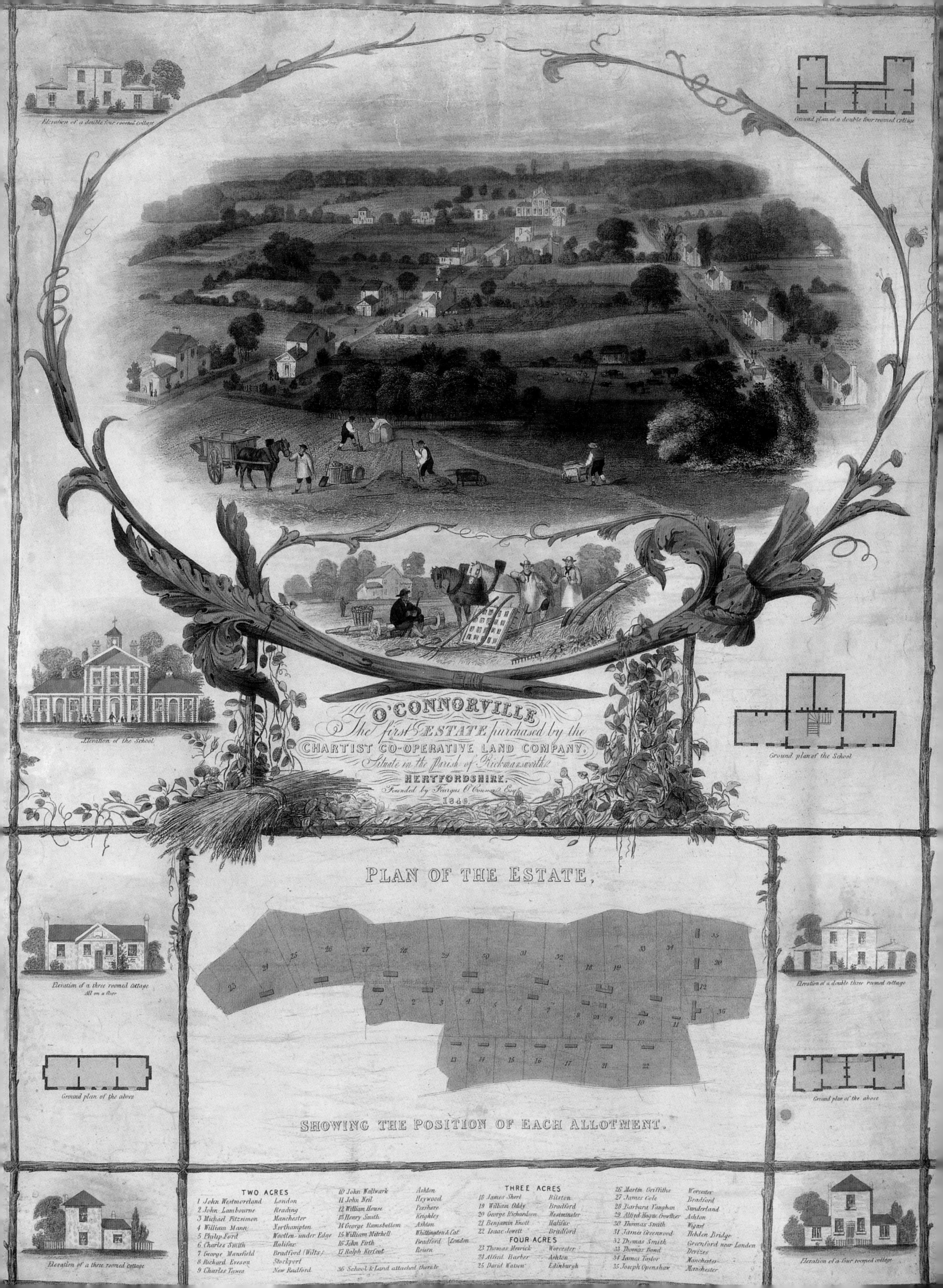
Elevation of a double four roomed cottage
Ground plan of a double four roomed cottage
Elevation of the School
O'CONNORVILLE
The first ESTATE purchased by the
CHARTIST CO-OPERATIVE LAND COMPANY,
Situate in the Parish of Rickmansworth
HERTFORDSHIRE,
Founded by Feargus O'Connor Esq.
1846.
Ground plan of the School
PLAN OF THE ESTATE,
Elevation of a three roomed cottage All on a floor
Elevation of a double three roomed cottage
Ground plan of the above
Ground plan of the above
SHOWING THE POSITION OF EACH ALLOTMENT.
TWO ACRES
1 John Westmoreland London
2 John Lambourne Reading
3 Michael Fitzsimon Manchester
4 William Mann Northampton
5 Philip Ford Wootten-under Edge
6 Charles Smith Halifax
7 George Mansfield Bradford (Wilts)
8 Richard Eveson Stockport
9 Charles Tawes New Radford
10 John Walwark Ashton
11 John Neil Heywood
12 William House Pershore
13 Henry Smith Keighley
14 George Ramsbottom Ashton
15 William Mitchell Whittington & Cat
16 John Firth Bradford [London
17 Ralph Kerfoot Rouen
36 School & Land attached thereto
THREE ACRES
18 James Short Bilston
19 William Oddy Bradford
20 George Richardson Westminster
21 Benjamin Knott Halifax
22 Isaac Jowett Bradford
FOUR ACRES
23 Thomas Merrick Worcester
24 Alfred Barber Ashton
25 David Watson Edinburgh
26 Martin Griffiths Worcester
27 James Cole Bradford
28 Barbara Vaughan Sunderland
29 Alfred Hague Crowther Ashton
30 Thomas Smith Wigan
31 James Greenwood Hebden Bridge
32 Thomas Smith Greenford near London
33 Thomas Bond Devizes
34 James Taylor Manchester
35 Joseph Openshaw Manchester
Elevation of a three roomed cottage
Elevation of a four roomed cottage

A map largely swathed in Imperial Pink to denote the British Empire provides practical and patriotic information for emigrants to the colonies.

Wider Still and Wider

Right Extract, showing Australia, from Smith Evans, *Emigration Map of the World.* **Following pages** The map in full. **See also** 1597 (p.134).

TODAY WHEN WE think of travelling we think of an airport's rolling display – tantalizing you with names of faraway places which are only a jet flight away. One hundred and fifty years ago this map provoked a similar response from Britons who were looking to expand Queen Victoria's empire.

Produced by Smith Evans it provides information for those who were to set sail for a new life in one of Britain's many colonies. At a time when Britain held sway over an empire spanning the four corners of the world, the map aptly colours Britain's processions in Imperial Pink. In this context, it is perhaps surprising that Britain is not placed centre stage. Instead it appears twice at each edge forming an imperial 'book-end' enveloping the rest of the world. Indeed, the map carries a decoration of a British ship sailing over the globe with the words 'Britain upon whose Empire the sun never sets'.

It gives information about what might await a traveller at the end of their journey. The approach is captured in its lengthy subtitle, 'Geographical and physical map of the world on Mercator's projection showing the British possessions and the date of the accession, population etc, all the existing steam navigation and the overland route to India with the proposed extension to Australia and also the route to Australia via Panama'.

So if you were planning to travel to Australia, it lists the principal towns of Australia and the distance from London. It also gives you the price of your ninety-five-day steamer journey – rising from £15 steerage to a £75 cabin – and allows you to plan how many acres of Australian land you might buy at £1 per acre.

Today when we travel our health concerns might include malaria pills and vaccinations – this map carries a stark warning for the Victorian traveller: 'the Anglo Saxon Race deteriorate between the 25th latitude north and south of the equator'.

Retailing at one shilling as a sheet, and up to two shillings on linen with a pamphlet, this map connected Britons with their imperial reach. Today, looking again at this 150-year-old-map, we can see how our ancestors viewed the world and saw their overarching and indeed unquestioned place within it.

RCTIC OCEAN
OCEAN
Nova Zemlia
Sea of Kara
North Cape
Lapland
White Sea
Archangel
Onega
St. Petersburg
Moscow
Riga
Kiev
Odessa
BLACK SEA
Constantinople
ASIATIC TURKEY
Smyrna
Angora
Trebizond
Jerusalem
Alexandria
Cairo
Egypt
Suez
Akaba
Bussora
Bagdad
Ispahan
Teheran
Shiraz
Lahsa
Medina
Mekka
Muskat
Sana
Aden
Adel
Dongola
Sennaar
Abyssinia
Gondar
Brava
of the MOON
Gold dust Ivory Palm Oil
SIBERIA
RUSSIAN EMPIRE
ASIA
Berezow
Ob R.
Tobolsk
Tomsk
Omsk
Krasnoiarsk
Irkoutsk
Touroukhansk
Yenissey R.
Khantaiskia
R. Lena
Gigansk
Olekminsk
Yakutsk
Okhotsk
Sea of Okhotsk
R. Kolyma
Gijiginsk
KAMTCHATKA
Sea of Kamtcha
Kotelnoi I.
New Siberia
Linghoff I.
Oustie Olenskoie
C. Shalagskoy
Sea of Aral
Sandy Desert
Rainless District
TARTARY
CHINESE EMPIRE
Khiva
Bokhara
Kashgar
Caubul
Kandahar
Attock
Cashmere
Lahore
Delhi
BELOOCHISTAN
AFGHANISTAN
SIDNE
Hyderabad
THIBET
Lassa
Berhampooter R.
R. Ganges
HINDOOSTAN
Calcutta
Bombay
Poonah
Goa
Madras
Pondicherry
Laccadive Is
Seringapatam
Ceylon
Candy
Spice, Coffee & Cocoa Nut Oil
Maldive Islands
ARABIAN SEA
Bay of Bengal
Aden to Ceylon 1970
Pekin
Great Wall
Nankin
Canton
Amoy
Formosa
Hainan
Cambodia
Gulf of Siam
Malacca
Sumatra
Borneo
Java
Phillippine Islands
Samar
Mindanao
Palawan
Sea of Japan
Jesso
Niphon
JAPAN
Trade with the Dutch only
Japan ware, Pearls &c.
Area 240,000 Square Miles
Population 30,000,000
Bonin Is
Mariana or Ladrone Is
Caroline Islands
Desierta
5050 Miles
55 Days sail
SPERM WHALE FISHERY
Sandwich Is
Constant drift
NORTH
EQUATOR
New Guinea
Solomon's Isles
New Hebrides
N. Caledonia
Norfolk I.
INDIAN OCEAN
Christmas I.
Keeling I.
Mozambique
Quillimane
Zanzibar
Rodriguez
Mauritius
Mauritius to Swan R. 3400
Juan de los
Port Natal
Delagoa B.
NEW HOLLAND
AUSTRALIA
Tropical Productions
WESTERN AUSTRALIA
SOUTH AUSTRALIA
Perth
Swan R.
Moreton Bay
Sydney
VICTORIA
Melbourne
Port Phillip
Bass Strait
Van Diemens Land
Launceston
Hobart Tn
Amsterdam I.
St Paul's Is
NEW ZEALAND
Auckland
Ship Alfred Capt Flint Outward course 1842.
Pr. Edward's Is
Marion
Crozets
Possession I.
Kerguelen's I.
Westerley Winds prevail in these Latitudes
50

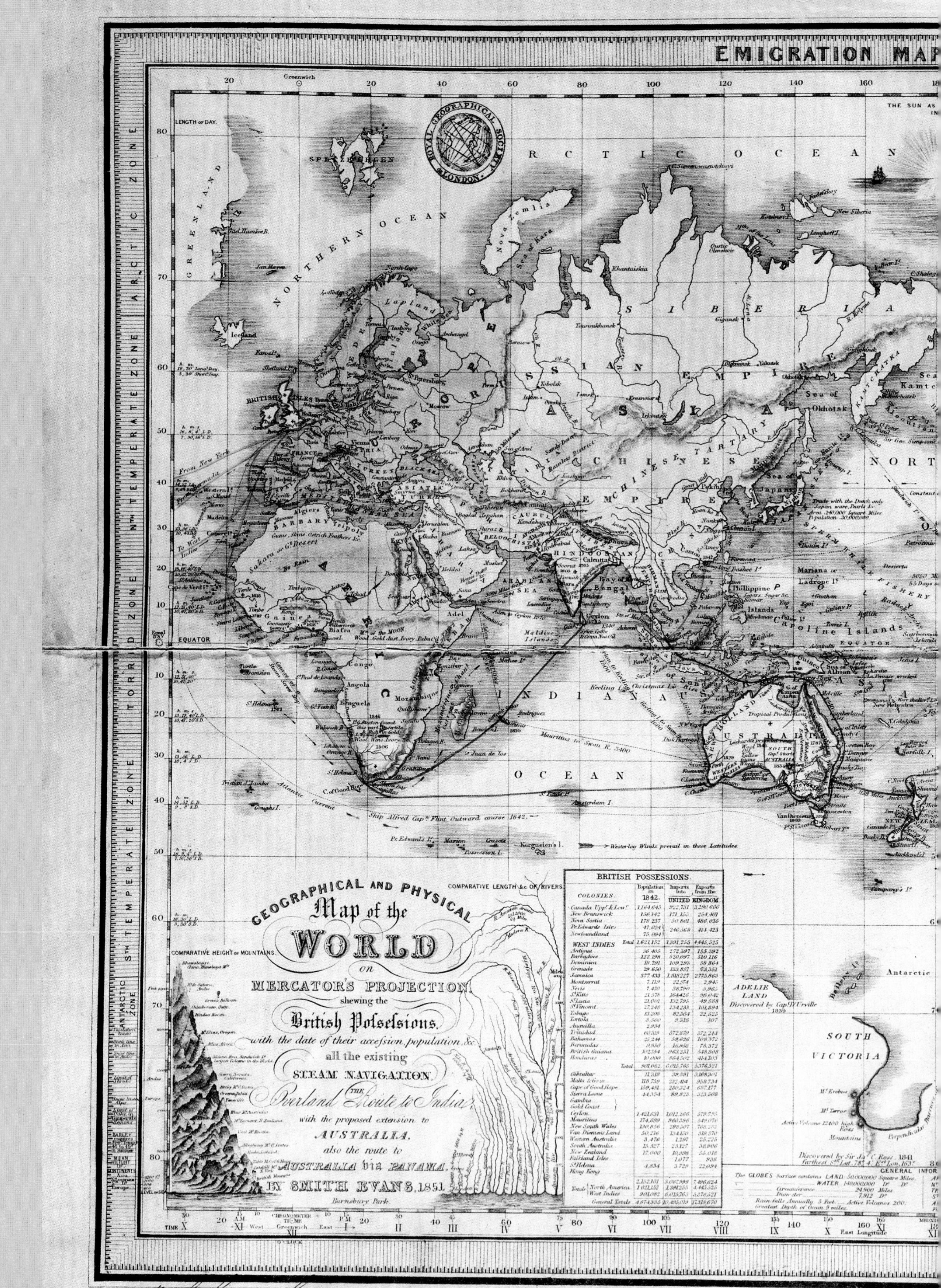

BRITISH POSSESSIONS.

COLONIES.	Population in 1842.	Imports into UNITED	Exports from the KINGDOM.
Canada Uppr. & Lowr.	1,164,645	922,731	3,290,666
New Brunswick	156,142	171,155	254,401
Nova Scotia	178,237	50,801	486,035
Pr. Edwards Isles	47,034	246,568	414,423
Newfoundland	75,094		
Total	1,621,152	1,391,255	4,445,525
WEST INDIES			
Antigua	36,405	272,397	155,392
Barbadoes	122,198	520,097	510,116
Dominica	18,291	109,298	58,864
Grenada	29,650	133,857	73,351
Jamaica	377,433	1,818,227	2,775,863
Montserrat	7,119	22,374	2,945
Nevis	7,470	38,790	5,965
St. Kitts	21,578	164,426	98,042
St. Lucia	21,001	132,795	49,588
St. Vincent	27,248	234,233	101,894
Tobago	13,208	82,564	22,525
Tortola	8,500	9,316	107
Anguilla	2,934		
Trinidad	60,319	572,879	372,214
Bahamas	25,244	59,626	108,372
Bermudas	9,930	16,958	78,372
British Guiana	102,354	963,231	548,808
Honduras	10,000	864,502	414,103
Total	901,082	6,015,765	5,376,521
Gibraltar	11,318	39,891	3,168,301
Malta & Gozo	118,759	232,484	958,734
Cape of Good Hope	159,451	280,324	687,177
Sierra Leone	44,334	89,823	323,508
Gambia			
Gold Coast			
Ceylon	1,421,631	1,012,266	579,795
Mauritius	174,699	960,396	549,076
New South Wales	130,856	298,507	768,282
Van Diemens Land	50,216	134,150	319,570
Western Australia	3,476	1,297	25,225
South Australia	15,527	23,127	38,906
New Zealand	17,000	10,998	55,018
Falkland Isles		1,077	938
St. Helena	4,834	3,729	22,094
Hong Kong			
	2,152,101	3,087,999	7,496,624
Totals North America	1,621,152	1,391,255	4,445,525
Totals West Indies	901,082	6,015,765	5,376,521
General Totals	4,674,335	10,495,019	17,318,670

With the Author's Compts.

THE WORLD
Greenwich
ARCTIC HIGH LANDS
PARRY or North Georgian Isl^ds discovered 1819
POLAR SEA
BAFFINS BAY
GREENLAND
SPITZBERGEN
NORTHERN OCEAN
RUSSIAN POSSESSIONS
Arctic Circle
Hudsons Bay
Labrador
Iceland
Lapland
Archangel
St Petersburg
Moscow
NORTH ATLANTIC OCEAN
BRITISH ISLES
Gulph Stream
Cod Fishery
FRANCE
Madrid
Algiers
BARBARY
Sahara or Gt. Desert
Rainless District
Timbuctoo
Guinea
Biafra
Congo
Angola
Benguela
Mozambique
St Helena
Ascension I.
Cape de Verd Is.
Madeira
Canary Is.
Western Is.
Gulf of Mexico
Caribbean Sea
Tropic of Cancer
Sandwich Isles
PACIFIC OCEAN
EQUATOR
PACIFIC 13.25 HIGHER THAN ATLANTIC OCEAN
Marquesas
Disappointment I.
Tropic of Capricorn
SOUTH SEA
SOUTH ATLANTIC OCEAN
Rio Janeiro
Buenos Ayres
Cape Horn
Falkland Is.
Tristian d'Acunha
Gough I.
Atlantic Current
Pacific Current
Outward course
Pr. Edward's Is.
Indian Ocean
New South Shetland
New Orkneys
Sandwich Land
Adelaide Is.
GRAHAMS LAND
One Minute of Time in this Latitude is 3,299 Miles
One Minute of Time on EQUATOR 17,280
Ship Alfred Capt. S. Flint Homeward course 1843
INFORMATION FOR EMIGRANTS.
COUNTRIES. | Principal Towns. | Miles from LONDON. | TIME taken by Sailing Ships &c. | EXTENT in Square Miles. | TEMPERATURE average Fahr. Therm. | PHYSICAL DESCRIPTION.
AUSTRALIA. Adults DIE annually 14 in every 1000. | SYDNEY, Port Philip, Adelaide, Swan River | 14,000. | 95 to 120 days | 3,000,000 | Nearly as large as EUROPE, with a Sea Coast of 3000 miles in extent
VAN DIEMANS LAND. 14 in 1000. | HOBART TOWN, Launceston | D^o. 570., D^o. 480. | 27,192 | The size of IRELAND. The Climate more temperate than AUSTRALIA
NEW ZEALAND. 15 in 1000. | AUCKLAND, Wellington, Nelson, Otago | 15,000. | 63,000 | As large as ENGLAND & SCOTLAND
C. OF GOOD HOPE. 13.7 in 1000. 9.8 East^n Coast. | CAPE TOWN, Algoa Bay, Port Natal | 7,000. | 45 to 70 days | 200,000 | The CAPE DISTRICT is a land of Steppes with numerous mountain streams
CANADA. 16.1 in 1000. | QUEBEC, Montreal, Toronto | 3,500. | 25 to 45 days | 390,000 | CANADA in connection with the Hudson's Bay Comp^y Territories
NOVA SCOTIA & NEW BRUNSWICK. 14.7 in 1000. | HALIFAX, St John | 2600. | 20 to 35 days | 50,000 | HALIFAX is the finest Harbour in the World
UNITED STATES. 18 in 1000. | NEW YORK | 2820. | 25 to 40 days | 2,300,000 | This Great Republic is composed of 30 States
TEXAS. 28 in 1000. | GALVESTON, Red River | 5,500. | 40 to 60 days | 310,000 | Lately annexed to the United States
CALIFORNIA. 20 in 1000. | SAN FRANCISCO via Cape Horn, Panama | 18,000., 8,500., 4,500. | 45 to 50 days | 50,000 | The Valleys of the Sacramento & St. Juan are fertile
The Anglo Saxon Race deteriorate usually in residing between 25° Lat. N^th & S^th of the Equator. SOVEREIGNS are usually worth 24s to 25s in the Colonies.
BRITISH POSSESSIONS.
EXPLANATION.
British Possessions marked thus +
Steam Navigation
D^o. contemplated
The Route of the India & Australia Royal Mail Steam Packet Company.
Aden to Sydney. | Miles. | Total London to Sydney.
Via Ceylon, Sincapore, P^t Essington, & Torres Str. | 7730 | 11,710 East^n Pass^e
P^t Essington, & Swan R. | 9308 | 13,288 West^n Pass^e
Christmas I. & Swan R. | 7758 | 11,738.
Keeling I. & Swan R. | 7388 | 11,368.
Mauritius, & Swan R. | 8118 | 12,098.
AUSTRALIA & CALIFORNIA.
London to Chagres 5057.
Chagres Panama 45.
Panama to San Francisco 3440 New Zealand 6662.
New Zealand Sydney 1138.
To California 8542. To Australia 12,902.
Sydney via Cape of Good Hope 12,870.
D^o. Cape Horn 13,257.
D^o. to Calcutta 6430.
D^o. to Hongkong 4500.
CHRONOMETER TIME
West Greenwich East
O'CLOCK
ENTERED AT STATIONERS HALL
Engraved by B.R. Davies.
...change, London.

A map created to counter the growing threat of nationalism plots the ethnic complexity of the Austrian Empire in unsurpassed detail and with a genuine attempt at impartiality.

Putting Peoples on Paper

AS THE BABYLONIAN world map demonstrates, maps have contained ethnic information from the earliest times. For ethnic maps to portray the distribution of peoples accurately, however, geometrically precise maps, an agreed definition of 'ethnic', reliable statistics and advanced printing technology are needed. Ethnic maps as generally understood first appeared in the late sixteenth century. It was only in the course of the nineteenth century that they began to proliferate, a reflection of technological advance, the coming of sophisticated statistics gathering and the growing importance of nationalism.

Nowhere were these factors more in evidence than in the dominions ruled by the Habsburg dynasty and covering most of central Europe. The Monarchy, as it was generally called, embraced numerous nationalities, generally but not invariably with their own distinct languages, and was administered by a large and self-confident, if also cumbersome and self-satisfied, bureaucracy. Awakened among the subject peoples and particularly the Germans, Hungarians and Czechs at the turn of the nineteenth century by the example of revolutionary France, by the middle of the nineteenth century nationalism was imposing enormous strains on the Monarchy and infiltrating almost all aspects of its existence. In order to control their peoples, the Habsburg authorities had to identify and pinpoint the location and numbers of the different nationalities. Maps were the simplest means of doing so.

This map, by a loyal servant of the Habsburgs, Karl von Czoernig (1804–89), is one of the greatest of all ethnic maps. The fruit of fourteen years of research, it was derived from meticulously gathered official information as well as field research in the most racially mixed parts of Hungary. The map captures small pockets of peoples, particularly Germans, living in 'islands' hundreds of miles east of their western heartlands, and narrow rings of one nationality surrounding town centres dominated by another.

By identifying nationality purely through language, and failing to differentiate between dialects and distinct languages, Czoernig missed out some significant groups like the Jews who spoke a variety of languages. Nevertheless his map can be regarded as the highwater mark of attempts to portray such emotive factors objectively before ethnic mapping, for all the geometric accuracy of the base maps, became dominated from about 1890 by politics and racism.

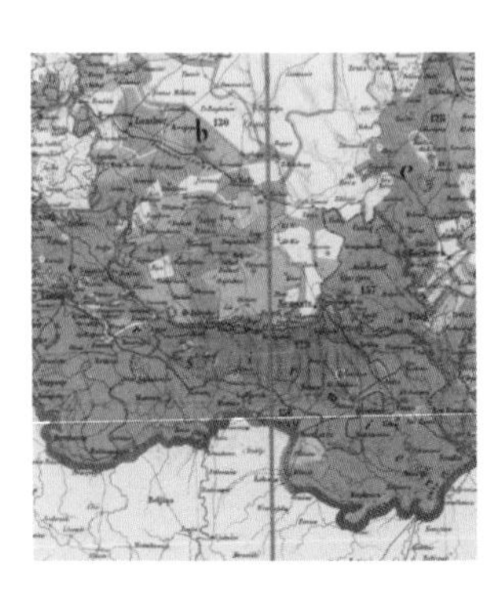

Right Karl von Czoernig, *Ethnographische Karte der Oesterreichischen Monarchie.* **Above** Detail showing Hungarian borderlands. **See also** 600 BC.

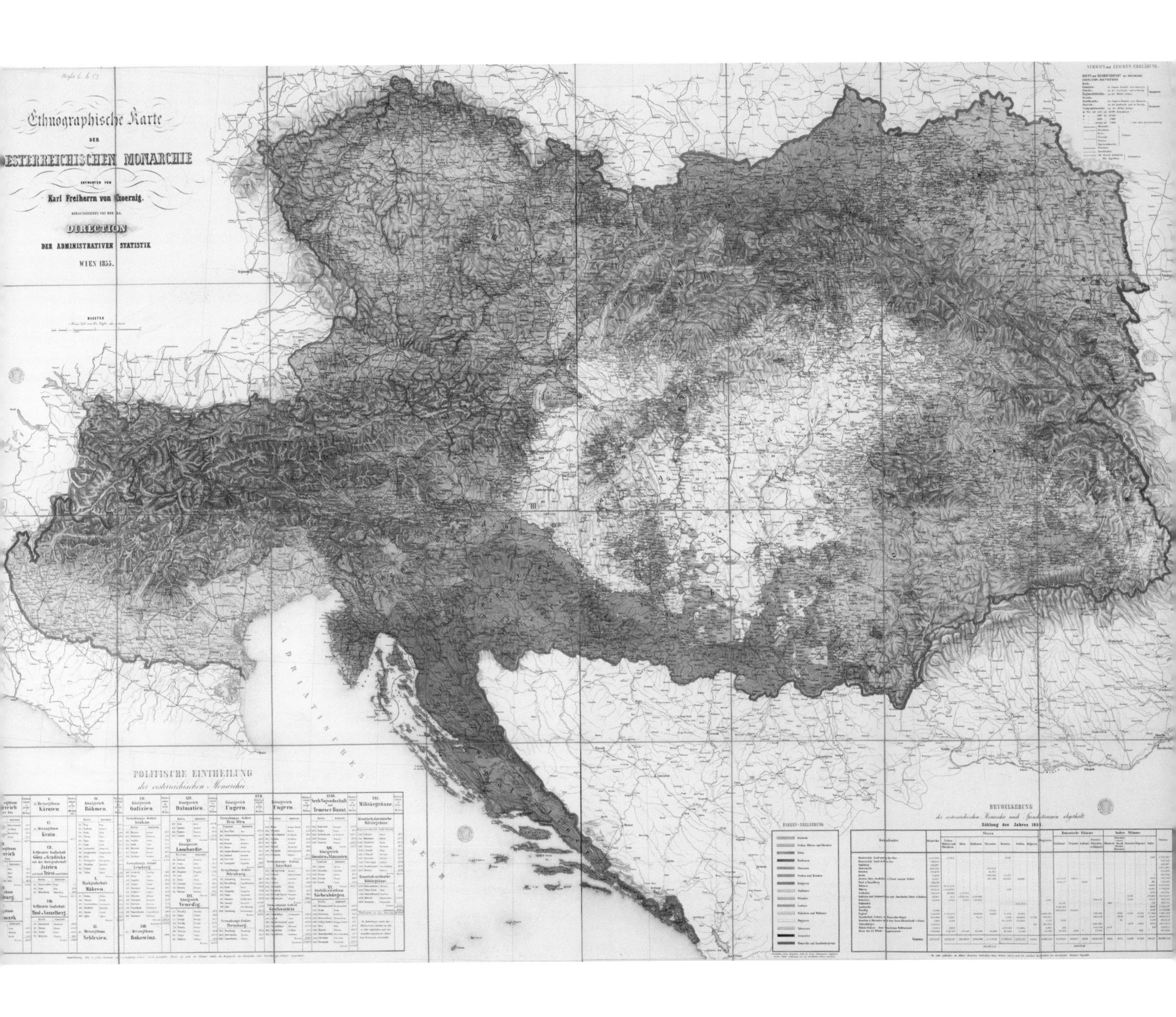
Ethnographische Karte
DER
ESTERREICHISCHEN MONARCHIE
Karl Freiherrn von Czoernig.
DIRECTION
DER ADMINISTRATIVEN STATISTIK
WIEN 1855.
MASSTAB
POLITISCHE EINTHEILUNG
BEVOELKERUNG

'Geographical spies', or Pundits, used by the British in areas closed to their industrious cartographers, helped to create the first map to name Mount Everest.

Geographical Spies

Right The first map of India to name Mount Everest. See also 1619, 1782.

TODAY A HAND-HELD Global Positioning System (GPS) – the size of a mobile phone – can give you your location anywhere in the world at a push of a button. GPS technologies can also allow us to forget the huge energies and arduous work that went into the production of maps throughout history.

Such endeavours are manifest in this map of the Himalayas. It was produced as a result of the Great Trigonometrical Survey of India and is the first map to name peak VX (as it was known) as Mount Everest. The map appeared in an article by Colonel A. S. Waugh in 1858 in the Proceedings of the Royal Geographical Society.

The Survey of India was a massive cartographic undertaking, lasting decades. It was undertaken by the British in order to record their colonial possession. A huge theodolite was hauled across India, with groups of men clearing forests, dragging survey chains and recording the almost countless survey calculations. But what of the areas – such as Nepal, which was a closed country – where the British could not enter? To gain access the British trained a group of Indian 'geographical spies' – the Pundits. These men crossed the border and took secret surveys of Nepal and the Himalayas. They carried adapted rosary beads which could be flicked over to record the number of their paces, recorded altitude by timing how long it took for water to boil, and kept their survey information in prayer wheels to avoid detection. The most famous of the Pundits was Nain Singh – on whom the Kipling story of Kim was based. Singh's contribution to the survey is highlighted in a letter from Colonel H. Yule to the Society which stated, 'I have a strong opinion that Singh's great merits cannot be fully recognised by anything short of one of the Society's gold medals. Either of his great journeys in Tibet would have brought this reward to any European explorer.' So today – sandwiched between H. M. Stanley and Sir George Nares is Singh's name as the RGS's Gold Medal winner of 1877, 'for his great journeys and surveys in Tibet and along the Upper Brahmaputra, during which he determined the position of Lhasa and added largely to our knowledge of the map of Asia'.

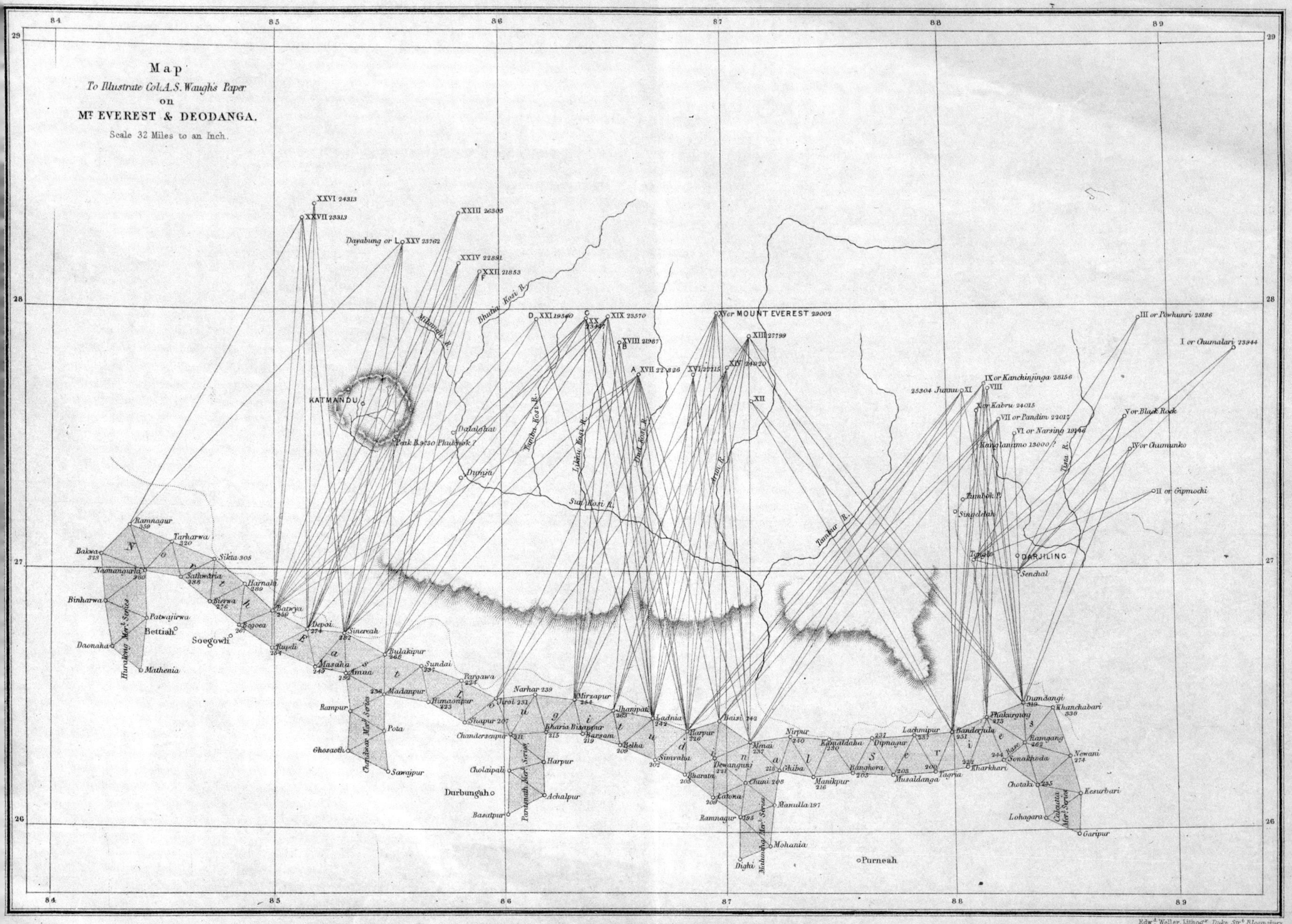
Map
To Illustrate Col. A. S. Waugh's Paper
on
Mt EVEREST & DEODANGA.
Scale 32 Miles to an Inch.
84
85
86
87
88
89
29
28
27
26
XXVI 24313
XXVII 23313
XXIII 26305
Dayabung or L XXV 23762
XXIV 22881
XXII 21853
F
Bhutia Kosi R.
Milamchi R.
D XXI 19540
C
XX
XIX 23570
XVIII 21987
B
A XVII 22826
XVI 22715
XV or MOUNT EVEREST 29002
XIII 27799
XIV 24020
XII
KATMANDU
Dalalghat
Dunja
Tamba Kosi R.
Likhu Kosi R.
Dud Kosi R.
Arun R.
Sun Kosi R.
Tambur R.
Tista R.
III or Powhunri 23186
I or Chumalari 23944
IX or Kanchinjinga 28156
25304 Junnu XI
VIII
X or Kabru 24015
VII or Pandim 22017
VI or Narsing 19146
Kanglanamo 13000?
V or Black Rock
IV or Chumunko
II or Gipmochi
Tumbok P.
Singelelah
Tonglo
DARJILING
Senchal
Ramnagur 359
Yarharwa 320
Bakwa 323
Naumangurla 360
Sikta 305
Sathwaria 288
Harnali 289
Binharwa
Bierwa 276
Batwya
Patwajirwa
Bettiah
Soegowli
Begoea
Depoi
Sinereah
Daonaha
Hurdlong Merl. Series
Rupdi
Mathenia
Masaha
Amua
Bulakipur
Sundai
Pargawa
Madanpur
Himaonpur
Rampur
Chandwar Merl. Series
Pota
Ghosaoth
Sawajpur
North East Longl. Series
Jirol 231
Narhar 239
Mirzapur 254
Shapur 207
Chanderseupa
Bharia
Bisanpur
Barsam
Jhanipati
Ladnia
Harpur
Baisi
Belha
Simrahi
Bharatn
Dewangunj
Chuni
Latona
Ramnagur
Malumba Merl. Series
Mohania
Dighi
Purneah
Menai
Nirpur
Ghiba
Manikpur
Kamaldaha
Dipnagur
Ranghora
Lachmipur
Musaldanga
Tagria
Banderjula
Kharkhari
Phakurgunj
Dumdangi
Khanchabari
Ramgang
Sonakhoda
Newani
Chotaki
Kesurbari
Lohagara
Calcutta Merl. Series
Garipur
Cholaipali
Harpur
Parisnath Merl. Series
Achalpur
Basatpur
Durbungah
Published for the Royal Geographical Society, London 1858.
Edwd. Weller, Lithogr. Duke Str. Bloomsbury.

Union engineers map unfamiliar ground as precisely as possible to overcome home advantage during the American Civil War.

Battling Through the Unknown

THIS MAP, DRAWN by Union (northern) engineers, incorporates a segment of the Bull Run or Manassas, Virginia, battlefield. In July 1861 and August 1862 the South won decisive victories here.

Despite their deadly serious purpose, American Civil War maps were often disconcertingly lovely. The meticulous and ornate lettering, the delicate watercolours, and the quite cheery subject matter – the farms, fence lines, taverns, cornfields, orchards – make these among the most charming of all military documents.

Form and function, however, remained the predominant concern of the topographical engineers who created the maps. Officers in the field who depended on these maps were often studying them by the light of a guttering lard candle, a sinking campfire, or a burning barn. The watercolours ensured that they didn't mistake a stream for a turnpike, a forest for a field. The precise lettering assured legibility. The names of each resident along a line of march served as route markers on the nondescript roads of the South which could be nothing more than a dry streambed or a hodgepodge of farm lanes. Distinctive stands of pine trees were carefully drawn in to serve as natural landmarks amid the semi-wilderness conditions of mid-nineteenth-century America.

A wonderful feature of the maps is the memoir-like comments that inform, warn or query the soldiers who will be using them. Typical notations culled from the maps require interpretation. 'New Cut Road is 50 years old', lets the commander know that the road is old and therefore packed down and practicable despite the name. 'Yarnell's house ... a strong rebel ... a good subject to bleed', is a Union engineer's ominous annotation. Other comments are unintentionally amusing: 'Unable to locate Lost Mt.', or alarming, 'Guess at what you don't know ... the way that I do'. 'Information wanted for this part', asked anyone with local knowledge to supply it to the mapmaker.

The necessities of the case in this sectional Civil War meant that Union armies operated on unfamiliar ground amid the unfriendly inhabitants of the South. Their need for maps was thus desperate while the map needs of the Confederates were merely pressing. The initial, dramatic successes of the Rebels reflect this topographical advantage. The ultimate Union victory came about when generals emerged who adapted to and surmounted the immense logistical and geographical challenges of this colossal war.

Right *Map of the Battlefield of Bull Run.* Thousands died here in July 1861 and August 1862. Both battles were decisive Confederate victories. The site is in Virginia, some 25 miles (40 kilometres) west of Abraham Lincoln's White House and Washington, DC.
Above Union topographical engineer Washington Roebling.
See also 1684, 1746, 1777, 1815.

Map
of the
Battlefield of Bull Run
August 29th & 30th
1862.
Surveyed by order of Maj. Gen. Sigel
and under direction of Maj. Kappner
by G. Stengel and U. de Fonvielle.
8 inches to 1 mile.
November 1862.
abandoned railroad partially graded
Position of enemy's artillery
Groveton
Turnpike to Centerville
to Sudley Spring
Gen. Pope's H.Q.
Stone House
Robinson
J. Dogan
Gen. Sigel's H.Q.
Henry
Chin
N.

A supposed 'Utopian' map of the time of Emperor Napoleon III conceals a core of traditional French nationalistic prejudices.

A European Union

Right Victor Marlet, *Carte Utopique de l'Europe Pacifiée.*
See also 1596, 1846, 1998.

EUROPEAN CONFEDERATION. Respect for human life. General disarmament. The rights of all substituted for the violence of some. Universal peace!

Thus resound the French slogans on this 'Utopian' map which dates from the time when Emperor Napoleon III of France was trying to reinvent himself as a liberal, almost democratic, ruler. The idealized vision of a United States of Europe embraces a Danubian Federation of the Czechs, Slovaks, Hungarians and Balkan peoples, a Gallic Confederation, an Iberia consisting of Spain and Portugal, Great Britain, a united Scandinavia, Italy and Greece and a resurrected and enormous Poland in one big, happy and peaceful family of European nations. Flashpoints like Palestine and the Dardanelles are neutralized while Switzerland and the Netherlands retain their traditional neutrality.

But a closer look at the map reveals the price. European peace and prosperity are to be paid for by the French, Italian and Spanish colonization of Northern Africa (the Germans, Greeks, Danubian and Mediterranean peoples being left only with Libya), and by the exploitation of sub-Saharan Africa. From their North African colonies the Europeans are exhorted to cross the Sahara, sustained by water from wells and shaded by palm trees, to the unexplored lands and enormous natural wealth of 'Nigrittia or Sudan' to their south 'which could be called the African Indies because they are as rich as the Asian Indies, but much closer to Europe'.

Nor is everything too innocent elsewhere. Inside Europe, the French Confederation includes much of western Germany. Britain loses Gibraltar and any chance of winning African colonies. It seems implied that the Habsburg dynasty, a hereditary enemy of the French, would not necessarily continue as rulers of the Danubian Federation. The Russians, another traditional rival, dismissed as an Asiatic people, are deprived of Poland and Finland and, having been forced to abandon their 'impossible dream of European domination' are redirected to Asia where, introducing the arts and sciences of Europe, they will be free to expand at will.

The Utopian façade barely conceals traditional French objectives.

CARTE UTOPIQUE DE L'EUROPE PACIFIÉE

LES NATIONALITÉS SOLIDARISÉES DANS UN LIEN FÉDÉRATIF

CONFÉDÉRATION EUROPÉENNE

LE DROIT DE TOUS SUBSTITUÉ A LA VIOLENCE DE QUELQUES UNS

RÈGNE DE LA JUSTICE

OCÉAN GLACIAL ARCTIQUE

Respect de la vie humaine.
Désarmement général.
Economie annuelle pour toute l'Europe 2 milliards de francs.

Paix Universelle.
L'humanité emploie toutes ses forces au travail productif.
Augmentation indéf.ie de la production Min.um 4 milliards de francs.

OCÉAN ATLANTIQUE

SCANDINAVIE

RUSSIE

D'origine Asiatique, la Russie a restitué la Finlande et la Pologne et abandonné son rêve impossible de domination sur l'Europe.
C'est désormais sur l'Asie qu'elle agit en la civilisant.

ASIE CENTRALE

Livrée à la libre expansion de la Russie qui la civilise en y introduisant les arts et les sciences de l'Europe.

GRANDE BRETAGNE

MER DU NORD OU D'ALLEMAGNE

GERMANIE OU ALLEMAGNE

POLOGNE

CONFÉDÉRATION FRANCE GAULOISE

CONFÉD DANUBIENNE

IBÉRIE

ITALIE

GRÈCE

TURQUIE

MER NOIRE

MER CASPIENNE

PERSE

SYRIE

ARABIE

MER MÉDITERRANÉE

MAROQUIE OU IBÉRIE AFRICAINE

ALGÉRIE OU FRANCE AFRICAINE

TUNISIE OU ITALIE AFRICAINE

COLONIES DES GERMAINS, DES GRECS, DES DANUBIENS ET AUTRES PEUPLES MÉDITERRANÉENS

ÉGYPTE

MER ROUGE OU GOLFE ARABIQUE

GOLFE PERSIQUE

MER D'OMAN

GRAND DÉSERT DU SAHARAH

Les Européens y tracent des routes parsemées de puits artésiens et ombragées de palmiers. Ils pénètrent dans la Nigritie ou Soudan, région inexplorée, d'où ils tirent les plus riches produits et qu'on pourrait appeler les INDES AFRICAINES.

NIGRITIE OU SOUDAN
OU
INDES AFRICAINES

Aussi riches que les Indes Asiatiques et beaucoup plus voisines de l'Europe.

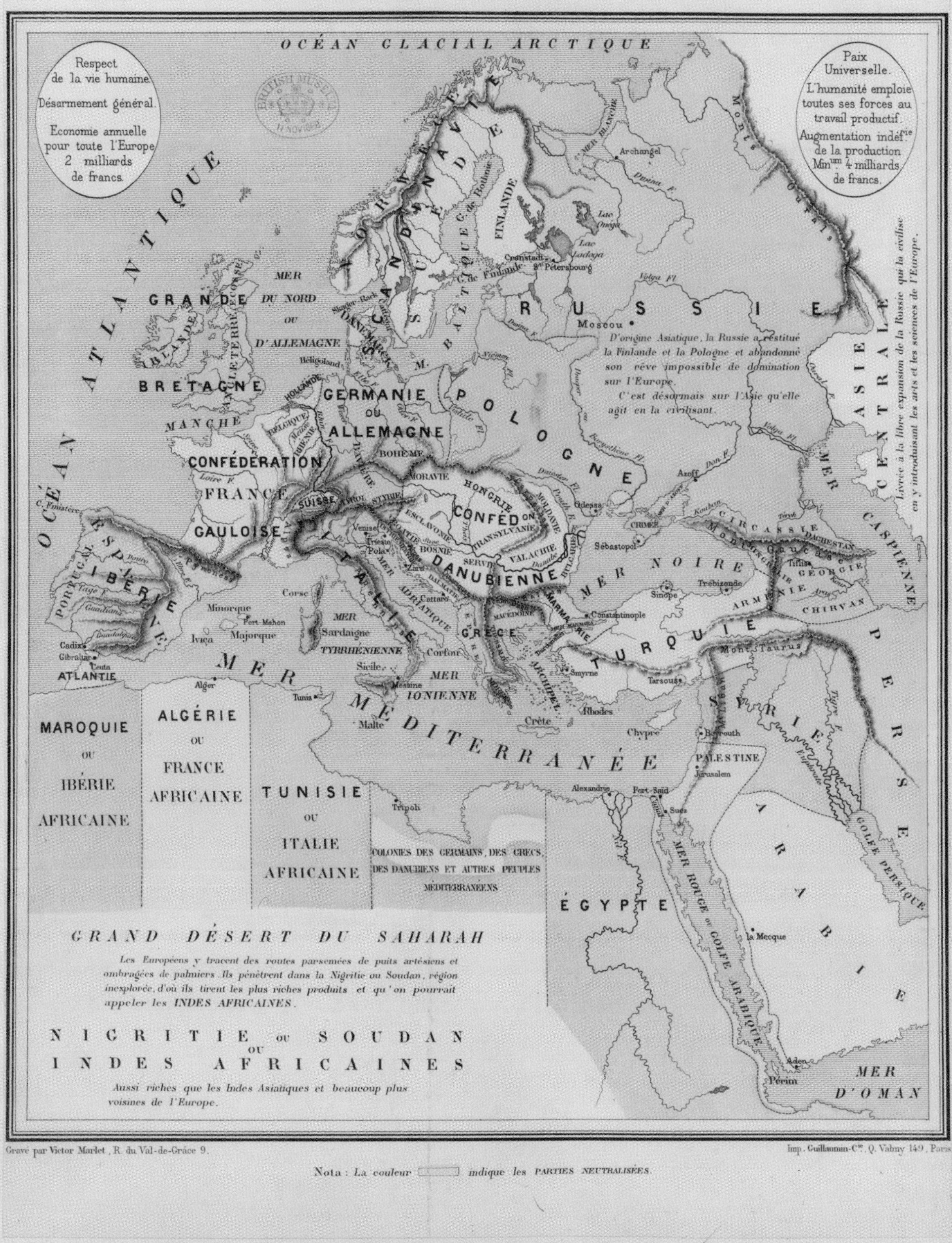

Gravé par Victor Marlet, R. du Val-de-Grâce 9.
Imp. Guillaumin-Cie, Q. Valmy 149, Paris

Nota : *La couleur* [ ] *indique les* PARTIES NEUTRALISÉES.

Two Africans saved from destruction Dr Livingstone's survey material and the fruits of his last expedition.

Dr Livingstone's Map

Right *Dr Livingstone's Map: A map of a portion of Central Africa by Dr Livingstone from his own surveys, drawings and observations between the years 1866 and 1873.*
See also 1867, 1899 (p.294).

THIS MAP SHOWS a portion of Central Africa by Dr Livingstone from his own surveys, drawings and observations between the years 1866 and 1873. It highlights Livingstone's achievements as one of the winners of the Royal Geographical Society Founders Medal in 1855, 'for his recent explorations in Africa', spending many years in East Africa as a missionary, explorer and geographer. His surveys, of previously uncharted lands, caused major revelations in Britain, bringing the Victorian public such sights as Victoria Falls and the Rwenzoris – the fabled 'Mountains of the Moon'. His fame was further enhanced by the famous 'discovery' of Livingstone by the Welsh-born American explorer H. M. Stanley in Ujiji in 1871.

As well as his geographical work Livingstone also worked as a missionary and campaigned against slavery. He once freed a group of slaves from a raiding party and later, on his return to Britain, dramatically crashed down the shackles during a speech denouncing slavery.

Yet we could so easily have been deprived of Livingstone's work. When Livingstone died in East Africa, he was surrounded by his papers, maps, sextant and other equipment. It would have been so easy for these invaluable records to have been lost to the forest forever. However, it is to the African members of his expedition that we owe a debt. It was they who safely and transported the materials, and Livingstone's body, to the coast. The contribution of two men – James Chuma and Abdulah Susi, a freed slave and woodcutter respectively – in particular were recognized for this work. They both subsequently came to Britain and spent this time with Horace Waller – who edited Livingstone's papers. As well as helping Waller, Chuma and Susi were also recognized by the RGS. They were RGS medallists in 1874 and the Society President connected their contribution to Livingstone's work in declaring 'let us never forget what has been done for geography by the faithful band who restore to us all that it was in their power to bring – our lost friend and who rescued his priceless writings and maps from destruction'.

Dr. LIVINGSTONE'S MAP

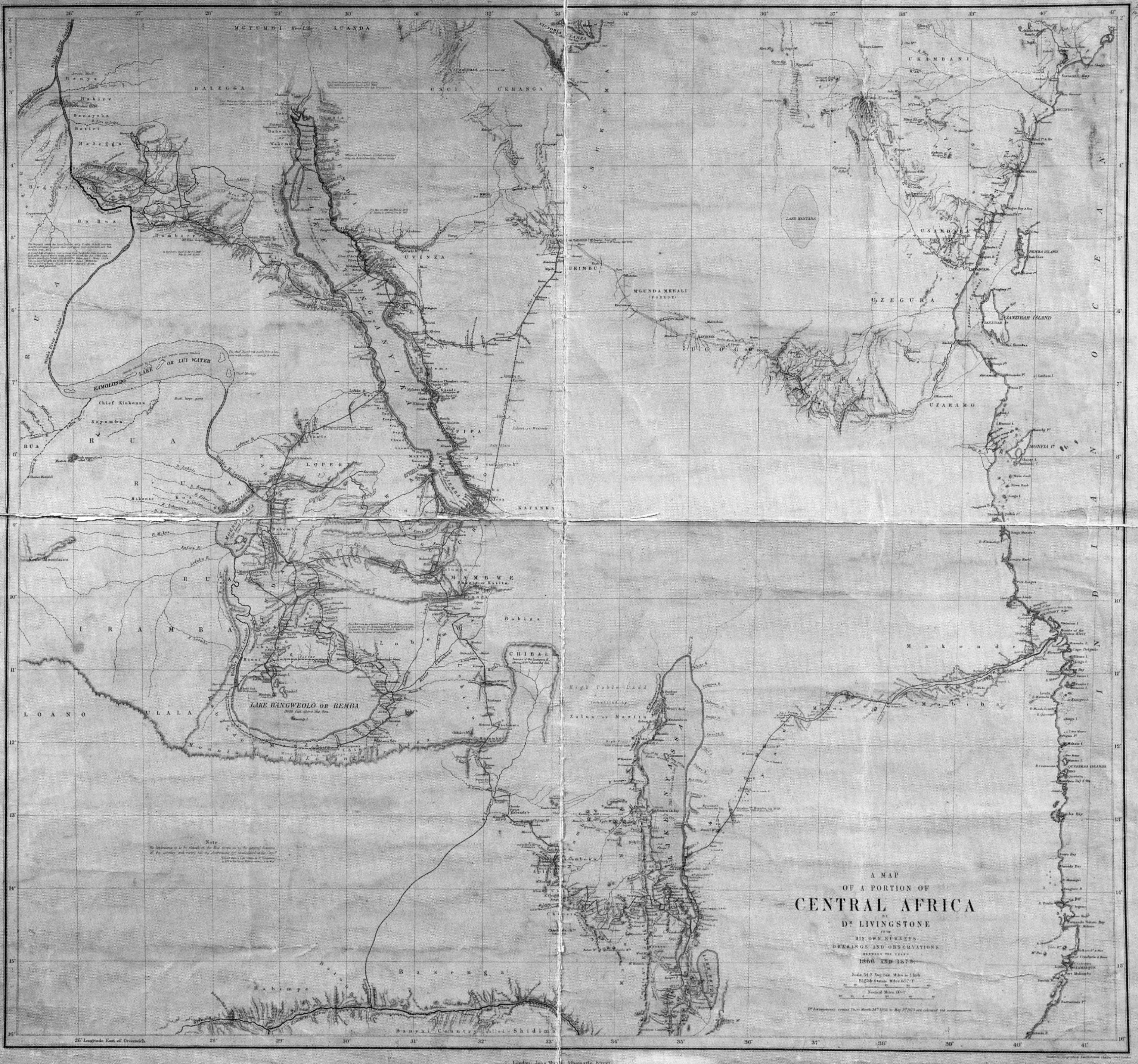

A MAP
OF A PORTION OF
CENTRAL AFRICA
BY
Dr. LIVINGSTONE
FROM
HIS OWN SURVEYS
DRAWINGS AND OBSERVATIONS
BETWEEN THE YEARS
1866 AND 1873.

Scale 34·3 Eng. Stat. Miles to 1 Inch
English Statute Miles 69·1
Nautical Miles 60·1

Dr. Livingstone's routes from March 24th 1866 to May 1st 1873 are coloured red

MUTUMBI
LUANDA
BALEGGA
USUI
UKHANGA
UKAMBANI
LAKE MANYARA
USAMBARA
UVINZA
UKIMBU
MGUNDA MKHALI (FOREST)
UZEGURA
ZANZIBAR ISLAND
PEMBA ISLAND
KAMOLONDO LAKE OR LUI WATER
Chief Kinkonza
Kayumba
RUA
FIPA
LOPERE
UZARAMO
MONFIA I.
KATANGA
MAMBWE
IRAMBA
Babisa
CHIBALE
LAKE BANGWEOLO OR BEMBA
ULALA
LOANO
Makonde
Mabiha
High Table Land inhabited by Zulus or Mazitu
Basenga
Babimpe
Banyai Country called Shidima

Note

26° Longitude East of Greenwich

London: John Murray, Albemarle Street

Maps & Design

Design is essential to a map's effectiveness, however precise the measurements and reliable the information that it contains. Design is dependent on purpose. The simplest form of map is a sketch map or diagram consisting only of a few lines and a few letters, with all the inessentials pared away (see AD 500). More often there is a multiplicity of information which needs disciplining if it is not to cancel itself out in a morass of illegibility, as a detailed map of Bohemia published by Peter Schenk in 1745 on the basis of a much larger-scale survey by J. Müller demonstrates. In these cases the mapmaker has to prioritize. Usually this presents little problem since surveyed settlements can easily be ranked in order of importance – be they county towns, market towns, villages or mere hamlets – and this is reflected in the size and style of lettering used, with bold capitals, for instance, being used for the most important, plain capitals for the less important and small italic for the least important categories. On small-scale maps of large areas, the mapmaker has a more difficult time because he has to take decisions about the relative importance of small settlements of which he may know little or nothing.

Economic and geographical considerations are not the only ones affecting the design of maps, however. On thematic maps, such as Czoernig's linguistic map of the Austrian Empire (1855), the selection of colours for the different phenomena is all important. Choose colours that are too closely related and the user will be confused. Some mapmakers, particularly in the late 1930s under the Third Reich turned this to political advantage by using similar colours for German and for supposed Polish dialects inside Poland with a completely different colour for standard Polish, to give the impression that the so-called 'Polish Corridor' that then separated Germany proper from East Prussia was effectively German-speaking and should again form part of the Reich.

In other cases again, purely aesthetic considerations have come into play. This particularly relates to commercially produced maps and those intended for presentation and/or for public display. In these cases the intended information inherent to the map has to be conveyed in a way that pleases the eye and flatters the vanity of the purchaser or dedicatee. An outstanding example is the Gallery of Maps in the Vatican (1580) where the Pope was to be flattered and the visitor (particularly the Italian visitor) awed. But even on a relatively workaday plan prepared for Henry VIII (1544, p.100), the mapmaker inserted irrelevant but decorative miniaturized town-views. If, at the other end of the scale, the map was intended for a local audience, then local susceptibilities had to be pandered to by appropriate decoration, such as a panorama in the case of the Homann heirs' 1730 map of London (see p. 196).

Frequently a decision needed to be taken on the map's artistic form. At a commercial level, map publishers soon realized that an appealing design surrounding a title would help to sell the map and these often combined aesthetic with political considerations (1777). Even if such designs are often outdated in terms of the latest fashions of the time, an English estate surveyor had to take into account that fashion had moved on from the figurative bravura of the baroque to the twirls and shell of the Rococo or the austere but elegant pillars and letterforms of neoclassicism (see 1720). For much of the

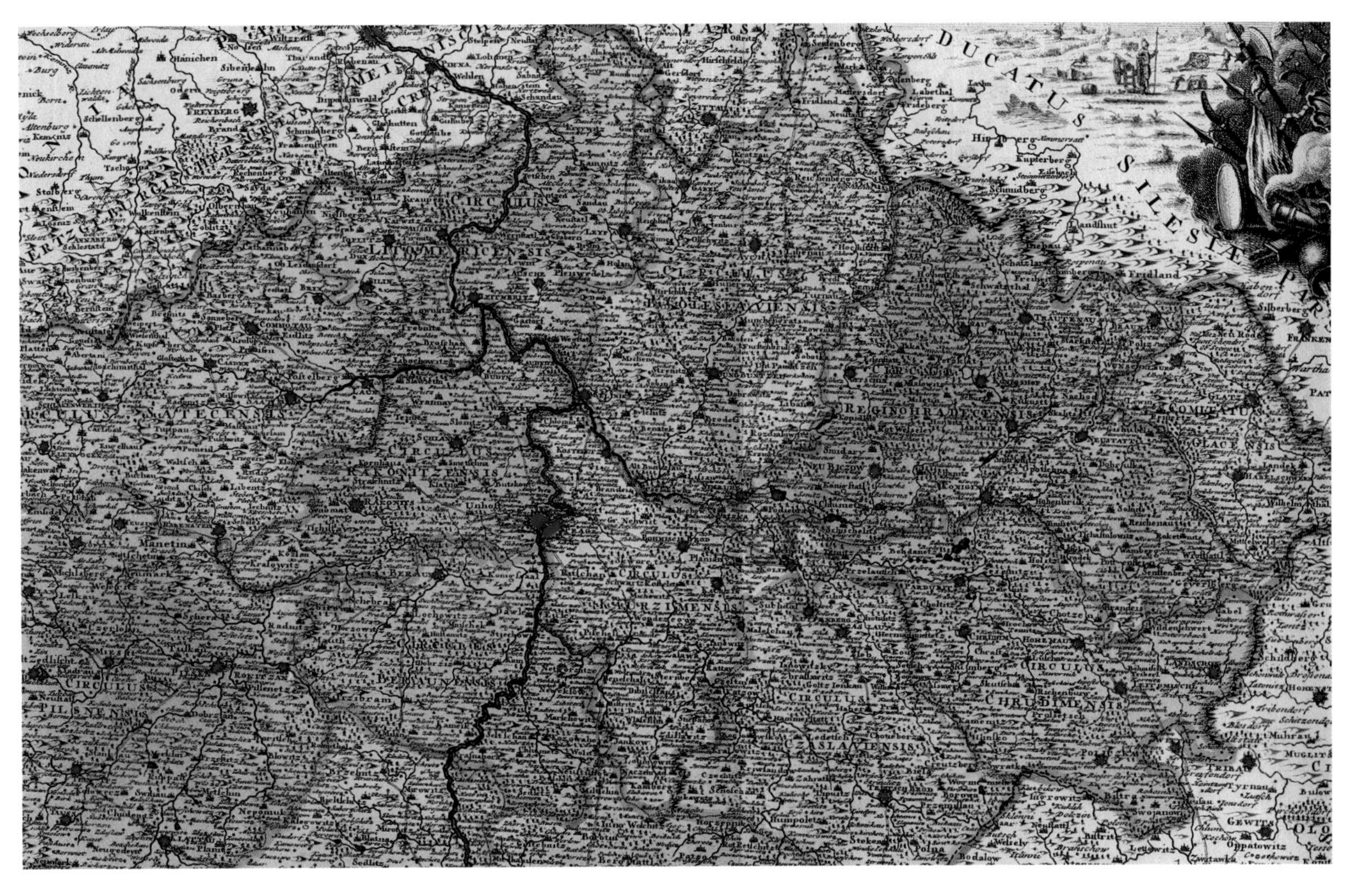

Detail from Peter Schenk's map of Bohemia (1745).

period between 1480 and 1750, and again for reasons of artistic fashion, the bird's-eye view was the preferred means of portraying cities and battles, even if they were based on measured surveys (1521, 1574, 1684). Sometimes the link between maps and art was more direct since several leading artists also practised as mapmakers. Paul Sandby, who is often acclaimed as the founder of the English school of watercolourists, not only drew the maps of the first detailed survey of Scotland between 1747 and 1755 but also taught drawing at the principal British military academy at Woolwich. It is not surprising, then, to find his theories of 'tinted drawing', or the application of watercolour over grey washes, reflected in the delicate colouring of an encampment plan of 1780 (see p. 132).

These considerations remain as relevant as ever in the computer age. It is the person at the screen who agonizes over the appropriate colours, from the supposedly natural to the wildly unnatural, to use when downloading the widely differing phenomena captured by remote sensing or when presenting the differing layers of data displayed on a Geographical Information System. When designing a cyclists' map for the Scottish borders in 2001, the creator was careful when depicting physical relief to choose colours that he considered to be pleasing as well as a material for the map that was waterproof. Paul Sandby would have applauded.

A cartoon map personifies the various countries of Europe and is used to illustrate the danger posed by the mighty octopus that is Russia.

The Russian Octopus and World Domination

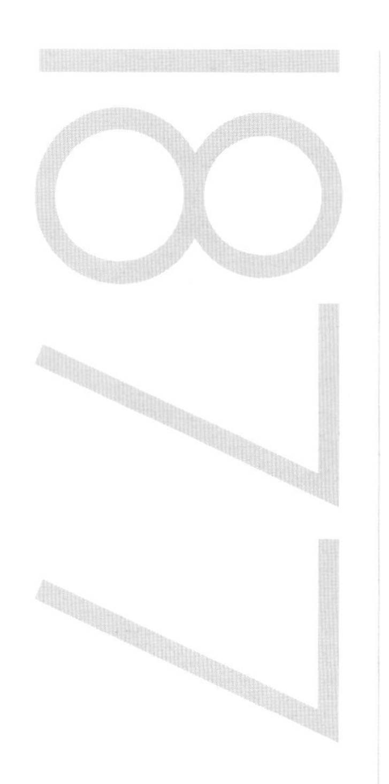

Right Fred W. Rose, *Serio-Comic War Map for the Year 1877. Revised Edition.*

IN THE AUTUMN of 1877, Europe was poised on the brink of war. Russia threatened to invade the enfeebled Ottoman Empire in support of its fellow Christian Bulgarians who had been victims of a Turkish massacre. Britain and Germany, concerned at the potential growth in Russian power, were determined that it should not conquer Constantinople, and with it direct access for its fleet to the Mediterranean.

The crisis inspired what is generally considered the British graphic artist Fred Rose's cartographic masterpiece. This is the less well-known and slightly revised version. Russia, with its two capitals of St Petersburg and Moscow, is portrayed as an octopus. Its tentacles throttle Poland, nearly strangle Finland, which both then formed part of the Russian Empire, and threaten Persia, central Asia, Christian Armenia, the Holy Land and, by way of Bulgaria, Constantinople, which is shown as the Sultan's gold watch. Rose acknowledges the Turks' crimes in Bulgaria, by way of the skull, but Greece alone, portrayed as a crab, is ready to join Russia in attacking the Turks. Hungary wants to intervene, but is restrained by the Austrian part of the Austro-Hungarian Empire. Wilhelm I, Emperor of the recently unified German Reich, is pushing back the Octopus and shepherding his plentiful ammunition. England and Scotland look on anxiously. The other powers are concerned with their own affairs. Spain, portrayed through its newly restored king, Alfonso XII, is resting after decades of civil wars, France, led by Marshal MacMahon and hoping to avenge its defeat in 1870–1, is training a machine gun on Germany. Recently unified Italy is a young girl, while Belgium is represented by its cruel and avaricious monarch, Leopold II, who is counting his money. Ireland is a monk with 'home rule' on the brain. Britain emerged as the principal gainer when peace was signed in Berlin in 1878, acquiring Cyprus. But Rose did not show it.

The cartoon belongs to a tradition stretching back to the 1330s when Opicinus de Canistris made moral points through the allegorical portrayal of Mediterranean coastlines as people. Rose's particular models go back to at least the 1850s, but Rose was the acknowledged master. His octopus had a long life: twenty-five years later Japanese propagandists were using octopus maps to win European support against Russia during the Russo-Japanese War.

REVISED EDITION
SERIO-COMIC WAR MAP
FOR THE YEAR 1877.
BY F. W. ROSE.

ERKLÄRUNGEN.
Der nördliche Kolosz—Russland—ist repräsen-
rt in der Form eines wild aussehenden Octopus,
essen Kopf den grösseren Theil des europäischen
ussland's einnimmt, während er mit seinen aus-
estreckten Krallen sich wunderbar nach allen
ichtungen ausdehnt und bereits verschiedene
änder festhält.
Die Türkei liegt hingestreckt unter ihm. Der
opf und die Brust eines Türken umfassen die
uropäische Türkei, während sein Unter-Körper
lein Asien darstellt. Der Bosphorus, das Marmora
leer, und die Dardanellen bilden zusammen einen
ürtel an der Figur, während der lüsterne Preis,
onstantinopel, als seine goldene Uhr repräsentirt
st. Griechenland, in der Gestalt eines Krebses
eunruhigt den Türken im Süden.
Eine Kralle des Octopus umringt Bulgarien,
nd scheint die umliegenden Distrikte zu bedro-
en. Eine Andere hat die Krimm umfaszt, welche
ber noch eine schlimme Wunde bei Sebastopol
eigt. Eine dritte Kralle hat den Fusz des Türken
rgriffen (Armenien); eine Vierte dehnt sich
veiter nach unten zu dem lange beneideten Ge-
obten Land; eine Fünfte umarmt den Schah von
ersien und eine Sechste umringt Khiva und die
nderen Eroberungen in Asien. Eine siebente
Kralle scheint Polen ganz erwürgt zu haben, weil
ie Achte, Finnland umfäszt, welches bei einem
euer das Wenige genieszt was Russland ihm
elassen.
Ungarn ist nur von seiner Schwester, Oester-
reich, davon zurückgehalten seinen Nachbar Russland anzugreifen.
England und Schottland beobachten aufmerk
sam die Scene, letzteres bewaffnet mit Dolch und
Schwert. Irland ist als Mönch abgebildet. Alle
Drei aber scheinen ganz entschieden wenigstens
des Türken goldene Uhr zu retten.
Frankreich ist der Marschall Mc Mahon mit
einer Mitrailleuse auf seinen Nachbar zielend.
Deutschland zeigt seinen Kaiser in Uniform,
von Kugeln und Kanonen umgeben, ein Zeichen
dass es sich für jeden Nothfall bereit hält.
Spanien ist der junge Alphonso welcher sich von
seinen Anstrengungen ausruht.
Italien ein junges Mähchen freut sich seiner
neu errungenen Freiheit. Den Papst erkennt man
in Rom. Der reiche König von Belgien bewacht
seine Schätze. Dänemark's Fahne zwar nur klein,
doch ist es stolz darauf. Sicilien ist als drei Wein
Fäszer dargestellt, und in Aegypten zeigt sich der
Kopf einer Figur, welche dem Khedive und der
Sphinx ähnlich sieht.

REFERENCE.
"The Northern Colossus—Russia—is represented in the form of a vicious-looking Octopus, the head of which occupies the greater portion of European Russia, while, with its outstretched arms, it is extending marvellously in every direction, and embracing many countries in its grasp.
Turkey lies prostrate beneath it. The head and bust of a Turk make up European Turkey, while the lower vestments stretch over Asia Minor. The Bosphorus, Sea of Marmora, and the Dardanelles, form together a girdle for the figure, whilst the coveted prize, Constantinople, is seen in the shape of a gold watch. Greece, shown as a crab, is annoying the Turk on the south.
One of the arms of the Octopus encircles Bulgaria, and seems threatening the surrounding districts. Another envelopes the Crimea, but this arm shows a bad wound at Sebastopol. A third arm has seized hold of the Turk's foot (Armenia). A fourth is stretching far down to the long-coveted Holy Land, while a fifth is giving the Shah of Persia a gentle embrace as it curls round his neck. A sixth is encircling Khiva and the other acquisitions in Asia. A seventh seems to have wrung all life out of Poland while the eighth arm passes round Finland, who is warming up and making the most of what little Russia has left him. Hungary is only prevented from attacking his neighbour Russia through being held back by his sister Austria.
England and Scotland are eagerly watching the scene, from afar, the latter armed with a dagger and claymore. Ireland is shown as a hooded monk, with an indication of "Home Rule" on the brain. All three look fully determined to save at least the Turk's watch.
France is Marshal MacMahon pointing a mitrailleuse towards his neighbour.
Germany is represented by her Emperor, in uniform, surrounded by shells, cannons, and shot, indicating her readiness for any emergency which may arise.
Spain is young Alfonso sleeping after his recent exertions.
Italy is a young girl rejoicing in her newly acquired liberty. The Pope's triple crown is seen at Rome. The wealthy King of the Belgians is taking care of his treasure. Denmark's flag is small, but she is evidently proud of it. Sicily is made up of three wine barrels, and in Egypt is seen a figure suggesting both the Khedive and the Sphinx.

RUSSIA
NORWAY
SWEDEN
FINLAND
G. OF BOTHNIA
G. OF FINLAND
BALTIC SEA
MOSCOW
SKAGERRACK
CATTEGAT
DENMARK
NORTH SEA
BRITISH ISLES
IRISH SEA
St GEORGES CHANNEL
ENGLISH CHANNEL
POLAND
Die heilige Schrift
BAY OF BISCAY
PORTUGAL
SPAIN
STRAIT OF GIBRALTAR
CORSICA
SARDINIA
SICILY
ADRIATIC SEA
AUSTRO HUNGARY
MEDITERRANEAN SEA
BULGARIA
CRIMEA
SEA OF AZOV
BLACK SEA
Constantinople
BOSPHORUS
SEA OF MARMORA
DARDANELLES
ARCHIPELAGO
GREECE
CANDIA
ASIA MINOR
OTTOMAN EMPIRE
CASPIAN SEA
PERSIA
Khokan
Samarkand
Khiva
Bokhara
MOUTHS OF THE NILE
EGYPT
ALL RIGHTS RESERVED
(COPYRIGHT) London, Published by G.W. Bacon & Co 127 Strand.

A native Shan artist depicts in remarkable detail the disputed border between his British-dominated state and Chinese territory.

Border Conflict in Asia

Right Shan map relating to a border dispute between Burma and China. **Following pages** Detail of the map.

EVEN BEFORE THE advent of the British, the Burmese Empire had developed a fairly sophisticated cartographic capability. Maps were prepared not only for general topographic intelligence, but also for such purposes as planning and memorializing military campaigns, guiding the construction of new capital cities and land revenue assessment. In the late eighteenth and nineteenth centuries British diplomatic representatives and, later, colonial authorities in Burma collected numerous existing indigenous maps and commissioned the others to serve various purposes. Such maps were made by both ethnic Burmans and members of various tribal groups, especially the Shans, a T'ai people straddling a large, hilly and mountainous region contested by Burma, Siam and China.

The work depicted here forms part of a remarkable collection of forty-seven – mainly Shan – maps bequeathed to Cambridge University in 1933 by the widow of James George Scott, who spent two decades as a colonial administrator in the Shan States. It is at once a topographic document and a guide for determining the then disputed border between the British Shan state of Möng Mäo and Chinese territory, shown in red and yellow respectively. Painted by an anonymous Shan artist in tempera on indigenously manufactured paper and drawn to a rather large – if not quite precise – scale, this remarkably accurate map covers an area of perhaps 47 square miles (75 square kilometres) along the Nam Mao (Burmese Shweli) River. Easily discernible on the map are the then existing stream meanders, while others, indicative of a previous river course, can easily be inferred. Named on the map are more than eighty villages and hamlets, shown by ovals, and the town of Namkhan, depicted by the irregularly bounded shape in green. Stylized mountains run the length of the south-western and the shorter north-eastern limits of the map. Although a great many Shan and Burmese maps are richly embellished with diverse vegetation symbols, such symbols are lacking in this example. The map text is in Chinese Shan, though Burmese notes have been added in pencil.

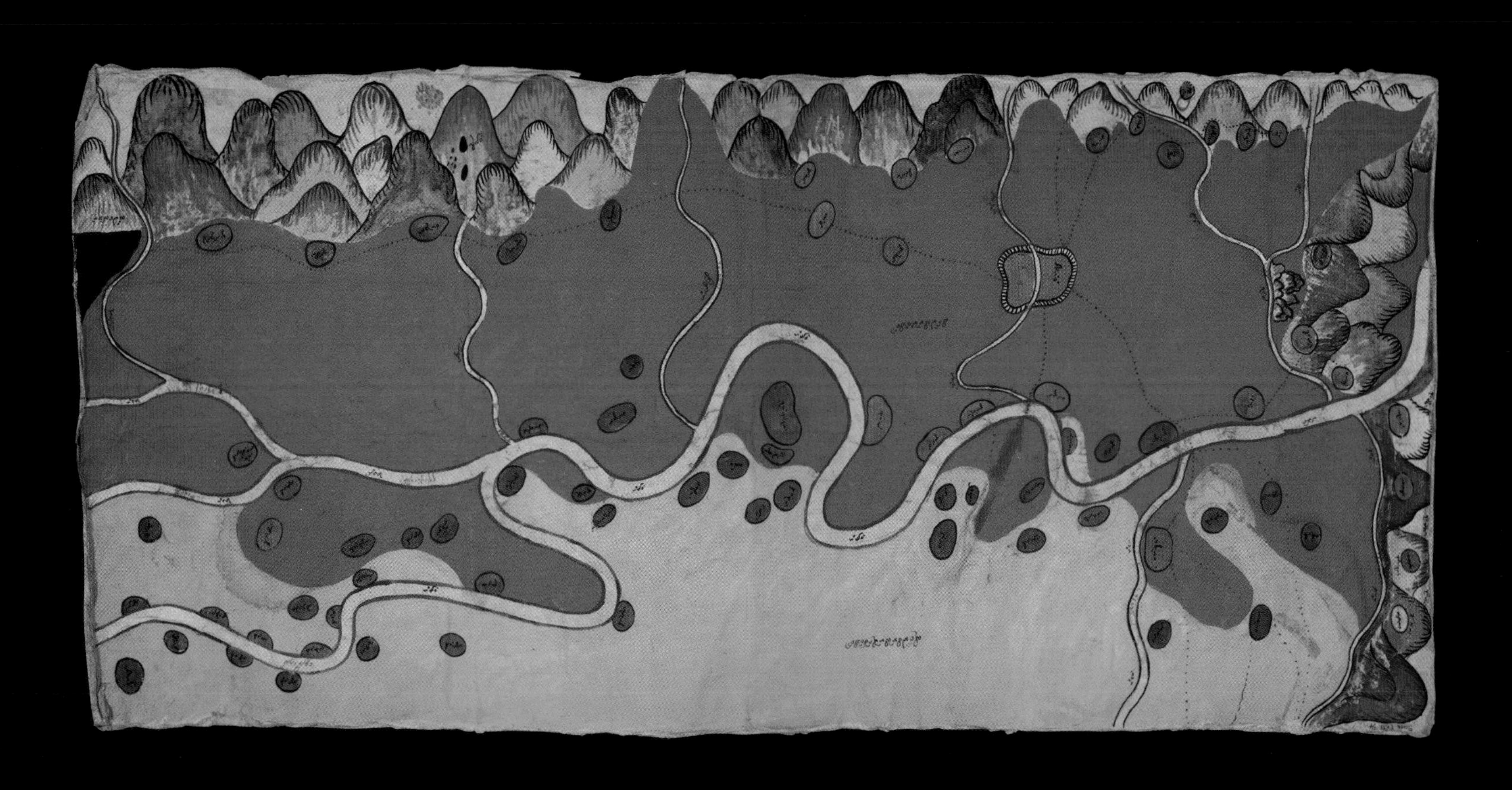

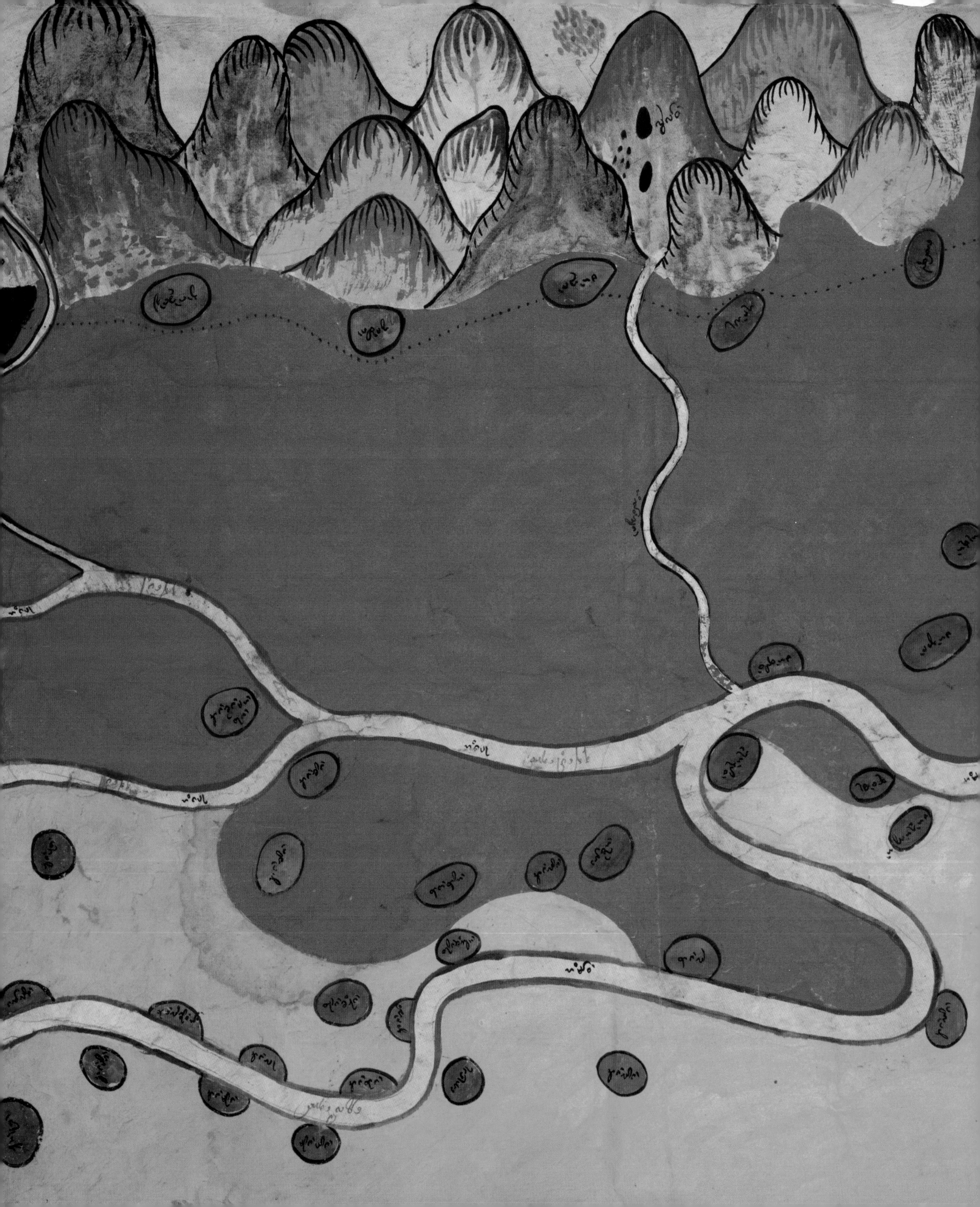

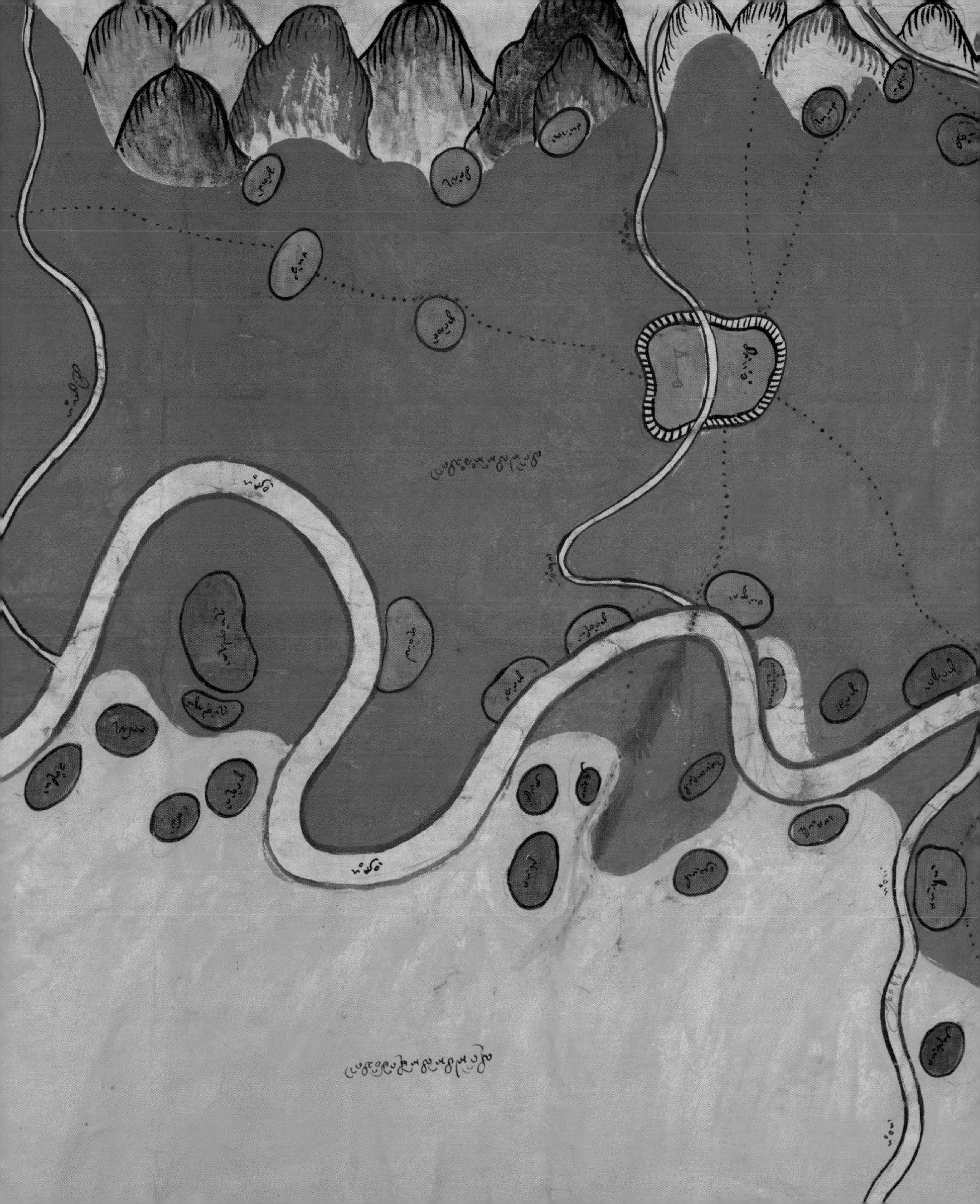

An early cycling map produced in the boom years of the bicycle recycles an old map to attract a new, and often female, audience.

Queen of the Road

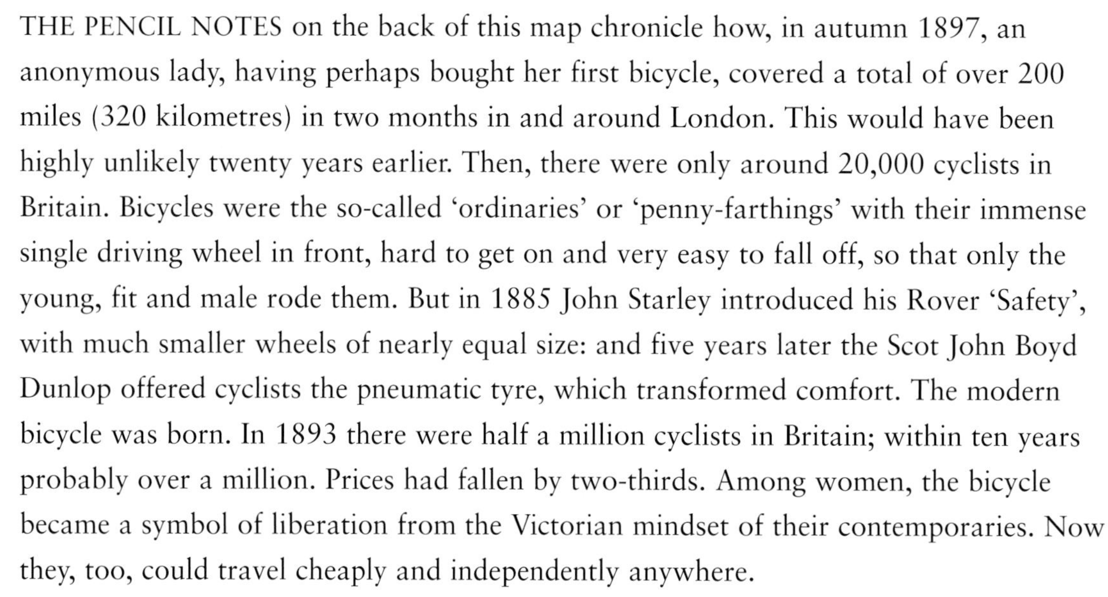

THE PENCIL NOTES on the back of this map chronicle how, in autumn 1897, an anonymous lady, having perhaps bought her first bicycle, covered a total of over 200 miles (320 kilometres) in two months in and around London. This would have been highly unlikely twenty years earlier. Then, there were only around 20,000 cyclists in Britain. Bicycles were the so-called 'ordinaries' or 'penny-farthings' with their immense single driving wheel in front, hard to get on and very easy to fall off, so that only the young, fit and male rode them. But in 1885 John Starley introduced his Rover 'Safety', with much smaller wheels of nearly equal size: and five years later the Scot John Boyd Dunlop offered cyclists the pneumatic tyre, which transformed comfort. The modern bicycle was born. In 1893 there were half a million cyclists in Britain; within ten years probably over a million. Prices had fallen by two-thirds. Among women, the bicycle became a symbol of liberation from the Victorian mindset of their contemporaries. Now they, too, could travel cheaply and independently anywhere.

The map shown was typical of the bottom end of the market, in being a cheap adaptation of a very old and once prestigious map made originally for travellers by horse and carriage. Created by the cartographer John Cary in 1794, it had then been Britain's most up-to-date sectional map – one that ran continuously across counties without breaks, at a uniform scale. The military adopted it for defence use: in France, ironically, it was also recommended to Napoleon for his projected invasion in 1804. In mid-century it began its long decline. The printing plates, bought by the mass-market publisher George Frederick Crutchley around 1850, next went to George Washington Bacon, another such entrepreneur, who from the 1880s aimed the map at cyclists and then, from the early 1900s, at motorists too. He would also print maps for anyone wanting to advertise their goods or services, such as, here, *The Cycle* magazine. Since urban cyclists used the train to transport their machines to and from the country, railways had to be kept up to date, but the roads they used had changed hardly at all since being overshadowed by the railways, so little else was altered and no attention was paid to cyclists' particular requirements. By the time the anonymous lady wrote up her notes, the map was a colourful, bargain-basement antique.

Right *The Cycle Road-Map of 50 Miles Round London.*
Above Detail showing roads.
See also 1785, 1790, 2001.

CYCLE
ROAD-MAP
OF
o Miles Round
LONDON
ONE SHILLING
PUBLISHED AT THE OFFICES OF "THE CYCLE"
106, FLEET ST. LONDON. E.C.
THE CYCLE ROA
BANBURY
BRACKLEY
BUCKINGHAM
DEDDINGTON
BICESTER
WOODSTOCK
WITNEY
OXFORD
ABINGDON
FARINGDON
LECHLADE
FAIRFORD
THAME
AYLESBURY
PRINCES RISBORO
WENDOVER
HIGH WYCOMBE
GREAT MARLOW
HENLEY
READING
NEWBURY
HUNGERFORD
MARLBOROUGH
WALLINGFORD
WANTAGE
WINSLOW
FENNY STRATFORD
WOBURN
DUNSTABLE
IVINGHOE
CHESHAM
AMERSHAM
BEACONSFIELD
UXBRIDGE
MAIDENHEAD
WINDSOR
STAINES
CHERTSEY
WOKINGHAM
BASINGSTOKE
ODIHAM
KINGSCLERE
HEMEL HEMPSTEAD
TOWCESTER
NEWPORT PAGNELL
STONY STRATFORD
BEDFORD

A series of maps commissioned by the millionaire Charles Booth charts the levels of relative wealth and poverty in late Victorian London.

Wealth and Poverty in Victorian London

1891

Right Charles Booth's *Descriptive Map of London Poverty.*

BY THE LATER nineteenth century, London, with a population of nearly 4 million, was the world's largest and richest city. The social geography of London, however, included large tracts of dire poverty, especially in the notorious slums of the East End. The questions of the extent, causes of, and solutions to poverty, coupled with a pervading fear of social disorder, were a major preoccupation of contemporary reformers and politicians.

Charles Booth (1840–1916), a wealthy businessman and owner of the Booth Shipping Line, was at first sight an unlikely candidate for a pioneering sociologist. Nevertheless, his Unitarian religious background and his Liberal family connections made him interested in issues of social reform, and he contributed a number of papers on statistical measures of poverty as well as participating in debates with radical and liberal thinkers.

In 1886 the *Pall Mall Gazette* published an article by members of the Marxist-oriented Social Democratic Federation which claimed that 25 per cent of London's population lived in poverty. Booth regarded this figure as wildly exaggerated and itself an incitement to disorder. In response he resolved to carry out a comprehensive survey combining statistical methods with direct observation. He recruited a team of volunteer researchers, including the socialist Beatrice Webb. They compiled a monumental analysis of London's social condition, incorporating reports of School Board visitors, Poor Law and census returns, in addition to extensive field visits and interviews with local police, clergy and employers. The result was the publication, in 1889, of *Life and Labour of the People*, which covered the East End and showed that 35 per cent, rather than 25 per cent, lived in poverty. A second volume, *Labour and Life of the People* (the title changed for copyright reasons), covered the rest of the city and appeared in 1891. To Booth's concern, this demonstrated that no less than 30 per cent of the city's total population could be classed as poor.

The 1891 volume included a detailed map with each street colour coded according to the degree of wealth of the inhabitants. The colours ranged from black ('Lowest class. Vicious, semi-criminal'), blue ('Very poor. Casual, chronic want'), pale blue ('Poor'), purple ('Mixed. Some comfortable, others poor') to pink ('Fairly comfortable'), red ('Well to do') and yellow ('Wealthy'). The extract displays Westminster, Pimlico and parts of Lambeth, and clearly shows the complex patterns of juxtaposed wealth and poverty which characterized the London urban landscape.

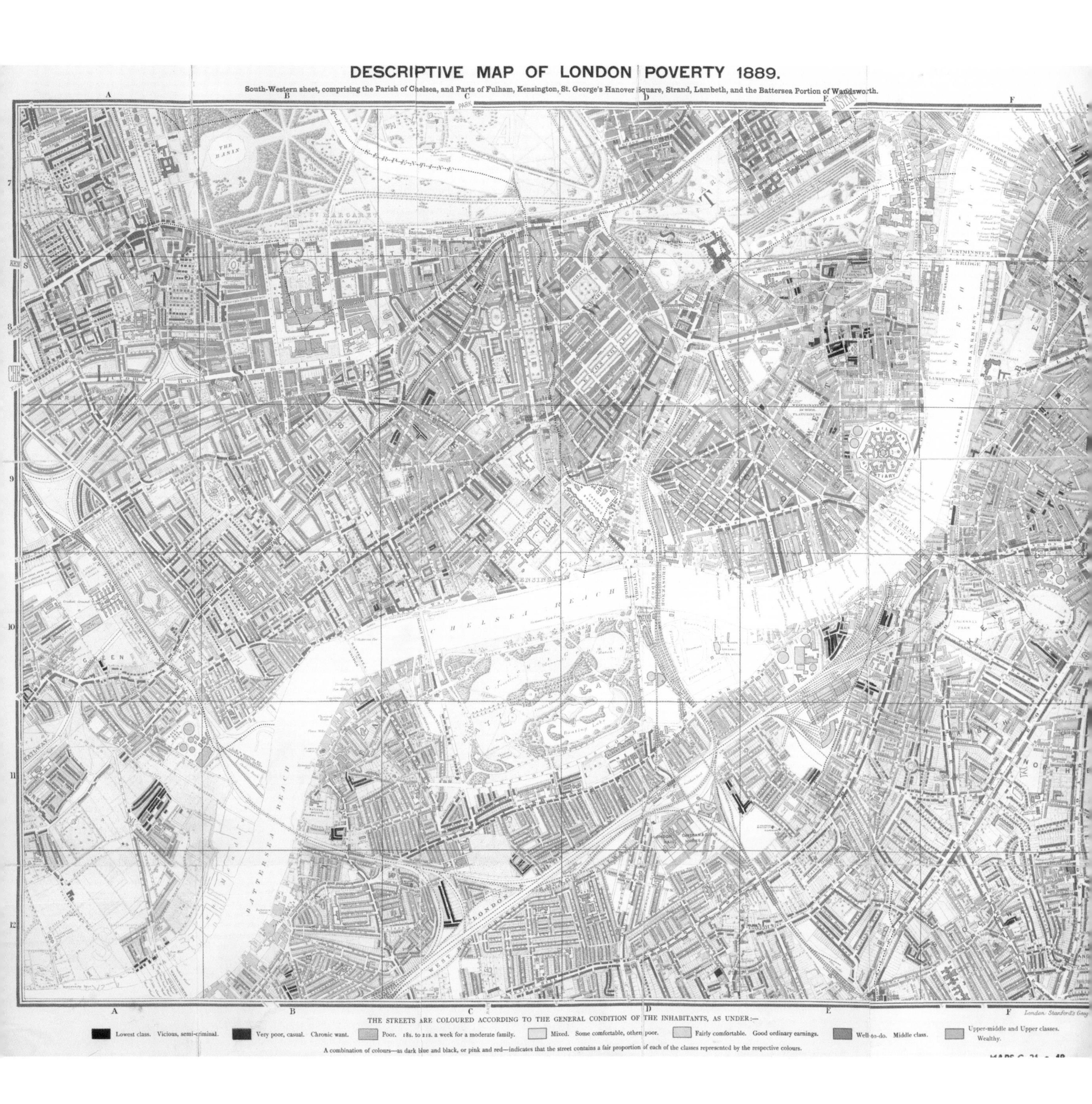
DESCRIPTIVE MAP OF LONDON POVERTY 1889.
South-Western sheet, comprising the Parish of Chelsea, and Parts of Fulham, Kensington, St. George's Hanover Square, Strand, Lambeth, and the Battersea Portion of Wandsworth.
CHELSEA REACH
BATTERSEA REACH
WESTMINSTER BRIDGE
LAMBETH BRIDGE
THE STREETS ARE COLOURED ACCORDING TO THE GENERAL CONDITION OF THE INHABITANTS, AS UNDER:—
Lowest class. Vicious, semi-criminal.
Very poor, casual. Chronic want.
Poor. 18s. to 21s. a week for a moderate family.
Mixed. Some comfortable, others poor.
Fairly comfortable. Good ordinary earnings.
Well-to-do. Middle class.
Upper-middle and Upper classes. Wealthy.
A combination of colours—as dark blue and black, or pink and red—indicates that the street contains a fair proportion of each of the classes represented by the respective colours.
London: Stanford's Geog

Joseph Conrad unveils the grim realities underlying the colours and neat lines of European maps of Africa.

A Place of Darkness

Right Top The interior of Africa explored: Stanley in Africa from 1867 to 1889, c.1890. **Bottom** The earlier image: *A New Map of Africa Exhibiting its Natural and Political Divisions, 1865.* **See also** 1867, 1873.

MARLOW'S TERRIFYING, illuminating voyage up the Congo in Joseph Conrad's *Heart of Darkness* (1899) really begins with a map. Sitting in a cruising yawl watching the sun fall on the Thames, Marlow – whose experiences draw heavily upon Conrad's own – tells his old friends:

> Now when I was a little chap I had a passion for maps. I would look for hours at South America, or Africa, or Australia, and lose myself in all the glories of exploration. At that time there were many blank spaces on the earth, and when I saw one that looked particularly inviting on a map (but they all look that) I would put my finger on it and say, When I grow up I will go there.

Once grown up he fulfils his wish, although the map of Africa he chances upon in a shop window in Fleet Street is no longer a 'blank space of delightful mystery'. On the contrary, it has 'become a place of darkness', the mighty Congo resembling 'an immense snake uncoiled'.

Marlow is charmed by the snake and investigates the possibility of piloting a steamboat for a trading concern. His temptations are irresistibly answered by another map at the morbid company offices. Marvelling at all the codified colours – 'a vast amount of red – good to see at any time, because one knows that some real work is done in there' – he is set on his course: 'I was going into the yellow. Dead in the centre.'

The naive Marlow undergoes a rude awakening indeed (as did Conrad during his own Congo journey of 1890), once he leaves the schematic security of paper geography for the real thing and discovers degradation, brutality and horror of the highest order. And herein lies the brilliance of Conrad's choice of symbol. Like everything else in the novel associated with the 'civilized' West (and especially the 'improving' agendas of the colonial powers) the map turns out to be a sham, a papering over of greed, brute force and all the darkness of the human heart.

And what is Conrad telling us through the tale that is at once Marlow's and his own? That the apparently rational and pleasingly ordered account of the world we find in our school atlases, spread out on the walls of our command centers or shoved in the seat pockets of our cars might be just so many lies.

STANLEY
IN
FRICA
1867 to 1889
EXPEDITIONS
1867-68. Abyssinian Expedition.
1871-72. The Finding of Livingstone.
1873-74. The Ashanti Campaign.
1874-78. The Great "New York Herald" and "Daily Telegraph" Expedition.
1879-82. The Founding of the Congo State: First Expedition.
1882-84. The Founding of the Congo State: Second Expedition.
1887-89. The Emin Pasha Relief Expedition.
Scale of Statute Miles
0 50 100 200 300 400 500
G. OF GUINEA
CAMEROONS
PROTECTORATE
CONGO
STATE
BRITISH
FRENCH
COMORO ISLANDS
VICTORIA NYANZA
UNIAMWESI
TURKAN
BAGIRMI
ADAMAUA
EQUATOR OR EQUINOCTIAL LINE
ANZICO OR MICOCO
VICTORIA NYANZA
LAKE TANGANYIKA
CONGO
LONDA
INDIA
OCEAN
ULF OF GUINEA
UTH
NTIC
EAN
DAMARA LAND

Alexander Supan's map of the ocean floors is the first to name features after the nearest major geographical masses.

Mapping the Ocean Floor

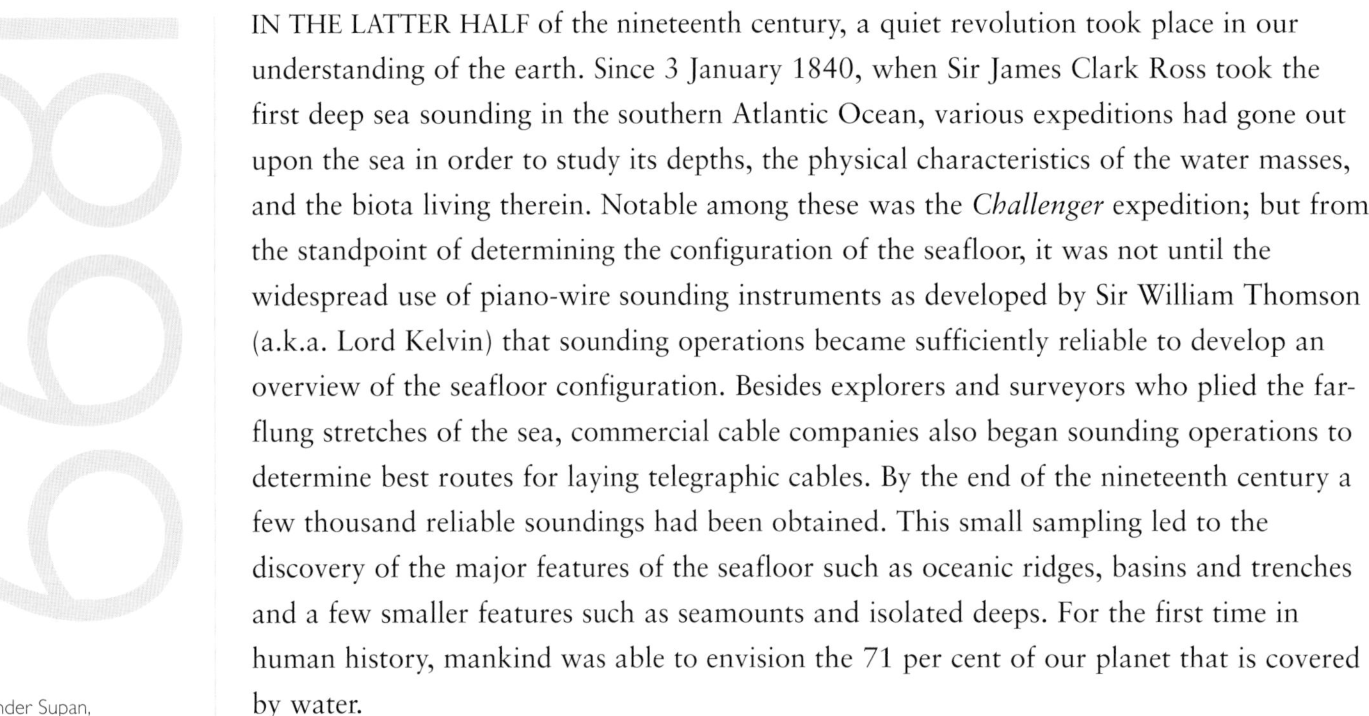

IN THE LATTER HALF of the nineteenth century, a quiet revolution took place in our understanding of the earth. Since 3 January 1840, when Sir James Clark Ross took the first deep sea sounding in the southern Atlantic Ocean, various expeditions had gone out upon the sea in order to study its depths, the physical characteristics of the water masses, and the biota living therein. Notable among these was the *Challenger* expedition; but from the standpoint of determining the configuration of the seafloor, it was not until the widespread use of piano-wire sounding instruments as developed by Sir William Thomson (a.k.a. Lord Kelvin) that sounding operations became sufficiently reliable to develop an overview of the seafloor configuration. Besides explorers and surveyors who plied the far-flung stretches of the sea, commercial cable companies also began sounding operations to determine best routes for laying telegraphic cables. By the end of the nineteenth century a few thousand reliable soundings had been obtained. This small sampling led to the discovery of the major features of the seafloor such as oceanic ridges, basins and trenches and a few smaller features such as seamounts and isolated deeps. For the first time in human history, mankind was able to envision the 71 per cent of our planet that is covered by water.

Right Alexander Supan, *Tiefenkarte des Weltmeeres.*

However, with envisioning the seafloor comes the need to order and name features. Two schools of thought were prevalent. The first was to name all features, including the major structural features of the seafloor, for famous ships and scientists; the competing viewpoint felt that the major features should be named for associated geographic features such as Argentine Basin while only the secondary features could receive ship or personal names. The map presented, by the influential German geographer Alexander Supan, was the first map to use the second viewpoint which is the accepted naming philosophy today. With the advent of acoustic sounding systems, improved navigation methods, and even satellite-altimetry systems from which depths can be derived, our ability to define seafloor features has been continually improving. John Murray of the *Challenger*, Alexander Supan, and the myriad scientists and surveyors of their era, would hardly recognize the bathymetric map of the world as we know it today. Significantly, the ability to better define the nature of the seafloor is what led directly to the great earth sciences revolution of the twentieth century, the theory of plate tectonics.

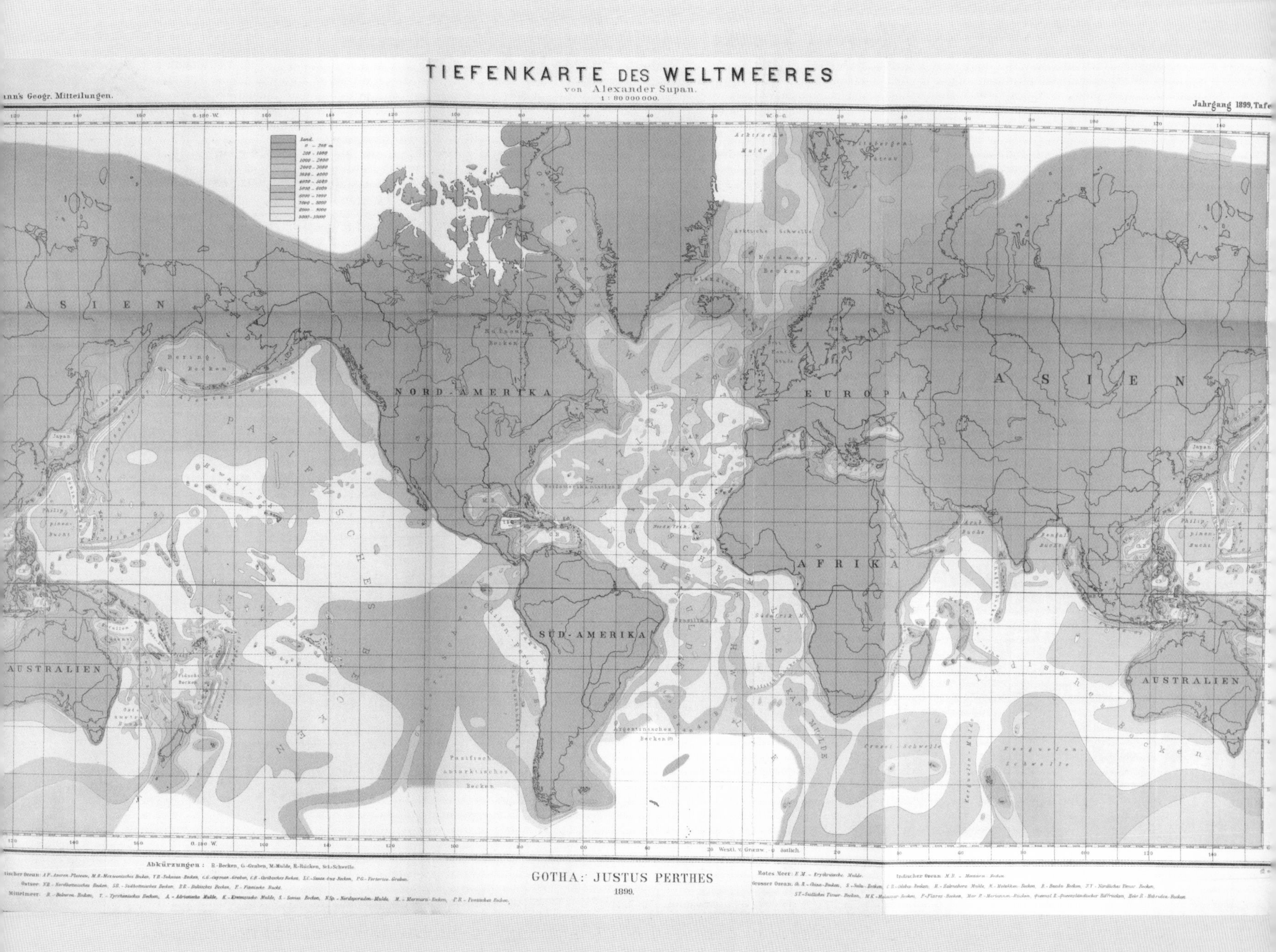
TIEFENKARTE DES WELTMEERES
von Alexander Supan.
1 : 80 000 000.
Jahrgang 1899, Tafe
NORD-AMERIKA
SÜD-AMERIKA
EUROPA
AFRIKA
ASIEN
AUSTRALIEN
Abkürzungen: B.-Becken, G.-Graben, M.-Mulde, R.-Rücken, Sch.-Schwelle
GOTHA: JUSTUS PERTHES
1899.

A map of a siege, during which the Boers beseiged the British, and a diamond mine which also reveals underlying racist attitudes.

War, Diamonds and Racism

THE CLOSE RELATIONSHIP between politics and commerce is nothing new. In October 1899 war broke out between the British-ruled Cape Colony and the Orange Free State and South African Republic (also known as Transvaal) both ruled by the Afrikaans-speaking Boers. In large part the dispute centred around the control of the gold and diamond mines. Gold, then the basis for the world economy, had been found in increasing amounts in Transvaal since 1886 and there was a danger that it would soon out-produce the British-owned gold mines.

As soon as war broke out Boer forces laid leisurely siege to Kimberley, the site of the De Beers' most important diamond mine, in Cape Colony. The British found it harder to relieve than they had expected, and their forces, predominantly consisting of Highlanders, received a particularly bad mauling at Magersfontein in December 1899. From then onwards conditions for the inhabitants of Kimberley deteriorated seriously before the town was relieved on 15 February by forces led by Sir John French.

This enormous map was drawn soon afterwards. Despite its title, it is as much a map of the mines as it is a map of the siege of Kimberley. While it shows redoubts, search lights, camps and hospitals, it also shows De Beers' mines, workshop, washing machines, debris heaps and reef tips. The mapmaker Claude Lucas was both a Cape government land surveyor and assistant surveyor to De Beers Consolidated Mines Ltd.

Doubtless in an attempt to avenge memories of Magersfontein, Lucas brutally satirized the Boers: at the top left-hand corner, two ragged Boers, grotesquely impaled on compass points, are fighting each other. More significant for the future is what the map tells us of the racist attitudes shared by the British and the Boers. The Apartheid mentality for which South Africa was later to become notorious is apparent from the (white) 'Kimberley Club', the 'native mission church', 'the native compound' next to a hospital, the 'coolie location', 'the native and malay cemetery' as well as 'native' and 'malay' military camps. Fifteen thousand black Africans are estimated to have died in a war which had nothing to do with them.

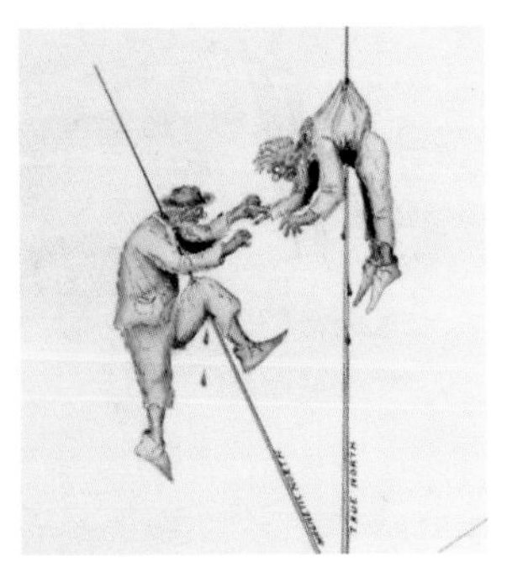

Right Claude O. Lucas under the supervision of Captain A. J. O'Meara. R. E., *Siege of Kimberley*. **Above** Detail showing two impaled Boers. **See also** 1684.

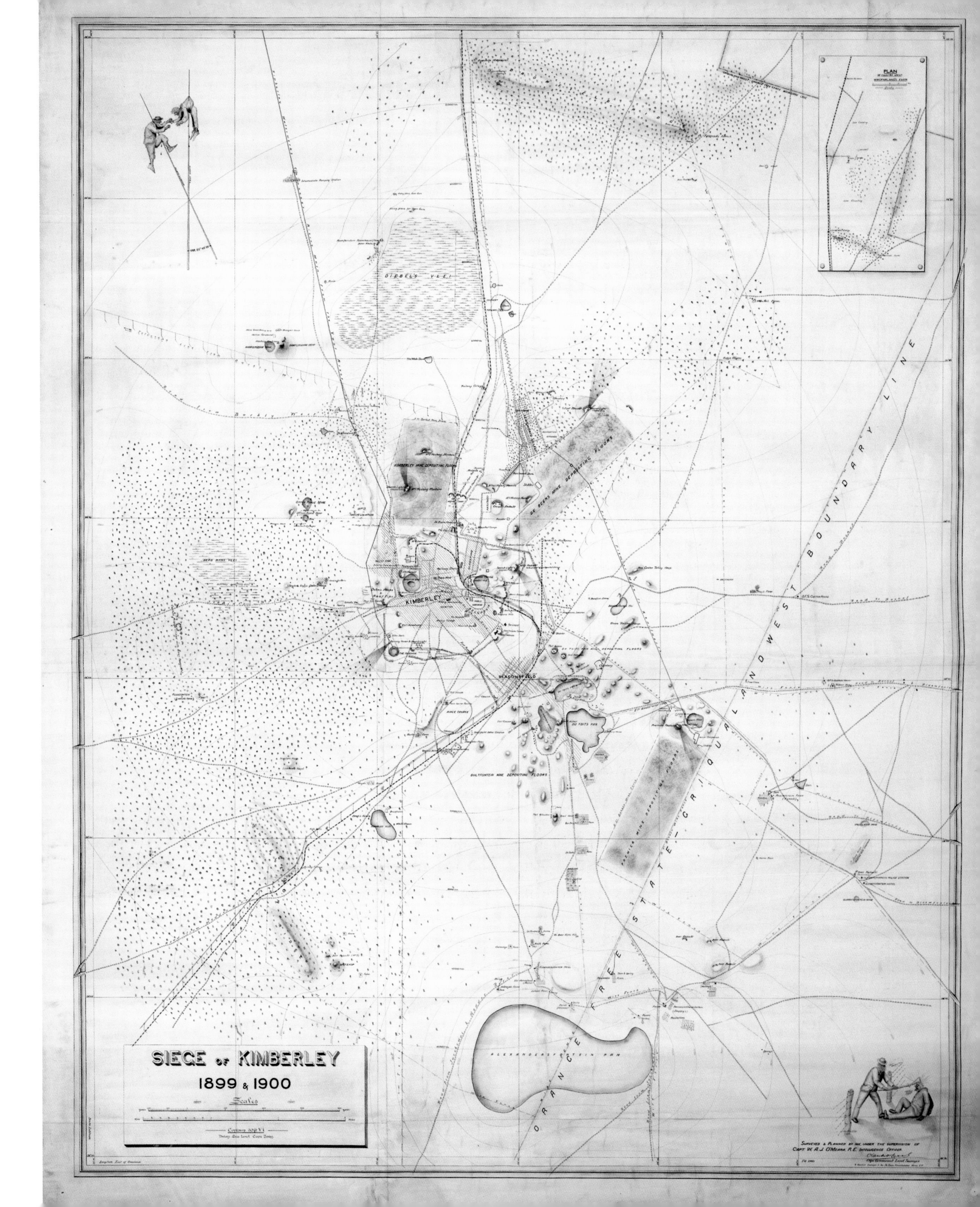

SIEGE of KIMBERLEY
1899 & 1900
Scales
Contours 30ft V.I.
KIMBERLEY
BEACONSFIELD
KIMBERLEY MINE DEPOSITING FLOORS
DE BEERS MINE DEPOSITING FLOORS
BULTFONTEIN MINE DEPOSITING FLOORS
PREMIER MINE DEPOSITING FLOORS
DU TOITS PAN
RACE COURSE
DIBBEL'S VLEI
ALEXANDERSFONTEIN PAN
ORANGE FREE STATE – GRIQUALANDWEST BOUNDARY LINE
PLAN
Surveyed & Planned by me under the supervision of
Capt. W. A. J. O'Meara R.E. Intelligence Officer

Fearful of Russian ambitions, British Army officers undertook the mapping of Anatolia in secret because the Ottomans were just as fearful of British Imperial ambitions.

Spying on One's Allies

1901

FOR MUCH OF THE nineteenth century the British government was concerned about Russian ambitions in central and western Asia. One of the main threats was to the Ottoman Empire, 'the sick man of Europe', since Ottoman control of the Bosporus prevented Russian warships having access to the Mediterranean. The Crimean War had been fought to prevent this in the 1850s, but the Russians had tried again in 1877. On that occasion, it had not been necessary to intervene militarily, but it was recognized that if Britain needed to intervene against the Russians, it would need maps of Ottoman territory, particular of Anatolia, then called 'Eastern Turkey in Asia', in case a Russian army invaded through the Caucasus.

Unfortunately, the Ottomans were just as fearful of British imperial ambitions as they were of Russia's. This meant that the mapping had to be carried out using covert methods. British Army officers, who had been sent on missions to reform the Ottoman gendarmerie, carried out some of the secret mapping. Other officers, such as Francis Maunsell, carried out surveys, often at great personal risk, while pretending to be carrying out archaeological explorations. To provide cover for this covert mapping, they borrowed surveying instruments from the Royal Geographical Society, rather than use army issue.

Francis Maunsell then compiled all the maps covering Eastern Turkey in Asia, using not only his own, and other officers' surveys, but whatever map information he could obtain. All the sheets of this series carefully list the sources used, providing a valuable insight into British activity in the area.

The Van-Bitlis sheet is the most attractive of this important series. It shows the area around Lake Bitlis in Turkish Armenia, including the impressive crater of Mount Nimrud to the west of the lake. In the absence of accurate height information, relief is shown using a technique called form-lines. The sophisticated use of form-lines on the Van-Bitlis sheet creates a good visualization of the relief.

Ironically, soon after Maunsell had finished his task, Russia had become an ally, and the Ottoman Empire had become a potential enemy.

Right Francis Maunsell, sheet showing Lake Bitlis from *Eastern Turkey in Asia*. **See also** 1877

EASTERN TURKEY IN ASIA

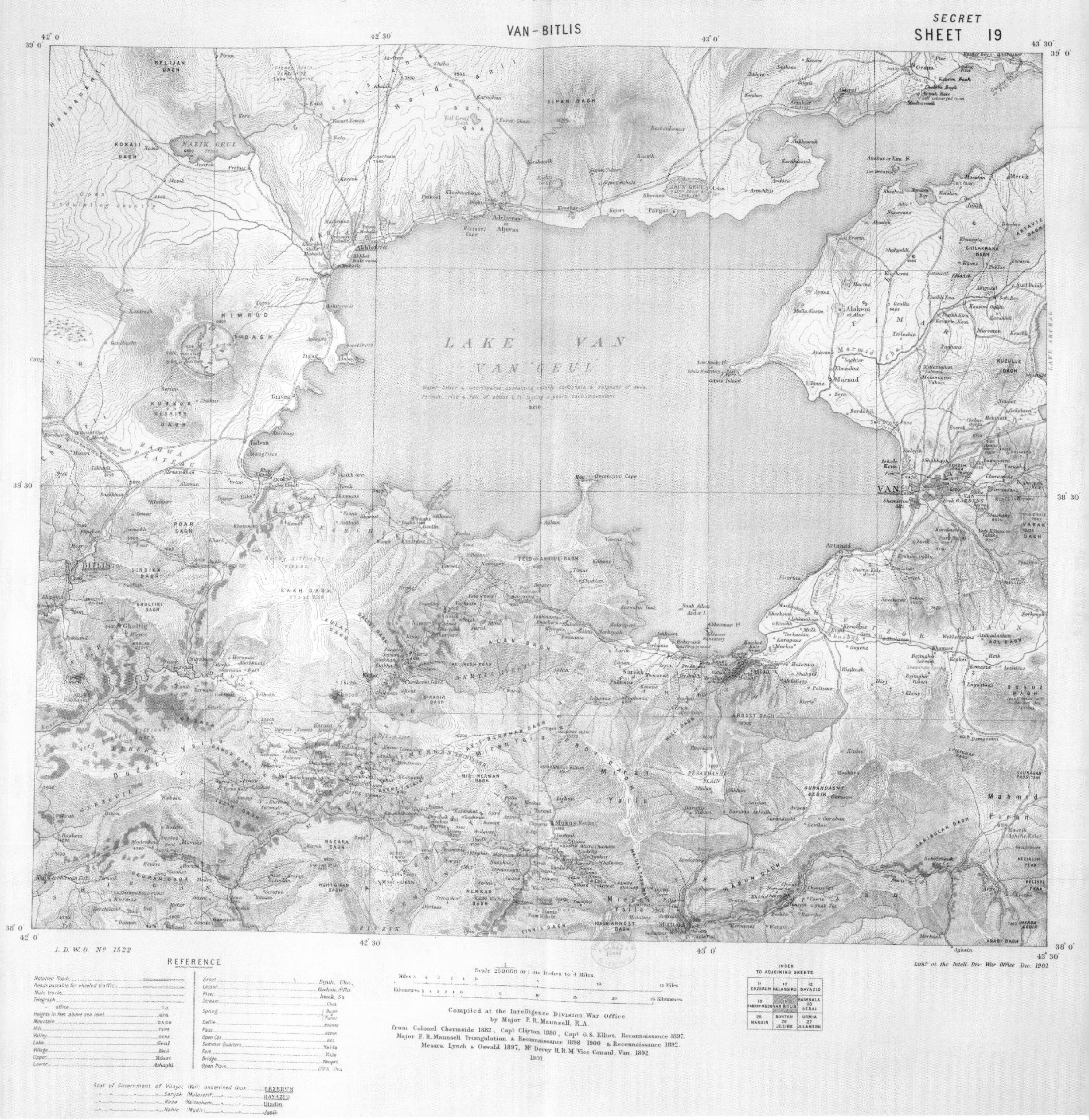

A young surveyor takes cartographic revenge, perpetuated in print on an Admiralty Chart, on his commanding officer.

The Human Element

THE BRITISH HYDROGRAPHIC Office came into existence in 1795 at about the same time as its terrestrial equivalent, the Ordnance Survey. Within a few decades its charts had become accepted as amongst the best available – not surprisingly at a time when the British Navy was the most powerful in the world. Periodic surveys were often followed by decades of minor changes with the same copperplates being used. Sometimes, however, a fresh survey had to be undertaken.

At the beginning of the twentieth century the Aegean was a bone of contention between Greece and the Ottoman Empire. With the distinct possibility that the British fleet might have to see action in those waters (as was to happen at Gallipoli in 1915), it was decided to refresh the mapping of the region.

As a consequence of this policy, Captain Corry of the Royal Navy sent a junior officer to survey a group of hills near the south-west of Mudros Bay, on the island of Lemnos which though mainly Greek-speaking was still ruled by the Turks. It was perhaps a hot day. Or perhaps the flies were just very aggressive. At all events, the young man did not care for the task. He named the hills Yam Hill, Yrroc Hill, Eb Hill and Denmad Hill. Perhaps he was counting on the inability of most Englishmen to understand any language other than English, and assumed that in this case and quite genuinely, 'it would be all Greek' to the revisers. If so he was right. No one doubted the authenticity of the names which were included in the next edition of the Admiralty Chart in 1920 (the year in which Lemnos formally became part of Greece).

And so his resentment towards his commanding officer was publicized and perpetuated. In reverse the hill names read, in good English, 'May Corry be damned'. Mapmakers, after all, are only human and occasionally their maps, for all their technical excellence, will reflect this.

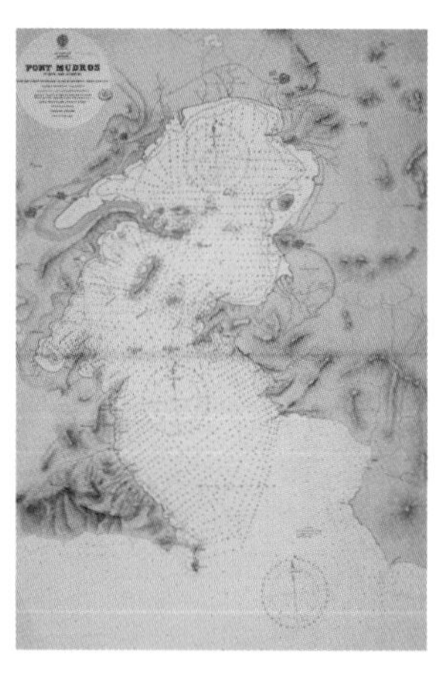

Right Extract from Admiralty Chart, Port Mudros. **Above** The complete chart.

Square Tr.
Cultivated
Landing
ALAGO
WEST PASS
MIDDLE PASS
EAST PASS
Black Rk.
Meganoros Pt.
VRULIDHI BAY
Vrulidhi Pt.
Conspicuous houses
Chapel
Farm
Fuller Cove
Terrant Cove
Limni Pt.
Sandy Cove
Barren
Foul Cove
Buda Pt.
SHOAL BAY
Freshwater Bay
Spring
Delias Pt.
C. Malathria
MAGNETIC
Penfold Cove
Durham Cove
Williamson Hill
Paterson Range
Crawford Cove
Hut
Wilkinson Beach
Well
BROWELL VALLEY
(Cultivated in places)
Greystone Hill
Undulating country
Redcliff Pt.
Cairn
Mt. Phako or San Antonio
Barren rocky country
Nulma Hill
Queen Hill
Yam Hill
Yrroc Hill
Eb Hill
Denmad Hill
Cone Hill
Coombs Cove
Hawkins Cove
Foul ground
Kombi Pt.
Dawkins Hill
Kombi Id.
boat passage
Kastro Id.
MUDROS BAY

The only known copy of one of the most sophisticated and beautiful relief maps to be produced by Ordnance Survey.

Ordnance Survey Quality

Right Ordnance Survey of England and Wales, sheet 144, Plymouth, one inch to the mile. Revised 1912–13, printed 1914. See also 1800 (p. 246).

BETWEEN 1801 AND 1897 the Ordnance Survey one-inch (2.5-centimetre) map of England and Wales was published as a black and white engraved map. The principal method of showing relief was by engraved hachures. The artistic skill and technical expertise of the Ordnance Survey engravers in this respect was internationally renowned.

The introduction of colour to the one-inch map came in 1897. The colour was printed by lithography, but used as a key the image generated by a transfer from the copperplate. The final printed map was, therefore, a marriage between copperplate engraving and lithography. This was, initially, an uneasy union because the sharpness of the engraved line did not transfer with the same clarity to the zinc plate or lithographic stone.

In 1911, the newly appointed Director General of Ordnance Survey, Colonel Sir Charles Close, began a programme to revolutionize the appearance of the one-inch map. It was to be produced in two versions. One was to be a simple, cheap contoured map (this became the Popular Edition), and the other was to be a luxury product – a sophisticated relief map.

Building on the ideas of the eminent map publisher, George Philip, for an ideal topographical map, Close decided on the innovative combination of a decreased contour interval (50 feet/15 metres instead of the existing 100 feet/30 metres), hachures, layers and hill shading. The first map to be produced was a tourist map of Killarney, in 1913. It was hailed as the most beautiful and complex system of hill representation yet adopted by the Ordnance Survey. A few other experimental one-inch maps on a similar specification followed, and, by April 1914, it was announced that several sheets of the new relief map of England and Wales were near publication. None of these appears to have survived. The advent of the First World War, and the financial retrenchment which followed it, brought about the demise of this lavish design which marked the zenith of Ordnance Survey colour printing techniques.

This sheet of the Plymouth area, produced for military use in 1914, is very similar to the Killarney style. It has almost imperceptible layer colouring, contours are drawn in black dots at 50-foot (15-metre) intervals, hachures are printed in a sunny brown, and a delicate purple hill shading is applied to south-eastern slopes. The existence of this map is a recent discovery, only one copy is known.

PLYMOUTH
Devonport
SALTASH
TORPOINT
PLYMPTON
Plympton Erle
Plymstock
Yealmpton
Brixton
Newton Ferrers
Noss Mayo
Wembury
Wembury Bay
Wembury Pt
Great Mew Stone
Mewstone
THE SOUND
The Breakwater
Drake's I.
Cawsand
Cawsand Bay
Kingsand
Rame
RAME HEAD
Penlee Point
Queener Point
Millbrook
Maker
Mt Edgcumbe
St John
St John's Lake
Antony
Landulph
Cargreen
Tamerton Foliot
Bickleigh
Shaugh Prior
Shaugh Moor
Cornwood
Sparkwell
Holbeton
Staddon Heights
Down Thomas
Mouthstone Point
Gara Point
Stoke Point
Netton Island
Beacon Point
Hooe
Turnchapel
Cattewater
Laira
Mutley
Stoke
Keyham
Egg Buckland
Crownhill
Roborough Down
Cann Wood
Lee Moor
Weirquay

A whimsical view of London transport and of Londoners by the brother of artist and typographer Eric Gill is suffused with jokes, puns and rhymes.

Cartographic Chaos

MOST MAPMAKERS HAVE to create their maps to strict specifications. But just occasionally the controls are loosened and the artist's whimsy is given free rein.

This is what has happened here. It may not look like it at first glance, but this is a map of the London Underground. In the early twentieth century a wide variety of artists were commissioned to design advertising posters for the Underground. In 1913 Macdonald Gill (brother of the artist Eric Gill) designed this map poster. The inscription around the image gives its official purpose: 'The heart of Britains empire * here is spread out for your view. it shows you many stations * &'bus routes not a few. You have not the time * to admire it all? Why not take a map home to pin on your wall.' There is an outsize standard symbol for Underground stations and if you look carefully you can see the lines connecting them. The major roads used by the trams and buses are also there. And some older forms of transport like horses and the sedan chair.

These are almost lost among the other elements. It is a tourist map as well as a transport map and most of London's most famous sights are shown. It is also sprinkled with extracts from nursery rhymes, pantomime characters, cartoons of London life with private jokes and some awful puns: 'Here, matey, lend a hand,' says a coalman on St Pancras Goods Yard (now the site of the British Library). 'Can't,' says the man standing near Euston, 'got a chill on my pancreas.' This is mapping anarchy.

The map poster gave pleasure to thousands of Londoners, but in the early 1920s when Macdonald Gill tackled the London Underground again, he produced a sober depiction of the lines and the stations, with virtually nothing else.

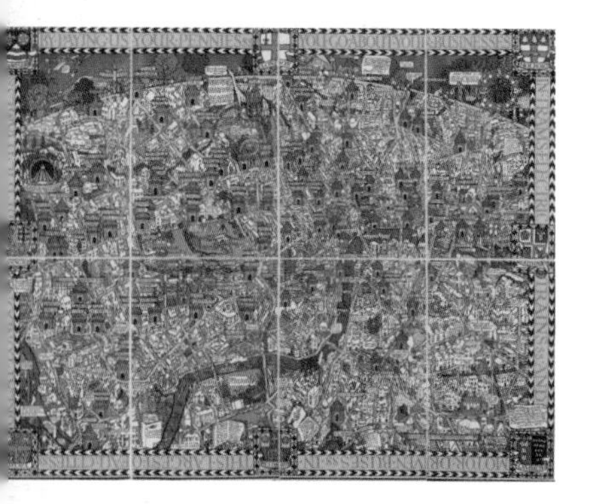

Right Extract from Macdonald Gill, *The Wonderground Map of London Town*. **Above** The complete map.

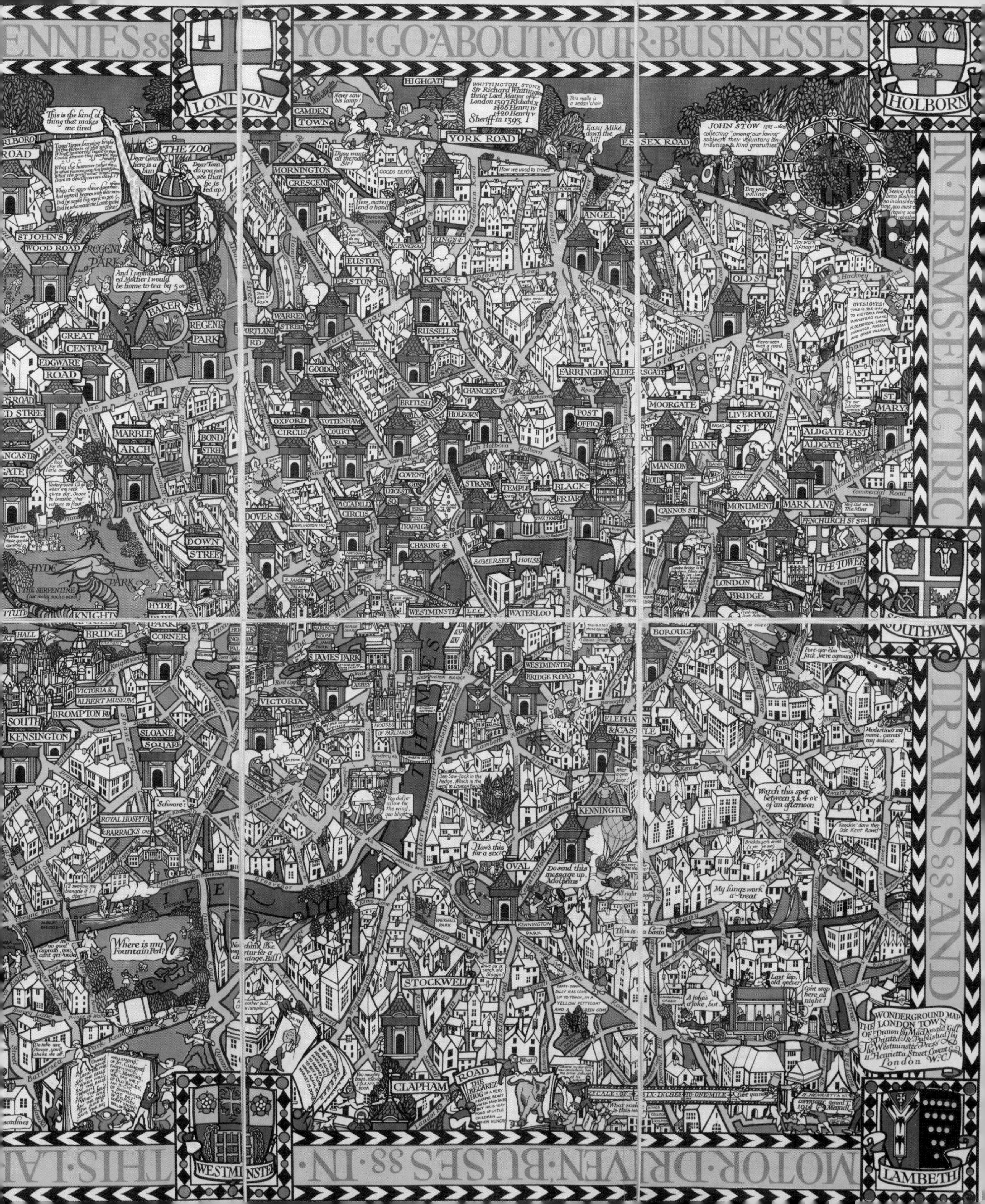
ENNIES ss YOU·GO·ABOUT·YOUR·BUSINESSES
IN·TRAMS·ELECTRIC·TRAINS ss AND
MOTOR·DRIVEN·BUSES ss IN
THIS·LA
LONDON
HOLBORN
SOUTHWARK
WESTMINSTER
LAMBETH
HIGHGATE
CAMDEN TOWN
YORK ROAD
ESSEX ROAD
WHITTINGTON STONE Sir Richard Whittington thrice Lord Mayor of London 1397 Richard II 1406 Henry IV 1420 Henry V Sheriff in 1393.
Never saw his lamp
This really is a sedan chair
Easy Mike, down the hill
JOHN STOW 1525–1605 collecting "among our loving subjects their voluntary contributions & kind gratuities"
This is the kind of thing that makes me tired
THE ZOO
Dear Giraffe here is a bun
Dear Tom do you not see that he is fed up
D'you want all the road Sir?
How we used to travel
MORNINGTON CRESCENT
Here matey lend a hand
ANGEL
CITY ROAD
ST JOHN'S WOOD ROAD
REGENTS PARK
And I promised Mother I would be home to tea by 5 o'c
BAKER ST
REGENTS PARK
GREAT CENTRAL
EDGWARE ROAD
EUSTON
EUSTON SQ.
KINGS +
WARREN STREET
RUSSELL SQ.
PORTLAND RD
GOODGE ST.
FARRINGDON
ALDERSGATE
OLD ST.
Hackney
OYES! OYES! This is the way to Victoria Park, Wanstead Flats, Woodford, Chigwell, Barking, Ilford, Romford, Harwich and other villages
Never seen such a road
MOORGATE
LIVERPOOL ST.
ST. MARY
ALDGATE EAST
ALDGATE
MARBLE ARCH
BOND STREET
OXFORD CIRCUS
TOTTENHAM COURT RD.
BRITISH MUSEUM
HOLBORN
CHANCERY LN
POST OFFICE
BANK
MANSION HOUSE
COVENT GARDEN
LEICESTER SQ
STRAND
TEMPLE
BLACK-FRIARS
CANNON ST.
MONUMENT
MARK LANE
FENCHURCH ST STA.
PICCADILLY CIRCUS
DOVER ST.
TRAFALGAR SQ.
CHARING +
SOMERSET HOUSE
THE TOWER
Underground I'll go when my neck gives out. Ozone to breathe, the vulture to flout.
DOWN STREET
HYDE PARK
THE SERPENTINE (nor really such a worm)
KNIGHTSBRIDGE
HYDE PARK CORNER
WESTMINSTER
L.C.C.
WATERLOO
LONDON BRIDGE
BOROUGH
Port-yer-'elm Jack, we're aground
S. JAMES PARK
WESTMINSTER BRIDGE ROAD
VICTORIA & ALBERT MUSEUM
BROMPTON RD
SOUTH KENSINGTON
VICTORIA
SLOANE SQUARE
HOUSES OF PARLIAMENT
THAMES
ELEPHANT & CASTLE
Modestine's my name, carrots my solace.
See-Saw Jack in the hedge. Which is the way to London Bridge?
Watch this spot between 3 & 4 o'c of an afternoon
ROYAL HOSPITAL
BARRACKS
Schware!
You didn't allow for the wind you blighter
KENNINGTON
Knockin' 'dem ther Ode Kent Road
How's this for a six?
OVAL
Do send this message up, Adolphus
My lungs work a-treat
RIVER
KENNINGTON PARK
This is a basin
Where is my Fountain Pen?
STOCKWELL
Last lap, old geezer
Can't stop here all night!
A joke's a joke, but...
WONDERGROUND MAP OF LONDON TOWN Drawn by MacDonald Gill Printed & Published by The Westminster Press 411 Henrietta Street Covent Garden London W.C.
CLAPHAM ROAD
THE BEGAREZ HOG IS A VERY FEARFUL BEAST WHICH FLIES VERY FOND OF LITTLE CHILDREN DRIVEN HUNGRY
SCALE OF SIX INCHES TO ONE MILE

One of the earliest aircraft accident reports incorporating a topographic map of the crash site records the tragic end of a pioneering Turkish venture.

An Early Aircraft Accident Report

IN THE EARLY YEARS of aviation, prior to the outbreak of the First World War, daring adventurers tried to establish flight records. They were part of man's self-challenge, pushing the human spirit's passion for achievement to the furthest possible limits.

In the last days of December 1913, two French airplanes executed the first successful landings in Palestine. The pilots took the three-continent route, from Paris to their end destination over the Pyramids in Cairo, via Istanbul, Damascus, Beirut and Jerusalem.

Meanwhile, the Ottomans grasped at any act that would enable them to restore their national pride. Early in 1914, the Ottoman authorities decided to stage long-distance flights to establish their own air presence in the Levant territories, Palestine and Egypt. The Ottoman pilots, having gained some experience during the Balkan Wars and encouraged by the French flights over their country, were compelled to take on the challenge. The most experienced Ottoman pioneer pilot was chosen for the first mission, accompanied by an observer. Everywhere on the way, proud officials and enthusiastic people greeted them and praised the brave aviators who were carrying the Ottoman flag over for the glory of the empire. On 27 February 1914, while on the way from Damascus to Jerusalem, the pilots faced an unknown problem and their Bleriot XI airplane crashed on the eastern flank of the Lake of Galilee in Palestine. They became the first martyrs of Turkish aviation.

While researching in Ankara we came across a bulletin of the Ottoman Police (*Polis Mecmuasi*) published two weeks after the crash, on 14 March 1914. It contained three official reports, constituting the first aircraft accident report, at least in that part of the world. Besides the police report and the forensic report, the geographic report included a 1:20,000 topographic map prepared by an unknown Ottoman surveyor on 28 February 1914. The map attempted to explain the scene of the event, the place where the wreckage of the aircraft was found about 2,214 steps (2 kilometres/1¼ miles) from the lakeshore, and the hazardous topography, by form-lines serving as contours to depict the deep slope from the Golan Heights (280 metres/920 feet above sea level) down to the Jordan Valley (200 metres/650 feet below sea level).

This map proved most valuable in providing the correct geographical information about the scene, as well as dispelling any doubts about the accident that took place, in a political environment infamous for alarmist, uncorroborated misinformation.

Right The official plan of the air crash of 27 February 1914. **Above** The aviators in Constantinople before take-off.

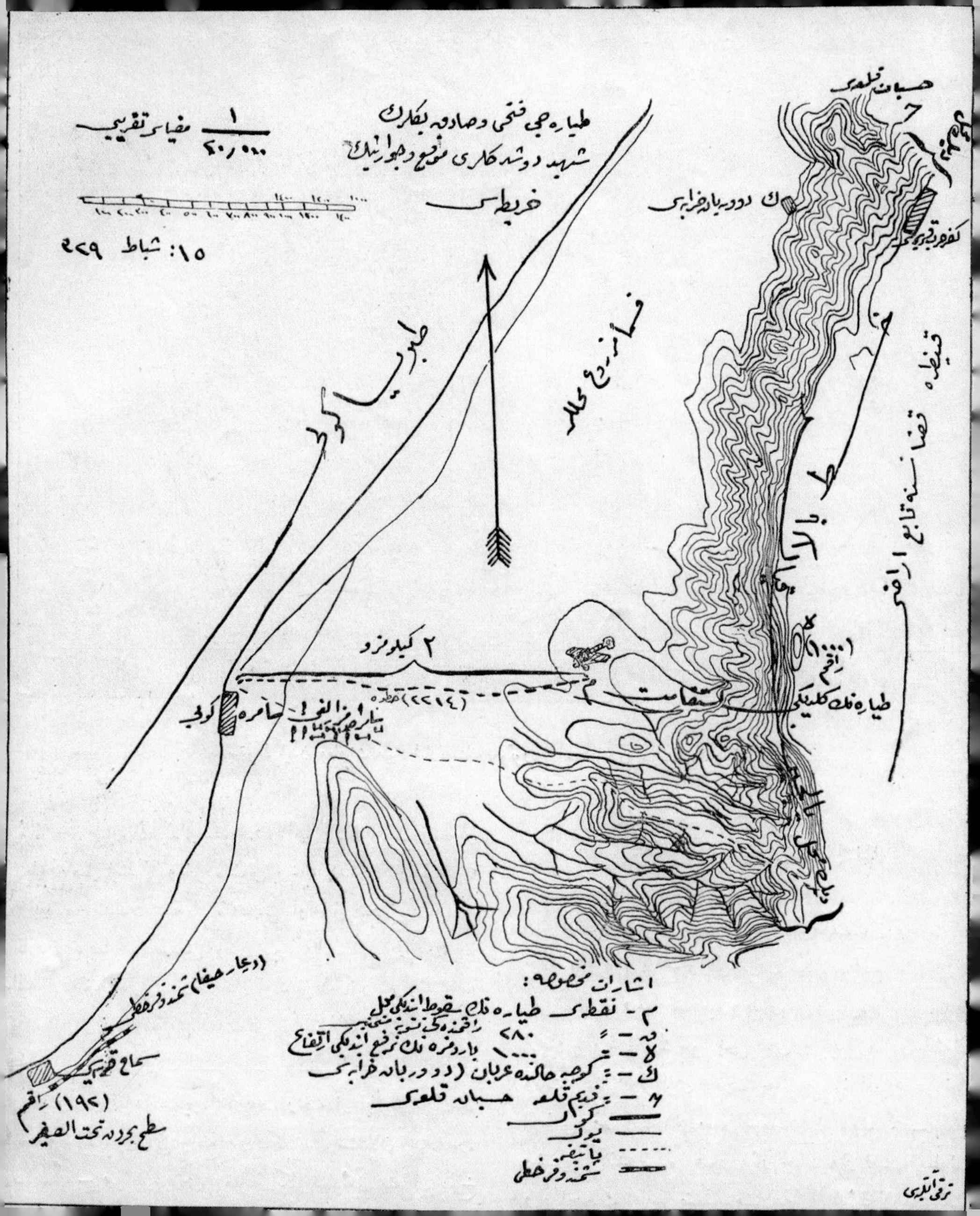

طیارەجی فتحی و صادق بکلرک
شهید دوشدکلری موقع و جوارینک
خریطه‌سی
مقیاس تقریبی ۱ / ۴۰,۰۰۰
۱۵ شباط ۳۲۹
طبریه کولی
قسم اعظمی مزروع محللر
۲ کیلومترو
سامره کویی
طیاره‌نک کلدیکی
حسبان قلعه‌سی
قنیطره قضاسنه تابع اراضی
(درعا - حیفا) شمندوفر خطی
سماخ قصبه‌سی
(۱۹۲) رقم
سطح بحرده تحت الصفر
اشارات مخصوصه:
م - نقطه‌سی طیاره‌نک سقوط ایتدیکی محل
کول
یول
پاتیقه
شمندوفر خطی

Sir Ernest Shackleton's sketches on the back of a menu card in the course of a dinner party establish his proposed route across Antarctica.

Ideas for Antartica

Right Sketch Map of Antarctica drawn by Sir Ernest Shackleton.

THIS MODEST PENCIL-drawn sketch map was a precursor to one of the most remarkable Antarctic survival stories. It was sketched by Sir Ernest Shackleton on 17 March 1914 on the back of his menu card for the London Devonian Association annual dinner. At the dinner Shackleton drew the sketch for his neighbour Clive Morison-Bell. The latter added the note 'presented to me by my neighbour at the dinner, in order to explain his prospective plans'.

The sketch map identifies Shackleton's route for his Imperial Trans-Antarctic Expedition which planned to be the first to cross the landmass of Antarctica. It is also accompanied by calculations showing how a fix could be made to establish 90 degrees south – the location of the South Pole.

Leaving South America and sailing in the ship *Endurance* through the Weddle Sea, Shackleton's plan was to travel to the South Pole and then continue across this continent finally arriving at the Ross Sea – a journey of 1,800 miles (2,900 kilometres). Shackleton had sent a separate party to the Ross Sea to lay supply depots which he would survive on after his team had crossed the Pole.

However, the *Endurance* became stuck and then crushed in the ice of the Weddle Sea and was abandoned on 27 October 1915. There then followed one of the most renowned chapters of Antarctic exploration. First, Shackleton brought his crew out of the pack ice in three open boats and landed them safely on Elephant Island. He then sailed across the southern ocean to South Georgia, crossed its uncharted glaciers and finally relieved his crew.

At the Devonian dinner Shackleton dined on tomato with tapioca, turbot dieppoise, kirsch sorbet and glace plombière – a far cry from the seal, penguin and pemmican that sustained his crew during their escape from Antarctica.

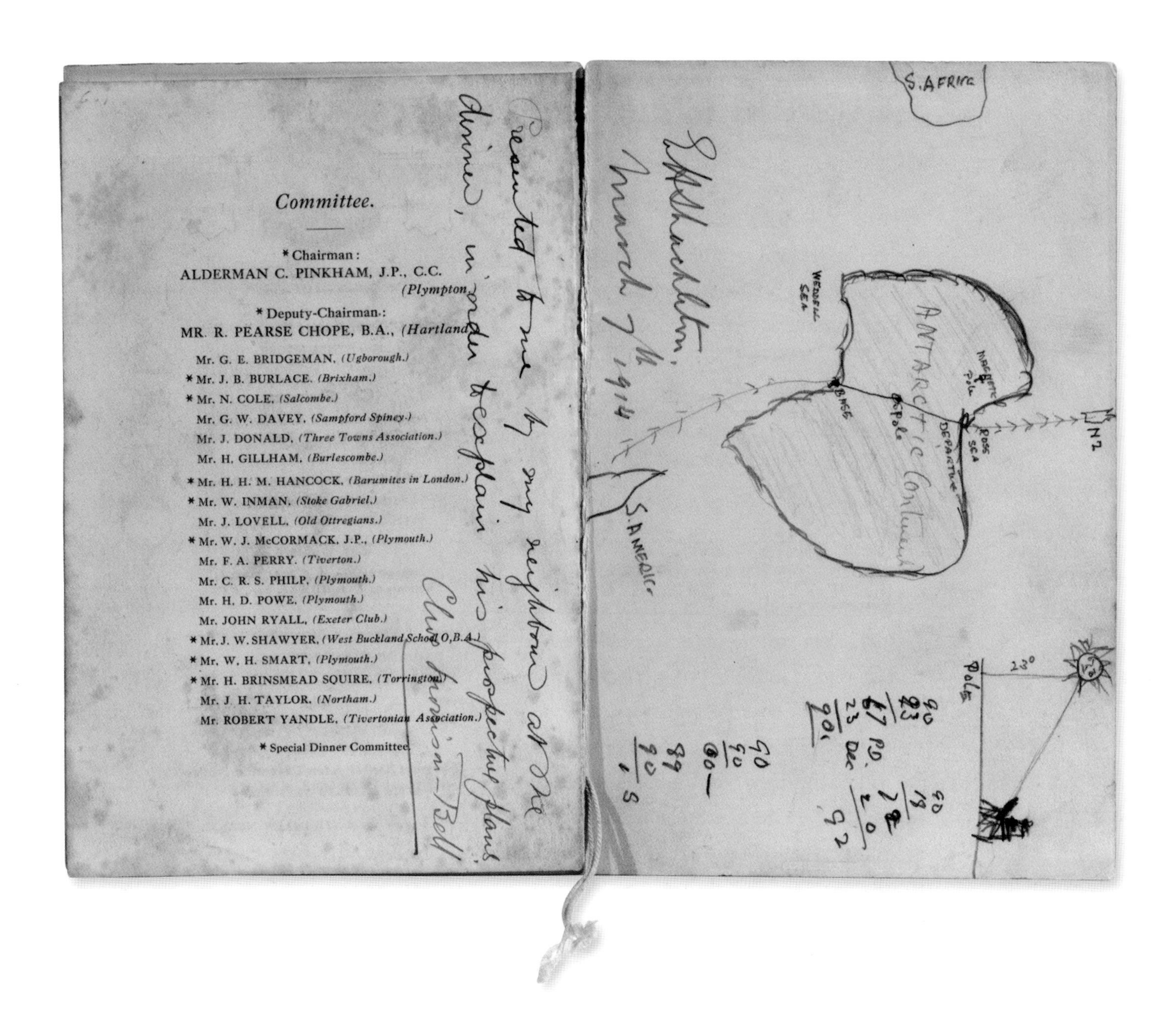

Committee.

* Chairman:
ALDERMAN C. PINKHAM, J.P., C.C. (Plympton.)

* Deputy-Chairman:
MR. R. PEARSE CHOPE, B.A., (Hartland.)

Mr. G. E. BRIDGEMAN, (Ugborough.)
* Mr. J. B. BURLACE, (Brixham.)
* Mr. N. COLE, (Salcombe.)
Mr. G. W. DAVEY, (Sampford Spiney.)
Mr. J. DONALD, (Three Towns Association.)
Mr. H. GILLHAM, (Burlescombe.)
* Mr. H. H. M. HANCOCK, (Barumites in London.)
* Mr. W. INMAN, (Stoke Gabriel.)
Mr. J. LOVELL, (Old Ottregians.)
* Mr. W. J. McCORMACK, J.P., (Plymouth.)
Mr. F. A. PERRY, (Tiverton.)
Mr. C. R. S. PHILP, (Plymouth.)
Mr. H. D. POWE, (Plymouth.)
Mr. JOHN RYALL, (Exeter Club.)
* Mr. J. W. SHAWYER, (West Buckland School O.B.A.)
* Mr. W. H. SMART, (Plymouth.)
* Mr. H. BRINSMEAD SQUIRE, (Torrington.)
Mr. J. H. TAYLOR, (Northam.)
Mr. ROBERT YANDLE, (Tivertonian Association.)

* Special Dinner Committee.

A recruitment poster for the Indian Army uses an image of India as a call to defend the motherland.

Defending Hearth and Home?

1914

Right Propaganda poster from India.

MAPS HAVE BEEN USED in propaganda for almost as long as there have been maps. Even maps which had been produced for other purposes could be used as propaganda. For example, presentation portolan charts could be used to enhance the prestige of both the giver and the recipient through the judicious use of decoration and dedications. However, for the widespread use of maps in propaganda, it was necessary to have methods of cheap reproduction, and, for maximum impact, they had to be in colour. In practice, this meant that the extensive use of maps in propaganda was a twentieth-century phenomenon.

Maps could be used in a variety of printed media, as independent maps in their own right, such as those scattered over enemy lines in an attempt to undermine morale, to accompany text in newspapers, books or pamphlets, on postage stamps, or on posters.

This recruitment poster for the Indian Army is interesting for a number of reasons. Firstly, it is a strong image, capable of conveying a message without any accompanying text. It is very much a twentieth-century image, clear and without any unnecessary detail. Graphically, it resembles the new style of advertising poster being produced just before the First World War for products such as washing powder.

The image shows an Indian soldier standing with gun held at the ready protecting India against external threats. India is shown in red, a dominant colour, which makes the subject depicted stand out, and was commonly used in British mapping to show the territories of the British Empire.

The choice of a map to represent India raises an interesting question, as it assumes that an Indian looking at the image will recognize the shape as being India. Whether Indian peasants, who were thought to make the best soldiers, would have recognized the map of India is open to question. The image also suggests that, in joining the army, Indian soldiers would be defending India against an unspecified enemy.

The accompanying text, in Urdu or Hindi, reinforces the message of the image with the claim that 'This soldier is defending India, he is guarding his home and his household. Therefore we are guarding your home. You have to join the army.' The reality was that the recruits were likely to be sent far away to East Africa, Mesopotamia, Palestine or the Western Front to fight on behalf of an imperial power.

यह सिपाही हिन्दुस्तान की हिफ़ाज़त कर रहा है। वह अपने घर और घरबारवालों की हिफ़ाज़त कर रहा है।।

अपने घरबारवालों की मदद करने का सब से अच्छा तरीक़ा यह है कि फ़ौज में भरती हो जाओ।।

A miniature atlas of the British Empire was specially commissioned by Stanford's, map publishers to the British establishment, for the library of Queen Mary's Dolls' House.

A Scaled-Down Atlas for a Scaled-Down Monarchy

THE LAST YEARS OF the First World War saw the fall of the Russian, German and Austrian emperors and numerous smaller German rulers. Most of the remaining west European monarchs realized that to survive they had to change their style from the grandiose to the more domestic.

This was not a problem for George V of Great Britain. Despite being head of the largest empire the world had ever seen, and titular Emperor of India, he had always been an unassuming individual. Since his accession in 1910 he and his consort, Queen Mary, had fostered a domestic, almost bourgeois image, underpinned by extensive charitable activity. In 1920, the King's cousin, Princess Marie Louise, had the inspired idea of creating a dolls' house for presentation 'as much [from] affection as ... respect and loyalty' to the Queen.

This was, however, to be a dolls' house with a difference. Actually a palace in miniature, it was designed by the most distinguished British architect of the time, Sir Edwin Lutyens, with miniaturized contributions from hundreds of leading craftsmen, inventors, artists and composers. Perhaps the most remarkable part of the Dolls' House was the library. One hundred and seventy-one living writers including Rudyard Kipling provided minute texts in exquisitely bound volumes. Among the atlases there was a specially prepared atlas of the British Empire. It was produced by Stanford's, map publishers to the British establishment, who had hitherto left such atlases to other publishers. It contained eight full-colour maps beginning with the British Isles and moving on to the colonies, each with their arms and for comparison – and to remind the reader of the enormous size of Britain's dominions beyond the seas – a still more miniaturized outline of the British Isles.

Right Stanford's *Atlas of the British Empire* made for Queen Mary's Dolls' House. **Above** Part of the library of Queen Mary's Dolls' House. Note the miniature globe. **See also** 1660 (p. 164).

The completed Dolls' House caught the imagination of the Queen's middle-class subjects, many of whom had bought or made dolls' houses for their own daughters. First displayed at the British Empire Exhibition at Wembley in 1924, it continues to attract a steady flow of visitors at Windsor Castle, with the income going to charity.

Stanford's presented a replica to the British Museum. Since then what may be the smallest atlas ever, a tribute to an ostensibly middle-class monarch, has been housed in the same building as the largest known atlas, presented some 260 years earlier to the flamboyantly baroque Charles II.

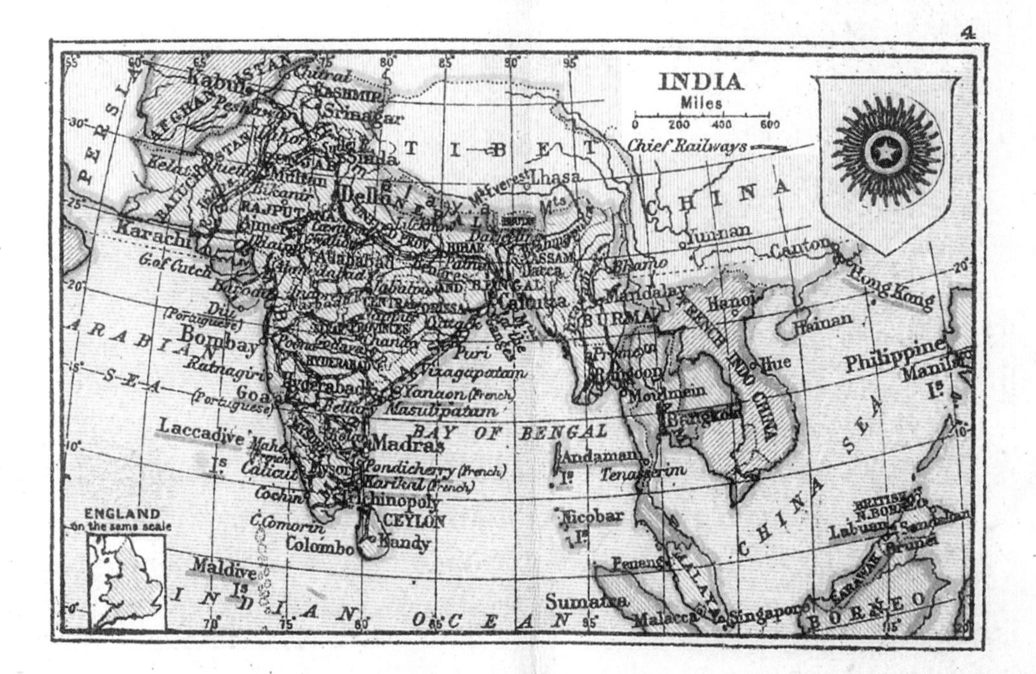

Contents

ATLAS
of the
BRITISH EMPIRE

This map illustrates the balance of economic as well as political power in inter-war Europe from a Marxist-Leninist perspective.

A Marxist-Leninist View of Europe

1926

THIS MAP PRESENTS a radical view of inter-war Europe. It is designed to explain the contemporary political and economic balance of power in terms of the Marxist-Leninist theory of imperialism. Lenin's theory holds that imperialism is not merely a question of establishing formal colonies such as the Belgian Congo, but more importantly involved economic and political influence over other nominally independent countries through overseas investment. It is here effectively represented in cartographical form.

The map comes from the pen of J. F. (Frank) Horrabin (1884–1962). Horrabin was not a cartographer but a cartoonist who cut his teeth on the provincial press and gained a reputation illustrating H. G. Wells's *Outline of History*. He was involved in many left-wing organizations, including the Plebs League founded in 1908 to provide a Marxist orientated education for trade unionists. For a short period he was a member of the Communist Party and later served as a Labour MP for his birthplace of Peterborough between 1929 and 1931. Despite his involvement in the working-class movement he contributed comic strips to various commercial Liberal newspapers throughout his career until his death. He was also a pioneer of news maps in the early years of television.

This map was originally published in a penny newspaper (*Lansbury's Labour Weekly*) so it was inevitable, given the economic and technological constraints of newspaper production, that it appeared in black and white. The same simple format is retained in *The Plebs Atlas* for reasons of economy. Designed for use by working-class students at the National Council of Labour Colleges a more lavish format would have forced up the price beyond the one shilling charged. The lack of colour is no deterrent to getting the message across.

Although it is arguable that the map is too simple as it fails to differentiate between the status of the French colonies in North Africa and those countries under French influence such as Romania and Turkey, any reader of the original newspaper article which this map illustrated or the accompanying caption in the atlas would have avoided any such misunderstanding.

Right *The New Map of Europe* from J. F. Horrabin's *The Plebs Atlas*. The original caption read: 'THE NEW MAP OF EUROPE – The workings of Imperialism are nowadays as clearly traceable in Europe as in the other continents; the Big Powers having reduced the smaller, nominally independent States to the status of colonies or spheres of influence. The map shows the countries under French, British, or Franco-British-American control.'

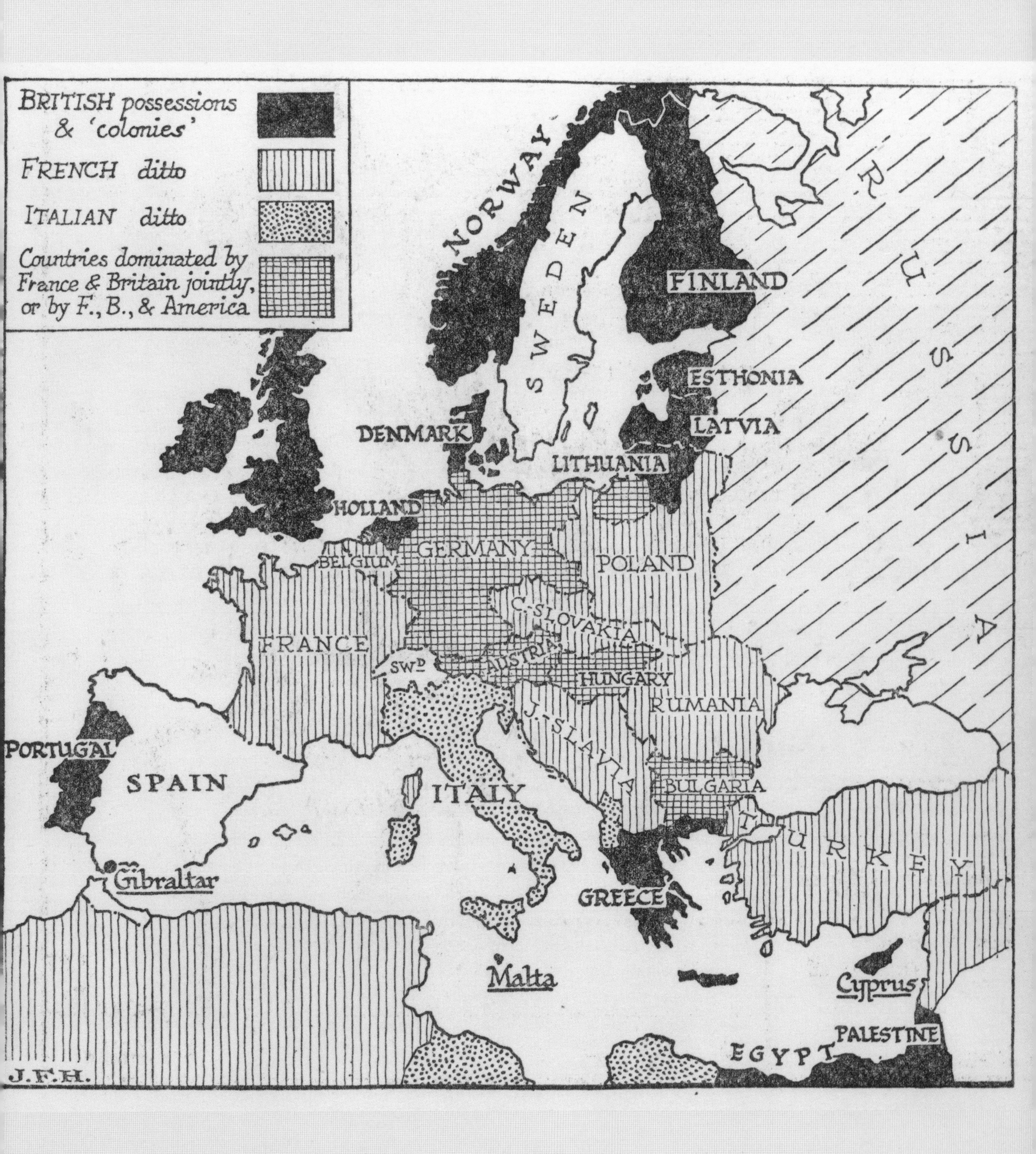
BRITISH possessions & 'colonies'
FRENCH ditto
ITALIAN ditto
Countries dominated by France & Britain jointly, or by F., B., & America
NORWAY
SWEDEN
FINLAND
RUSSIA
ESTHONIA
LATVIA
LITHUANIA
DENMARK
HOLLAND
BELGIUM
GERMANY
POLAND
C-SLOVAKIA
FRANCE
SWD
AUSTRIA
HUNGARY
RUMANIA
J.-SLAVIA
BULGARIA
PORTUGAL
SPAIN
ITALY
TURKEY
Gibraltar
GREECE
Malta
Cyprus
PALESTINE
EGYPT
J.F.H.

French commercial motoring maps based on a French Army map combined high quality mapping with stylish advertisements.

Selling to the Motorist

Right Extracts from *Paris-Sud: Carte Officielle du Service Géographique de l'Armée.*
See also 1941 (p. 332).

FRANCE AROUND THE turn of the nineteeth century was the cradle of the motor car. Frenchmen took it up, they developed it, and in the 1890s they were the first to turn occasional manufacture in two and threes into an industry. By the end of the 1920s one French person in twenty-seven had a car; a proportion bettered only in America, where nearly a quarter owned one.

Between the 1890s and the 1930s it was received wisdom in the advertising industry that maps sold other products advertised alongside them, but series of poster-format maps backed exclusively by colourful car-related advertising were something new in the 1920s. Copying a Belgian initiative, between 1928 and 1931 the Paris advertising agency EPOC supplied maps like the one shown to all kinds of businesses, among them oil companies and brandy distillers, which in turn gave them away as sales promotion. Most of the articles advertised on the EPOC maps were luxury goods – mostly Hotchkiss and Omega cars, Martell cognac, champagne, expensive tobacco and hotels. But whether the products were Hermès accessories, ordinary Shell petrol, or the Dunlop tyres advertised on the maps' protective slipcases, many of the advertisements were typical of commercial art in the spectacular new Art Deco style, which had been the height of fashion since the influential Exposition Internationale des Arts Décoratifs in Paris in 1925.

The maps themselves were of the highest quality. As with Britain's Ordnance Survey, the task of France's Service Géographique de l'Armée in the late nineteenth century was not to make money, but to create the best possible maps for the army's General Staff. As in Britain, however, commercial publishers used the official maps. The map shown was a reduction of the military map, but combined clarity with delicate depiction of hills, five colours, and detail down to mountain mule paths. As in Britain too, by the 1920s it had been decided that the official map must pay its way. The result was the series of regional maps described here, inspired jointly by the SGA, EPOC and the Touring-Club de France, an early champion of French motorists.

QUE DE L'ARMÉE

AT-MAJOR
IVERS ::
CIENNES

rvice Géographique
rincipaux Libraires
e. – On peut aussi
tement au Service
e de Grenelle, où
ouvert au Public.

s'adresser au Service
ercial : 136 bis, Rue de
ttré 77-50, 77-51, 77-52.
8. (Timbre pour réponse).

LA CARTE AU 200.000
FEUILLE
SUD

MELUN N° 25.
TOURS N° 32.
ORLÉANS N° 33.

FR. 50

SENTÉ AU RECTO DE CETTE
GÉOGRAPHIQUE AU PRIX DE
CS.

Harry Beck's famous diagram, first published in 1933, sacrificed geographical accuracy for clarity in a map of the complex London Underground system.

From Mental Map to Schematic Diagram

1933

HARRY BECK WAS A 29-year-old engineering draughtsman recently dismissed as a temporary employee by the Underground Railways when he conceived the idea of a rationalized Underground map which sacrificed geographical accuracy for clarity in what was a very complex system of interweaving, sometimes interconnecting, railway lines. Turning 'vermicelli into a diagram' is how he described his task. To solve the problem of the central area being densely crowded he imagined he was using a convex mirror to enlarge the more complex central section, so reducing the apparent length of the outlying lines. Submitted in 1931 to the Publicity Department it was at first rejected as being too 'revolutionary'. Resubmitted, it was accepted the following year and first printed in January 1933 as a card folder in an edition of 750,000 copies. The hand-lettering was replaced in the large-scale posters which followed by the distinctive and exclusive typeface 'Underground Railways Sans' designed by Edward Johnston in 1916. The diagram became an instant favourite with Underground railway commuters and was hailed as a design classic in its own right, reaching almost iconic status alongside images of the Tower of London, Big Ben and the reigning monarch. Beck continued to work on the diagram, searching always for simplicity and clarity in the face of continuous change in the Underground layout, until his last version of 1959. Thereafter, much to his chagrin, it was altered and kept up to date by other hands. His original arrangement with the London Passenger Transport Board by which, in return for a very modest fee, copyright was vested in the Board and Beck was to be called on to design all future amendments to the diagram, was never properly ratified. Subsequent versions of the diagram were attributed to other designers and Beck, an obsessive man posessive of his creation, remained embittered for the rest of his life.

Such was the impact of his diagram he was invited to design, in the late 1930s, a similar map for the Paris Metro. Drawings exist but were never used. Beck's basic concept, of overriding accuracy in favour of clarity, has been much copied since. Passed over in life, Beck's enduring fame has been recognized by a permanent display in a gallery named after him in London's Transport Museum.

Right Top: Map of the 'London Electric Railways' in 1906 displaying Beck's 'vermicelli'. Bottom: The card folder edition of 1933. **Above** H. C. 'Harry' Beck photographed in 1965 in front of a poster of his famous diagram and holding the original sketch made in 1931.

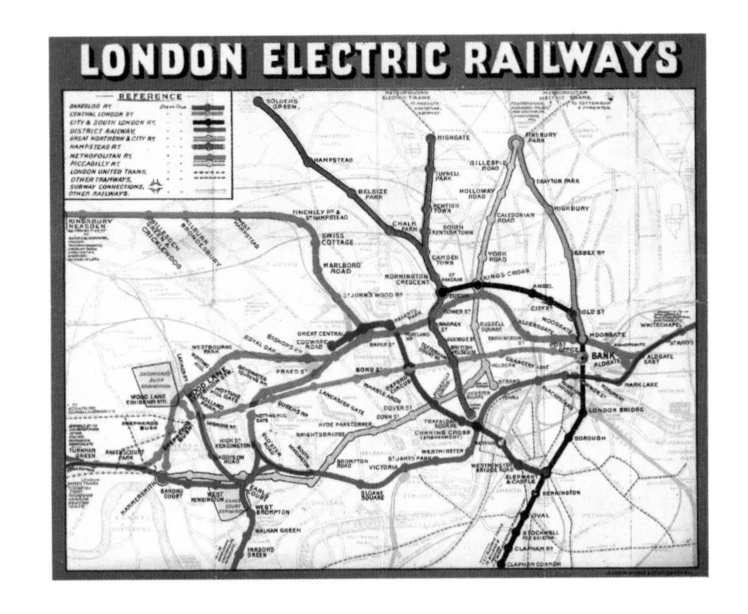
LONDON ELECTRIC RAILWAYS
REFERENCE

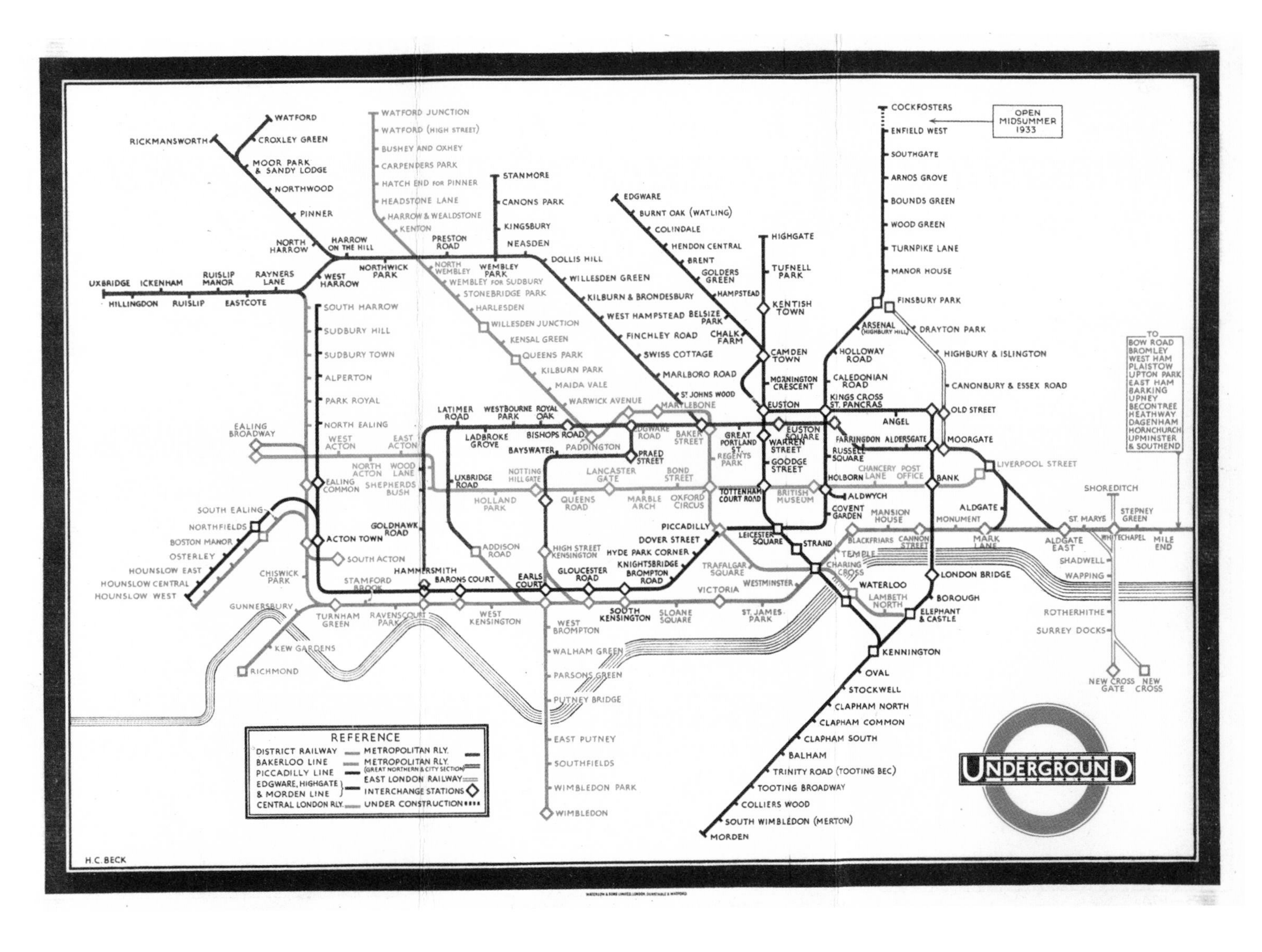
WATFORD
RICKMANSWORTH
CROXLEY GREEN
MOOR PARK & SANDY LODGE
NORTHWOOD
PINNER
NORTH HARROW
HARROW ON THE HILL
WATFORD JUNCTION
WATFORD (HIGH STREET)
BUSHEY AND OXHEY
CARPENDERS PARK
HATCH END FOR PINNER
HEADSTONE LANE
HARROW & WEALDSTONE
KENTON
STANMORE
CANONS PARK
KINGSBURY
NEASDEN
PRESTON ROAD
EDGWARE
BURNT OAK (WATLING)
COLINDALE
HENDON CENTRAL
BRENT
GOLDERS GREEN
HAMPSTEAD
BELSIZE PARK
CHALK FARM
HIGHGATE
TUFNELL PARK
KENTISH TOWN
CAMDEN TOWN
MORNINGTON CRESCENT
EUSTON
COCKFOSTERS
ENFIELD WEST
OPEN MIDSUMMER 1933
SOUTHGATE
ARNOS GROVE
BOUNDS GREEN
WOOD GREEN
TURNPIKE LANE
MANOR HOUSE
FINSBURY PARK
ARSENAL (HIGHBURY HILL)
DRAYTON PARK
HOLLOWAY ROAD
HIGHBURY & ISLINGTON
CALEDONIAN ROAD
CANONBURY & ESSEX ROAD
KINGS CROSS ST. PANCRAS
OLD STREET
ANGEL
UXBRIDGE
ICKENHAM
RUISLIP MANOR
RAYNERS LANE
HILLINGDON
RUISLIP
EASTCOTE
WEST HARROW
NORTHWICK PARK
NORTH WEMBLEY
WEMBLEY PARK
WEMBLEY FOR SUDBURY
STONEBRIDGE PARK
HARLESDEN
WILLESDEN JUNCTION
KENSAL GREEN
QUEENS PARK
KILBURN PARK
MAIDA VALE
WARWICK AVENUE
DOLLIS HILL
WILLESDEN GREEN
KILBURN & BRONDESBURY
WEST HAMPSTEAD
FINCHLEY ROAD
SWISS COTTAGE
MARLBORO ROAD
ST. JOHNS WOOD
MARYLEBONE
SOUTH HARROW
SUDBURY HILL
SUDBURY TOWN
ALPERTON
PARK ROYAL
NORTH EALING
EALING BROADWAY
WEST ACTON
EAST ACTON
NORTH ACTON
WOOD LANE
SHEPHERDS BUSH
EALING COMMON
LATIMER ROAD
WESTBOURNE PARK
ROYAL OAK
LADBROKE GROVE
BISHOPS ROAD
BAYSWATER
PADDINGTON
PRAED STREET
EDGWARE ROAD
BAKER STREET
GREAT PORTLAND ST.
WARREN STREET
EUSTON SQUARE
REGENTS PARK
GOODGE STREET
FARRINGDON
ALDERSGATE
MOORGATE
LIVERPOOL STREET
RUSSELL SQUARE
CHANCERY LANE
POST OFFICE
BANK
HOLBORN
ALDWYCH
BRITISH MUSEUM
TOTTENHAM COURT ROAD
UXBRIDGE ROAD
NOTTING HILL GATE
LANCASTER GATE
BOND STREET
MARBLE ARCH
OXFORD CIRCUS
HOLLAND PARK
QUEENS ROAD
COVENT GARDEN
MANSION HOUSE
MONUMENT
ALDGATE
SHOREDITCH
ST. MARYS
STEPNEY GREEN
ALDGATE EAST
WHITECHAPEL
MILE END
TO BOW ROAD BROMLEY WEST HAM PLAISTOW UPTON PARK EAST HAM BARKING UPNEY BECONTREE HEATHWAY DAGENHAM HORNCHURCH UPMINSTER & SOUTHEND
SOUTH EALING
NORTHFIELDS
BOSTON MANOR
OSTERLEY
HOUNSLOW EAST
HOUNSLOW CENTRAL
HOUNSLOW WEST
ACTON TOWN
GOLDHAWK ROAD
SOUTH ACTON
CHISWICK PARK
STAMFORD BROOK
HAMMERSMITH
BARONS COURT
ADDISON ROAD
EARLS COURT
HIGH STREET KENSINGTON
GLOUCESTER ROAD
PICCADILLY
DOVER STREET
HYDE PARK CORNER
KNIGHTSBRIDGE
BROMPTON ROAD
LEICESTER SQUARE
STRAND
TRAFALGAR SQUARE
CHARING CROSS
TEMPLE
BLACKFRIARS
CANNON STREET
MARK LANE
VICTORIA
WESTMINSTER
WATERLOO
LAMBETH NORTH
LONDON BRIDGE
BOROUGH
ELEPHANT & CASTLE
SHADWELL
WAPPING
ROTHERHITHE
SURREY DOCKS
NEW CROSS GATE
NEW CROSS
GUNNERSBURY
TURNHAM GREEN
RAVENSCOURT PARK
WEST KENSINGTON
WEST BROMPTON
SOUTH KENSINGTON
SLOANE SQUARE
ST. JAMES PARK
KEW GARDENS
RICHMOND
WALHAM GREEN
PARSONS GREEN
PUTNEY BRIDGE
EAST PUTNEY
SOUTHFIELDS
WIMBLEDON PARK
WIMBLEDON
KENNINGTON
OVAL
STOCKWELL
CLAPHAM NORTH
CLAPHAM COMMON
CLAPHAM SOUTH
BALHAM
TRINITY ROAD (TOOTING BEC)
TOOTING BROADWAY
COLLIERS WOOD
SOUTH WIMBLEDON (MERTON)
MORDEN
REFERENCE
DISTRICT RAILWAY
BAKERLOO LINE
PICCADILLY LINE
EDGWARE, HIGHGATE & MORDEN LINE
CENTRAL LONDON RLY.
METROPOLITAN RLY.
METROPOLITAN RLY. (GREAT NORTHERN & CITY SECTION)
EAST LONDON RAILWAY
INTERCHANGE STATIONS
UNDER CONSTRUCTION
UNDERGROUND
H.C. BECK

One people, one empire, one leader. The annexation of Austria into greater Germany in 1938 was celebrated in a map which combines both countries in a single landmass.

Propaganda Maps

1938

Right The map to celebrate the *Anschluss* of 1938 was printed on card for mass circulation.

ON THE MORNING of 12 March 1938, troops of the German Wehrmacht and the SS crossed the border into Austria to general acclaim. The next day Hitler announced in Linz, his birthplace, the legislation which would complete the annexation *(Anschluss)* of Austria into the German Reich. Unlike his subsequent forays into neighbouring Czechoslovakia, Poland and France, the annexation by Germany of a neighbouring country was welcomed by the population at large. In this instance, the political background was very different: after the Austro-Prussian War of 1866, when Chancellor Otto von Bismarck sought to bind the princely states into a reunified Germany, Austria was deliberately excluded, and following the collapse of the Austro-Hungarian Empire at the end of the First World War there was a move by the German-speaking majority in Austria to unite with the new German Republic. This was expressly forbidden by the draconian terms of the Treaty of Versailles in an attempt to forestall any future threat from an enlarged Germany. The Anschluss – literally the attachment or unification, rather than the show of force implied in 'annexation' – was therefore largely welcomed and the ratification by a popular vote under secret ballot thirty days after it was decreed was overwhelmingly approved.

To aid the resolve of the Austrian people the German government had resorted to overt propaganda in the years leading up to the *Anschluss*. The card shown here was part of a much wider attempt to present a unity of vision, a 'mythological aestheticism' which propagated the national destiny of the German people over all of Western civilization – the idea of the '1,000-year-Reich'. Propaganda took many forms including maps of racial characteristics, songs and film, exemplified by Leni Riefenstahl's 1934 film glorifying Hitler, *Triumph of the Will*, which also celebrated the notion of Das Volk, the Germanic people as one, as exemplified by this map of 'Greater Germany'. This sustained assault on an already susceptible population succeeded in convincing Austria to play its part in the historical destiny of the German 'race' rather than continue as a separate, sovereign nation.

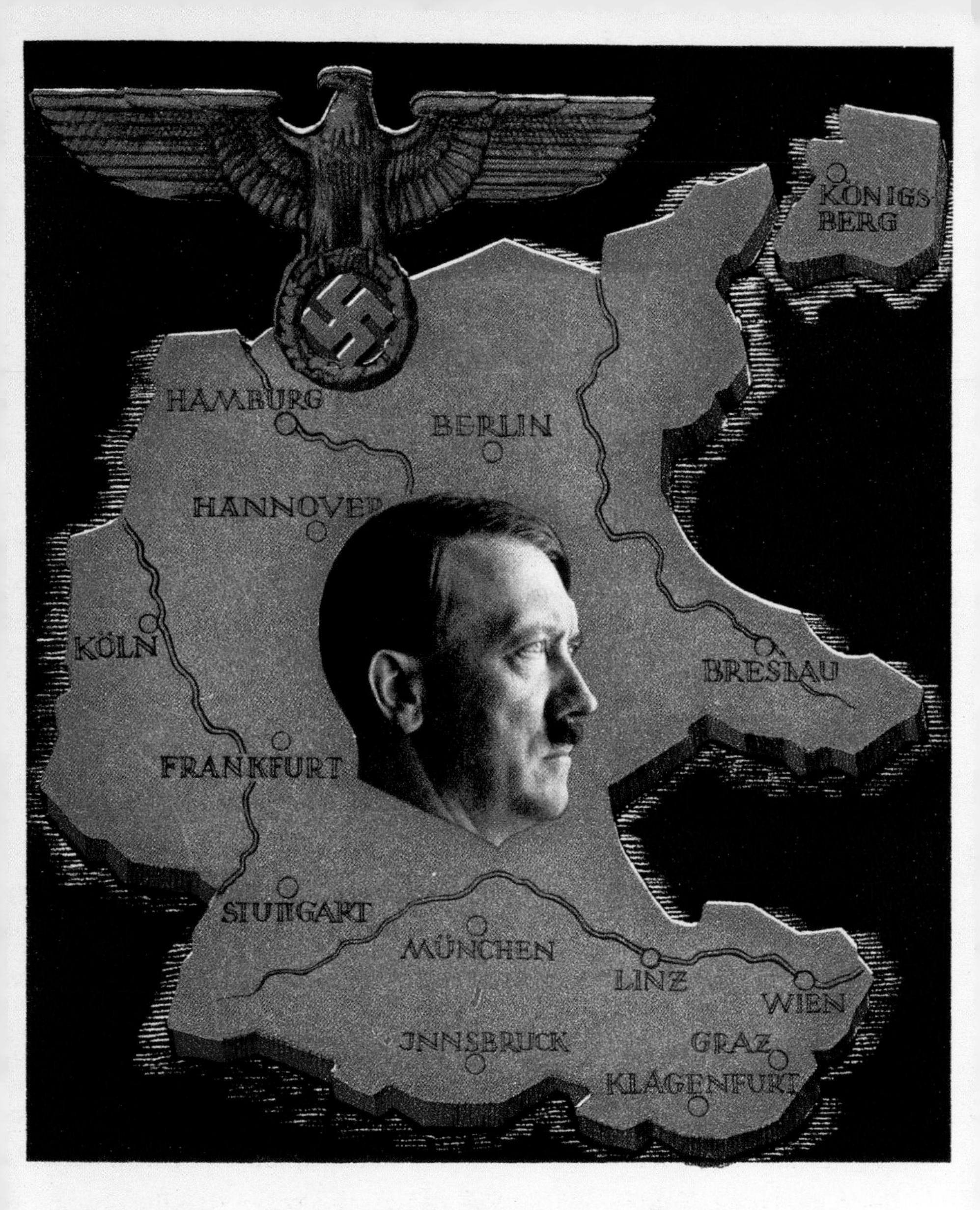

13·MÄRZ 1938
EIN VOLK EIN REICH
EIN FÜHRER

A map illustrating the detailed planning that underpinned the Nuremberg Rallies, the most important channel for Nazi propaganda.

Nazi Triumph

Right Planning map for the 1939 Nuremberg Rally: J. I. Lenke, *Aufmarsch der Politischen Leiter Reichsparteitag 1939 Zeppelinwiese*, overprinted on Leonhard Ammersdorfer, *Plan von Nürnberg-Fürth*.

EVERY SEPTEMBER IN the 1930s enormous Nazi Party rallies utilizing the latest technology took place in Nuremberg, the party's spiritual heart, to reinforce popular impressions of organized, monolithic invincibility. The parades needed detailed planning before they could be immortalized on film by producers of the calibre of Leni Riefenstahl.

The red and blue lines show how units of local party leaders were to march through the streets from the northern and southern parts respectively of the town, merging as they went, before converging on the Zeppelinwiese (or meadow named after another German nationalist hero, Ferdinand, Count Zeppelin, the inventor of the airship) and the Old Stadium in the south-east of the town. The numbers of participants and onlookers were so great that the narrow lanes of the romantic old town, with its half-timbered houses, medieval churches and antique bridges, so evocative of medieval Germany and the Meistersinger, had to be omitted from the marching routes in favour of the modern ring roads. The climax of the rally was the triumphalist parade of standards through the old stadium at the centre right of the plan, indicated by intermixed blue and red crosses. This preceded Hitler's speech. A set-piece of the Nazi year, it offered the Führer the opportunity to manipulate the German population and to menace his internal and external foes. Afterwards the contingents quietly made their way through the town to their camps, marked by dotted lines and tents.

All parts of the Reich were represented. By August 1939 these included Austria whose ancient provinces had already been reorganized as German regions or *Gaue* (e.g. Steiermark, Oberdonau, Wien) since its union with Germany, or *Anschluss*, in March 1938, and the mainly German-speaking Sudetenland of Czechoslovakia, which had been incorporated following the Munich Agreement of September 1938. Tendentiously the rally was also to include a contingent representing Danzig (Gdansk) whose cession Hitler had demanded only weeks before. This demand, however, provoked the outbreak of the Second World War in early September. As a result the 1939 Nuremberg Rally had to be cancelled at the last minute.

Plan von Nürnberg-Fürth.
Aufmarsch der Politischen Leiter
Reichsparteitag 1939 Zeppelinwiese
Koblenz-Trier
Mainfranken
Osthannover
Köln-Aachen
Württemberg-Hohenzollern
Hamburg
Essen
Kärnten
Salzburg
Sudetenland
Niederdonau
Danzig
Wien
Franken
Ostpreussen
Tirol-Vorarlberg
Steiermark
Oberdonau
Weser-Ems
Schleswig-Holstein
Saarpfalz
Ordensburgen
Bayr. Ostmark
Schwaben
Thüringen
München-Oberbayern
Kurhessen
Berlin
Westfalen-Nord
Westfalen-Süd
Schlesien
Sachsen
Pommern
Baden
Magdeburg-Anhalt
Düsseldorf
Mark Brandenburg
Halle-Merseburg
Südhannover-Braunschweig
Mecklenburg
Hessen-Nassau
ZEICHENERKLÄRUNG
Anmarschstrassen der Marschsäule „Nord"
Anmarschstrassen der Marschsäule „Süd"
Abmarsch Marschsäule Nord
Abmarsch Marschsäule Süd
Fahnen Anmarsch
Gaustandquartiere
angefertigt:
genehmigt:
Organisationsleitung der Reichsparteitage Nürnberg
FÜRTH
ZIRNDORF
ERLENSTEGEN
Märzfeld
Das Deutsche Stadion
SA-Lager
M=1:17500

This map was used to identify the states where Nazi money could most effectively be used to prevent the USA from entering the Second World War.

Recognizing One's Friends

THIS MAP LOOKS innocuous and highly academic. Yet it is marked at the top left '*Nur für den Dienstgebrauch!*' – 'Only for official use!' – in an effort to prevent it from falling into the wrong hands. And correctly so, since it is the key to one of the Third Reich's most controversial and secret political initiatives. Between September 1939 and December 1941 the German government spent large sums on a propaganda campaign aimed at trying to keep the USA out of the Second World War by mobilizing the residual feelings of loyalty of first and second generation people of German-speaking descent.

The data, based on statistics from the published US census of 1930, shows percentages of first and second generation white west and central European immigrants by state. It forms one of a series of maps showing, in Nazi fashion, the number of first and second generation white immigrants to the USA. The other maps illustrated overall white immigration as compared to total immigration and then immigration from the different parts of the world relative to total white immigration: one is devoted to northern Europe and predominantly British, Irish and Scandinavian immigration; another to eastern European and predominantly Czech, Slovak and Russian immigration; a third shows immigration from the Americas and finally from elsewhere in the world, such as Armenia. Reflecting Nazi racial theories, and possibly the course of the war, people of German, Austrian, Dutch and Belgian descent are all shown here, without distinction, as red.

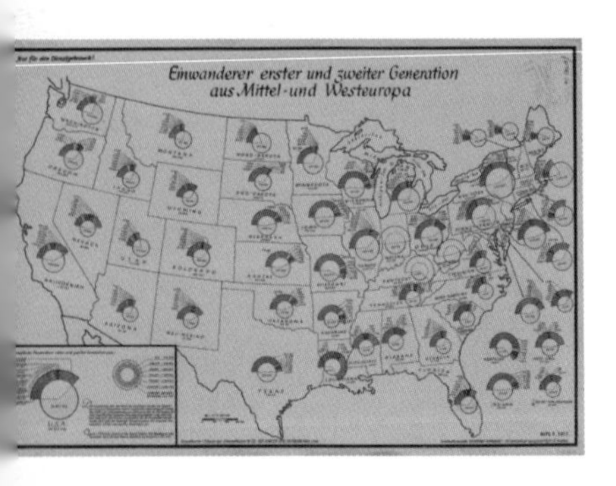

Right Detail of G. Hulbe, *Einwanderer erster und zweiter Generation aus Mittel- und Westeuropa*, Stuttgart and Hamburg. **Above** The complete map. **See also** 1590 (p. 130).

The maps enabled officials in Joseph Goebbel's Propaganda Ministry to identify significant concentrations of people with German-Austrian backgrounds at a glance and thus where isolationists and Nazi-supporters might most effectively use money from Berlin to place newspaper and cinema adverts and sponsor radio programmes. The maps suggest that German money would be best spent in Wisconsin, Michigan and other mid-western states as well as Washington, DC, and to least good effect in New England. Illinois, New York, New Jersey and possibly California, though they also had large numbers of people of Slav descent who were likely to be vociferously opposed to German propaganda, were also worth trying if funds permitted.

The campaign had a considerable impact on restraining President Roosevelt.

erster und zweiter Generation
ttel-und Westeuropa

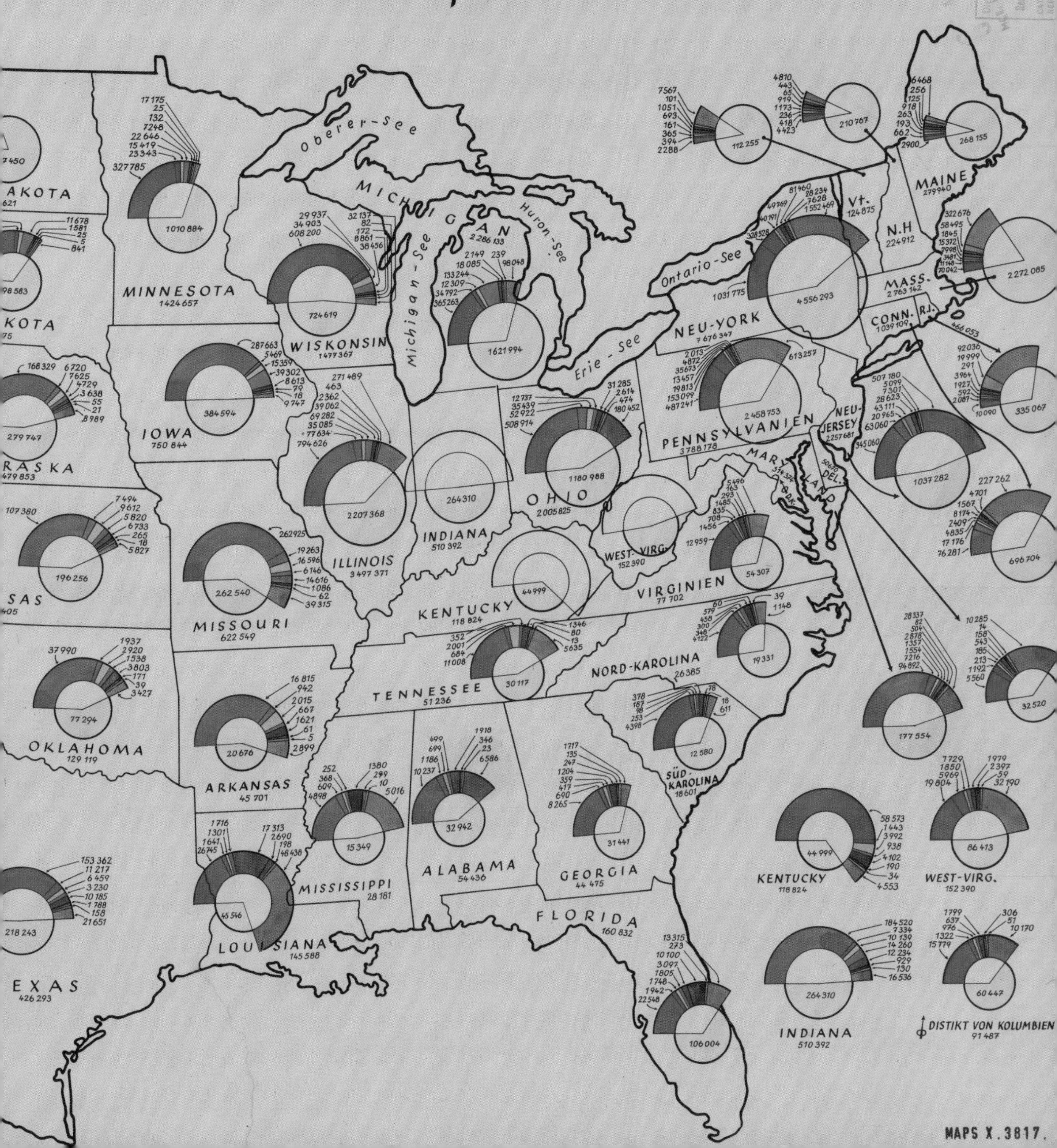

This German map, the targets coloured in a livid, luminous yellow, was specially created for use in a cockpit during a night raid on Manchester.

Bombs over Manchester

NIGHT BOMBING WAS a difficult business for Germans and the Allies alike. Getting hold of information about targets was much less of a problem than enabling the pilots and airmen to find them in conditions of darkness when any light in the cockpit could alert the enemy to your presence. The Germans gathered information about targets from large-scale pre-war mapping, air reconnaissance, espionage and such ephemeral material as picture postcards. Thick packets of such material, in brown mock-leather containers, were issued in large numbers to service personnel to accompany the up-to-date maps being published by the German military. They were intensively studied in the run-ups to the raids.

These maps were produced for use in the cockpit. They were printed on plastic-coated fabric so that they could be folded and manipulated without danger of tearing or being damaged by fluids. Only the most obvious ground features have been included (towns, railway lines, major roads, woodland) with a minimum of detail to aid observation from the air. The colouring imitates all that the pilots were likely to see of Manchester during a black-out. The targets (*Ziele*), however, are coloured in a livid yellow, intended to be luminous even in conditions of total darkness, with a red surround. The words '*Achtung. Deutsches Gefangenenlager in Oldham-Leeds*' calls the attention of Luftwaffe crews to a camp holding German prisoners of war – obviously an area to be avoided during the mission.

It is not possible to say how effective these minimalist maps were in practice. There is more than a suggestion, however, that in the closed conditions of the cockpit the fumes given off by the chemicals involved did long-term harm to the aircrews.

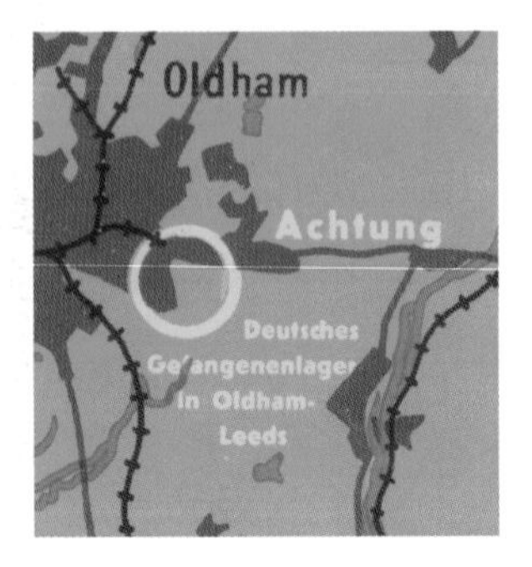

Right German bombing map for Manchester. **Above** Detail highlighting the Oldham prisoner-of-war camp.

Manchester

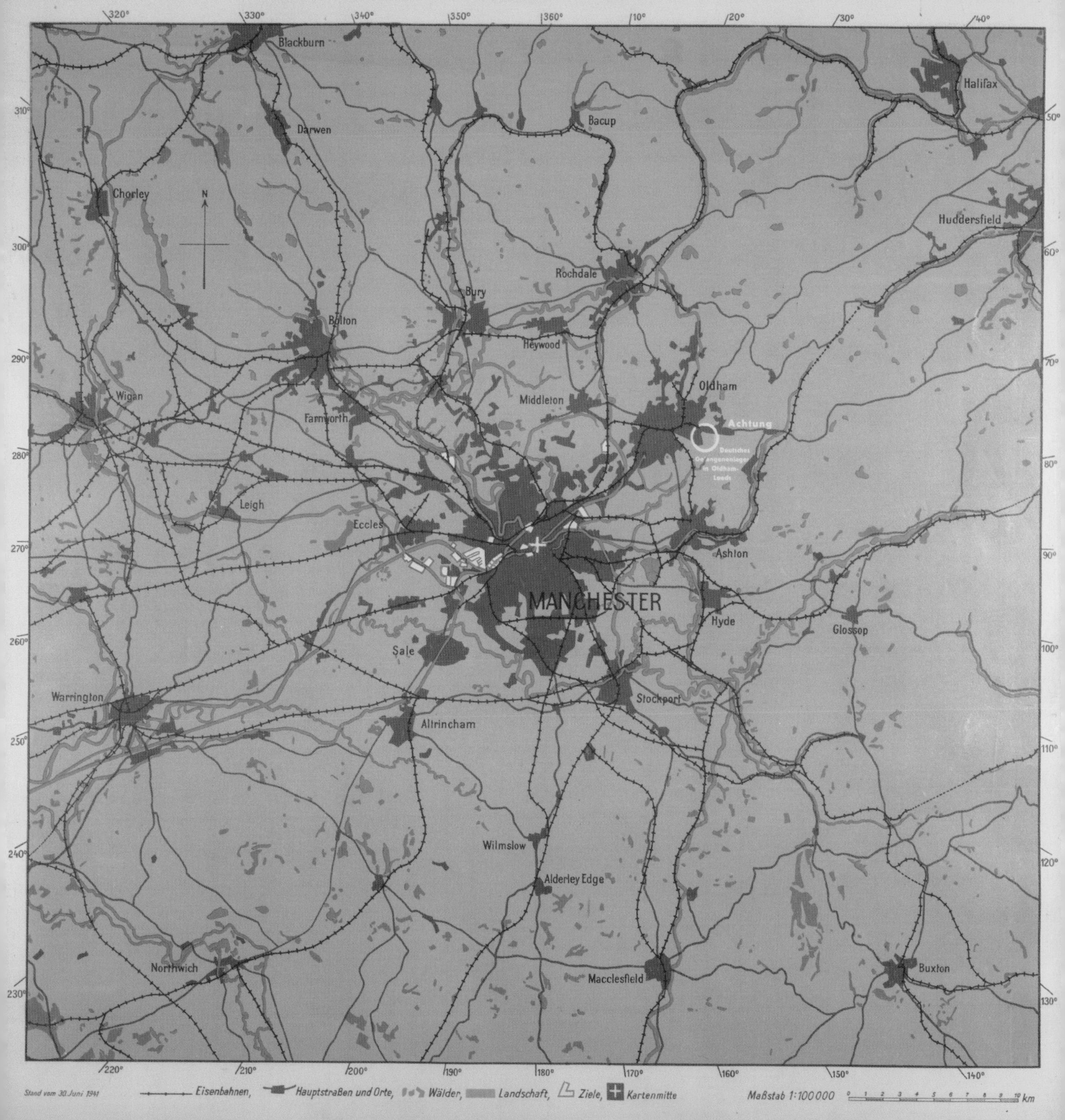
Blackburn
Halifax
Bacup
Darwen
Chorley
N
Huddersfield
Rochdale
Bury
Bolton
Heywood
Oldham
Wigan
Middleton
Farnworth
Achtung
Deutsches Gefangenenlager in Oldham-Leeds
Leigh
Eccles
Ashton
MANCHESTER
Hyde
Glossop
Sale
Warrington
Stockport
Altrincham
Wilmslow
Alderley Edge
Northwich
Macclesfield
Buxton
Stand vom 30. Juni 1941
Eisenbahnen,
Hauptstraßen und Orte,
Wälder,
Landschaft,
Ziele,
Kartenmitte
Maßstab 1:100 000
km

During the Second World War fabric escape maps were manufactured under conditions of great secrecy for the British government by Waddington Ltd.

The Great Escape

1941

THIS SILK MAP was designed as an escape map during the Second World War. It was one of a range of aids provided to Allied air crews to help them find their way to safety if they were shot down behind enemy lines. The project was the result of an unusual map creation venture during the early 1940s: Waddington Ltd., the British printing company best known for its games including Monopoly, printed escape maps on silk, rayon and tissue paper for military use. During the Second World War hundreds of thousands of these maps were produced by the British and later by the Americans. The idea was that a serviceman captured or shot down behind enemy lines should have a map to help him find his way to safety if he escaped or, better still, to evade capture in the first place.

When you see one of these maps, the unusual material is probably the first thing you notice. A map like this could be concealed in a small place (a cigarette packet, the hollow heel of a flying boot), did not rustle suspiciously if the captive was searched and, in the case of maps on cloth or mulberry leaf paper, could survive immersion in water and other wear and tear. Other escape aids such as tiny compasses and emergency food supplies were provided as well.

The idea appears to have been the brainchild of Christopher Clayton-Hutton, a creative and eccentric officer at the Ministry for Escape and Evasion (MI9). Waddington became involved because of their experience of printing theatre programmes on silk. A surviving archive of letters between Clayton-Hutton and staff at Waddington suggests a cloak and dagger atmosphere. Maps are referred to cryptically by various codes; parcels of maps were delivered to the left luggage office at Kings Cross station to be called for, rather than addressed to MI9. Supplying the maps directly to servicemen was only a start. Clayton-Hutton devised cunning methods of sending the maps into POW camps for those who had already been captured, by concealing them inside Waddington's more conventional products, such as board games and playing cards, and sending them to the camps in parcels from fictitious charities. The scheme was eventually expanded to include other items such as metal files, currency, compasses and even fake German Army uniforms. Escape maps on cloth continue to be produced for military use.

Right A letter from Captain Clayton-Hutton, 26 March 1941, with a folded silk escape map of Italy.

SECRET

c/o Room 321,
WAR OFFICE,
London, S.W.1.

26th March, 1941

Dear Mr. Watson,

Reference our conversation today. I am sending you, under separate cover, as many maps as I have in stock of the following:-

Norway and Sweden
Germany
Italy

I shall be glad if you will make me up games on the lines discussed today containing the maps as follows:-

<u>One game</u> must contain Norway, Sweden and Germany.
<u>One game</u> must contain N. France, Germany and frontiers.
<u>One game</u> must contain Italy.

I am also sending you a packet of small metal ...truments. I should be glad if in each game you could ...ge to secrete one of these.

... I want as varied an assortment containing these ... as possible. You had then better send to me ...ames on the straight.

...ose that are faked, you must give me some ... clue and also state what they contain.

...ove I also include, of course, packs of ... photographs and any other ideas you

...u in the next post some packs of cards ...cy boxes, etc. and insert in bridge

/I

From the golden era of American motoring come these strip maps combining maps of routes with views of tourist sites.

American Oil-Company Mapping

ADVERTISING HAS BECOME a major motivation for the publication and distribution of maps in the modern world, and nowhere more so than in the automobile-oriented United States. Private organizations, industrial concerns and state agencies gave away billions of promotional maps to American motorists over the course of the twentieth century, all in the interest of promoting automotive travel.

Roughly 5 billion maps alone were handed out to customers at gasoline service stations from the 1920s to the 1960s. Surviving copies are now treasured by Americans nostalgic for the golden era of American motoring, before the construction of the present superhighways. Most have beautifully illustrated covers (see left), but the design of the specially designed maps themselves was simple, which kept the product inexpensive for advertisers. Little attempt was made to represent topographic features. Even on maps of the mountainous West, roads appear to race across flat and featureless plains.

This example was published by the H. M. Gousha Company in 1941 for the Standard Oil Company of California, which had a market area encompassing most of the western United States, home of many of the country's most popular tourist attractions. Though the overall look of the map is typical of oil-company maps, it has several unusual features. Unlike most of its contemporaries, it was issued as a small atlas in four sheets. These comprise a strip map depicting the routes of the four major federal trunk highways in a corridor stretching from southern California to western Arkansas, including the famous Route 66. In order to fit these routes together on sheets that fall within the usual vertical dimensions of an American road map (about 9 inches/23 centimetres), the north–south scale has been compressed relative to the east–west scale. Small views of major natural features and monuments keyed to numbered locations on the map alert tourists along these popular routes to places of particular interest.

Though the United States was not yet at war, in hindsight it seems bittersweet and naïve that a map openly promoting pleasure travel and gasoline consumption should be published in 1941. Indeed, within a year of the publication of this map, gasoline rationing and official disapproval of extended leisure travel would be in place. Most oil companies suspended the issue of free road maps until peace ushered in the age of Disneyland, Holiday Inns and McDonald's.

Right An extract from one of the maps in the *Standard Oil Interstate Route Map, Southern Routes, US 60, 66, 70, 80.* **Above** Cover of the maps. **See also** 300, 1675, 1790, 1930.

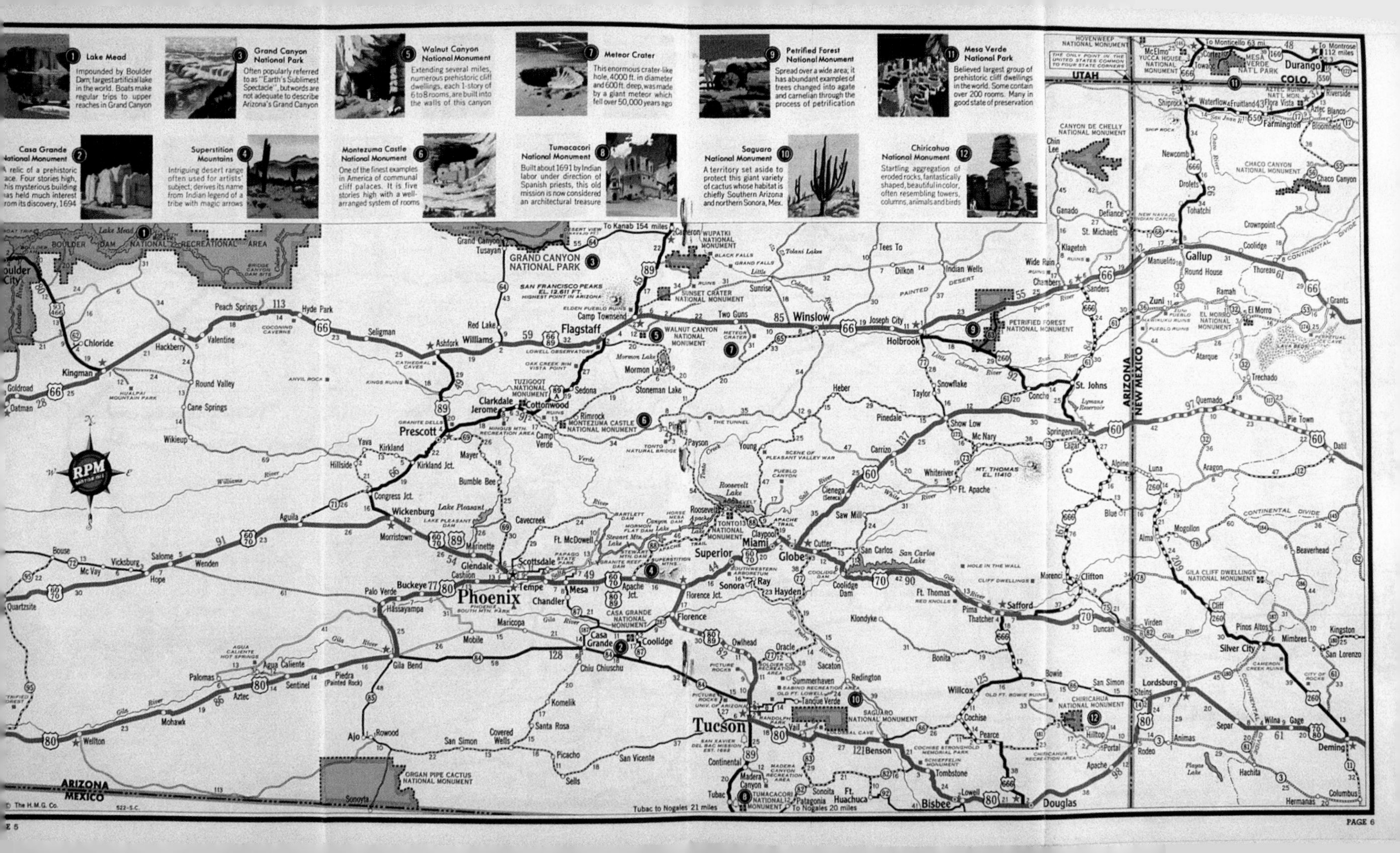

1 Lake Mead
Impounded by Boulder Dam; largest artificial lake in the world. Boats make regular trips to upper reaches in Grand Canyon
3 Grand Canyon National Park
Often popularly referred to as "Earth's Sublimest Spectacle", but words are not adequate to describe Arizona's Grand Canyon
5 Walnut Canyon National Monument
Extending several miles, numerous prehistoric cliff dwellings, each 1-story of 6 to 8 rooms, are built into the walls of this canyon
7 Meteor Crater
This enormous crater-like hole, 4000 ft. in diameter and 600 ft. deep, was made by a giant meteor which fell over 50,000 years ago
9 Petrified Forest National Monument
Spread over a wide area; it has abundant examples of trees changed into agate and carnelian through the process of petrification
11 Mesa Verde National Park
Believed largest group of prehistoric cliff dwellings in the world. Some contain over 200 rooms. Many in good state of preservation
2 Casa Grande National Monument
A relic of a prehistoric race. Four stories high, this mysterious building has held much interest from its discovery, 1694
4 Superstition Mountains
Intriguing desert range often used for artists' subject; derives its name from Indian legend of a tribe with magic arrows
6 Montezuma Castle National Monument
One of the finest examples in America of communal cliff palaces. It is five stories high with a well-arranged system of rooms
8 Tumacacori National Monument
Built about 1691 by Indian labor under direction of Spanish priests, this old mission is now considered an architectural treasure
10 Saguaro National Monument
A territory set aside to protect this giant variety of cactus whose habitat is chiefly Southern Arizona and northern Sonora, Mex.
12 Chiricahua National Monument
Startling aggregation of eroded rocks, fantastically shaped, beautiful in color, often resembling towers, columns, animals and birds
HOVENWEEP NATIONAL MONUMENT
THE ONLY POINT IN THE UNITED STATES COMMON TO FOUR STATE CORNERS
UTAH
COLO.
ARIZONA
NEW MEXICO
ARIZONA
MEXICO
To Monticello 63 mi.
To Montrose 112 miles
To Kanab 154 miles
McElmo
Yucca House National Monument
Cortez
Towaoc
Mesa Verde Nat'l Park
Durango
Aztec Ruins Nat'l Mon.
Shiprock
Waterflow
Fruitland
Flora Vista
Aztec
Farmington
Bloomfield
Riverside
Blanco
Canyon de Chelly National Monument
Chin Lee
Newcomb
Chaco Canyon National Monument
Chaco Canyon
Drolets
Tohatchi
Crownpoint
Coolidge
Continental Divide
Gallup
Manuelito
Round House
Thoreau
Grants
Zuni
El Morro National Monument
El Morro
Ramah
Atarque
Trechado
Quemado
Pie Town
Datil
Aragon
Luna
Alpine
Blue
Alma
Mogollon
Beaverhead
Gila Cliff Dwellings National Monument
Cliff
Pinos Altos
Silver City
Mimbres
Kingston
San Lorenzo
Virden
Lordsburg
Steins
Separ
Wilna
Gage
Deming
Animas
Hachita
Playas Lake
Rodeo
Columbus
Hermanas
Ganado
Ft. Defiance
St. Michaels
Klagetoh
Wide Ruin
Chambers
Sanders
Indian Wells
Painted Desert
Petrified Forest National Monument
Holbrook
Joseph City
Winslow
Two Guns
Sunrise
Tees To
Dilkon
Tolani Lakes
Wupatki National Monument
Black Falls
Grand Falls
Sunset Crater National Monument
Cameron
Grand Canyon
Tusayan
Grand Canyon National Park
San Francisco Peaks El. 12,611 ft. Highest point in Arizona
Elden Pueblo Ruins
Camp Townsend
Flagstaff
Walnut Canyon National Monument
Meteor Crater
Lowell Observatory
Mormon Lake
Stoneman Lake
Red Lake
Williams
Ashfork
Seligman
Peach Springs
Hyde Park
Coconino Caverns
Hackberry
Valentine
Boulder City
Boulder Dam National Recreational Area
Chloride
Kingman
Goldroad
Oatman
Hualpai Mountain Park
Round Valley
Cane Springs
Wikieup
Cathedral Caves
Kings Ruins
Anvil Rock
Tuzigoot National Monument
Clarkdale
Jerome
Cottonwood
Sedona
Oak Creek Rim Vista Point
Rimrock
Montezuma Castle National Monument
Camp Verde
Prescott
Granite Dells
Yava
Kirkland
Kirkland Jct.
Hillside
Mayer
Bumble Bee
Congress Jct.
Wickenburg
Lake Pleasant
Morristown
Aguila
Salome
Wenden
Hope
Vicksburg
Bouse
Mc Vay
Quartzsite
Pine
Payson
The Tunnel
Tonto Natural Bridge
Young
Scene of Pleasant Valley War
Heber
Pinedale
Taylor
Snowflake
Concho
St. Johns
Lyman Reservoir
Springerville
Eagar
Show Low
Mc Nary
Whiteriver
Ft. Apache
Mt. Thomas El. 11410
Carrizo
Cienega
Saw Mill
Roosevelt Lake
Roosevelt
Tonto National Monument
Apache Trail
Claypool
Miami
Globe
Cutter
San Carlos
San Carlos Lake
Hole in the Wall
Cliff Dwellings
Morenci
Clifton
Ft. Thomas
Pima
Thatcher
Safford
Duncan
Klondyke
Bonita
Bowie
San Simon
Willcox
Chiricahua National Monument
Hilltop
Portal
Apache
Cochise
Pearce
Dos Cabezas
Douglas
Bisbee
Lowell
Tombstone
Benson
Vail
Tanque Verde
Saguaro National Monument
Colossal Cave
Redington
Summerhaven
Sacaton
Oracle
Hayden
Ray
Sonora
Superior
Florence Jct.
Florence
Coolidge
Coolidge Dam
Apache Jct.
Superstition
Mesa
Chandler
Tempe
Scottsdale
Cavecreek
Ft. McDowell
Bartlett Dam
Horse Mesa Dam
Mormon Flat Dam
Stewart Mtn. Lake
Granite Reef Dam
Papago State Park
Marinette
Glendale
Cashion
Phoenix
Phoenix South Mtn. Park
Buckeye
Palo Verde
Hassayampa
Maricopa
Mobile
Casa Grande
Casa Grande National Monument
Chiu Chiuschu
Gila Bend
Agua Caliente
Agua Caliente Hot Springs
Palomas
Sentinel
Piedra (Painted Rock)
Aztec
Mohawk
Wellton
Ajo
Rowood
San Simon
Covered Wells
Komelik
Santa Rosa
Picacho
Sells
San Vicente
Organ Pipe Cactus National Monument
Sonoyta
Tucson
Univ. of Arizona
Picture Rocks
Owlhead
San Xavier del Bac Mission Est. 1692
Continental
Madera Canyon
Madera Canyon Recreation Area
Tubac
Tumacacori National Monument
Patagonia
Sonoita
Ft. Huachuca
Tubac to Nogales 21 miles
To Nogales 20 miles
Cochise Stronghold Memorial Park
Schieffelin Monument
Chiricahua Recreation Area
Old Ft. Bowie Ruins
Cameron Creek Ruins
City of Rocks
RPM Motor Oil
© The H.M.G. Co.
522-S.C.

A map which enabled planners to see at a glance the variety of ways in which the British countryside was used.

Maps for Planning

BRITISH REGIONAL LANDSCAPES were never so clearly shown on a map until Lawrence Dudley Stamp published the ten-mile (16-kilometre) to one-inch (2.5-centimetre) map of land utilization, generalizing the detailed pattern. It was intended to summarize the results of a survey of the whole of Britain carried out by schoolchildren.

Survival in wartime required agriculture to be supported to the hilt, but its maintenance in post-war conditions depended on a knowledge of how land was used. This map provided an essential tool for land planning in post-war Britain. Previously planning had focused upon towns and cities to the exclusion of agriculture. Without this map it would have been impossible for the Scott Report in 1942 to have undertaken its description of the countryside as it was. It became one of the building blocks of a series of maps for national planning.

Surveyed at 6 inches (15 centimetres) to one mile, land use was generalized onto 235 one-inch maps covering Great Britain. In creating the map of land utilization at one-tenth of the scale of the one-inch maps Stamp visualized its patterns as a key to reveal the way in which the British landscape was created. Without such a simplified map, one would almost literally not see the wood for the trees.

To be visible, the smallest single dab of colour represented 155 acres. We see combinations of dark green, brown, light green, yellow, red and purple, rather like an Impressionist painting. Cartographers combined blocks of the same kind retaining the proportion of six different land uses: woodland; arable; grassland; rough pasture; urban areas; orchards. Great Britain is covered by two sheets and makes a fine wall map and teaching aid.

Eastern England is seen as dominantly arable while the higher ground of the Pennines, Welsh mountains and hills of the south-west are largely rough pasture with improved pasture in sheltered valleys. Apart from the major urban centres of Lancashire, West Midlands and Greater London, the countryside presents subtle variations of grassland and arable ranging from the cultivated downlands of Wiltshire to the pastoral dairy lands of Midland clay vales. Woodland is mostly scattered but is noticeably significant in the Weald south of London. But the newly established forest on former heathland in Thetford Chase shows up as do the ancient Forest of Dean and the New Forest.

Right Ordnance Survey *Great Britain: Land Utilization*, Sheet 2.
See also 1800 (p. 246).

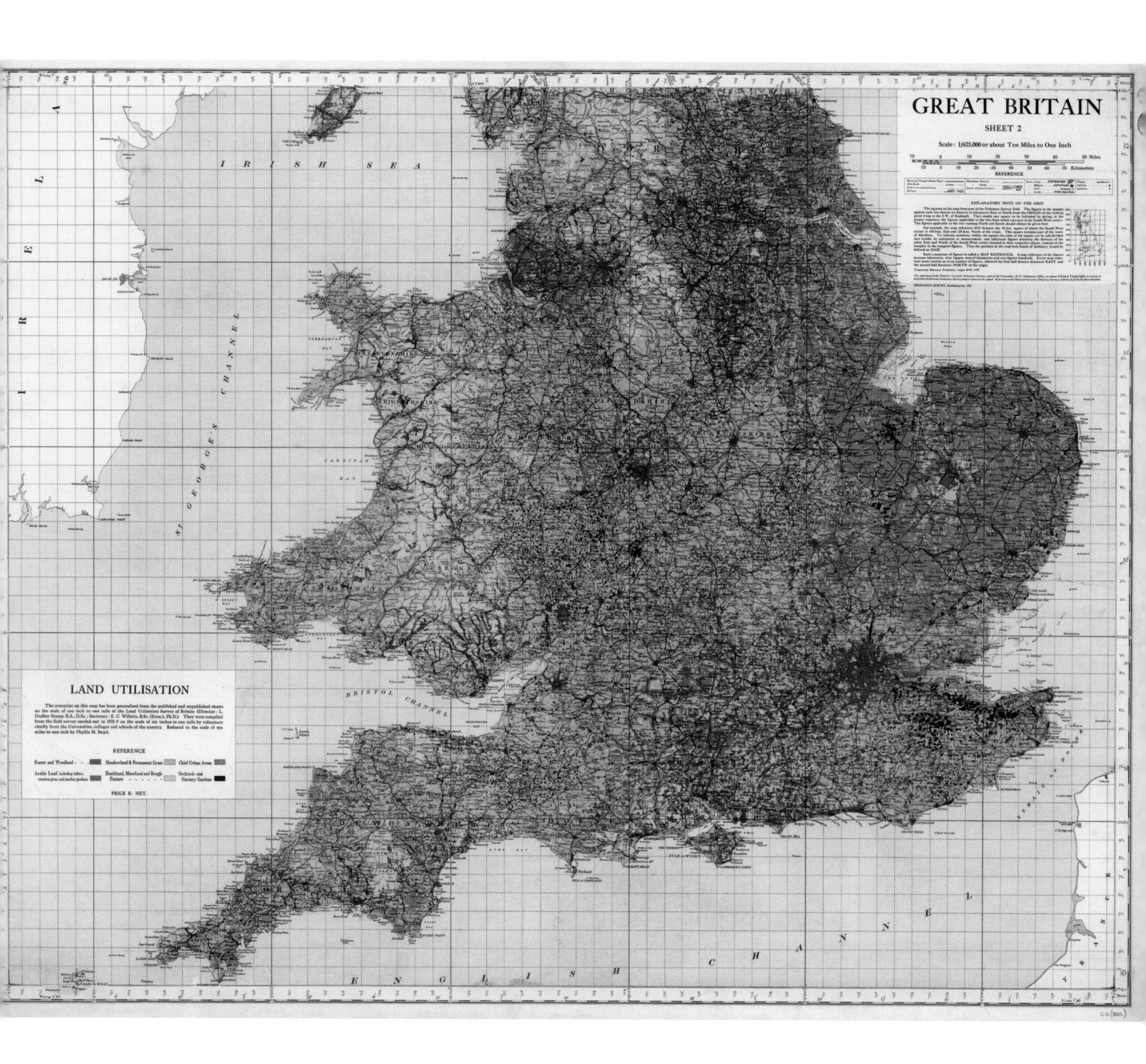
GREAT BRITAIN
SHEET 2
Scale: 1/625,000 or about Ten Miles to One Inch
IRISH SEA
ST GEORGE'S CHANNEL
BRISTOL CHANNEL
ENGLISH CHANNEL
LAND UTILISATION
The overprint on this map has been generalised from the published and unpublished sheets on the scale of one inch to one mile of the Land Utilisation Survey of Britain (Director: L. Dudley Stamp, B.A., D.Sc.; Secretary: E. C. Willatts, B.Sc. (Econ.), Ph.D.) They were compiled from the field survey carried out in 1931-9 on the scale of six inches to one mile by volunteers chiefly from the Universities, colleges and schools of the country. Reduced to the scale of ten miles to one inch by Phyllis M. Boyd.
REFERENCE
Forest and Woodland
Meadowland & Permanent Grass
Chief Urban Areas
Arable Land including fallow, rotation grass and market gardens
Heathland, Moorland and Rough Pasture
Orchards and Nursery Gardens
PRICE 5/- NET.

This map shows a village requisitioned by the British Army in 1943 that has remained closed to the public ever since.

'The Village that Died for England'

DURING THE SECOND World War, some 14.5 million acres of land (as well as vast quantities of industrial, storage and other premises) were requisitioned by the State. Requisitioned lands and buildings were put to an enormous range of uses, including airfields, army accommodation, underground shelters, factories, schools and hospitals. Land at Tyneham, Dorset, was taken over by the War Department in 1943 as a battle training site in preparation for D-Day. Tyneham House, long owned by the Bond family, was occupied and the Tyneham villagers were evacuated. After the war, the land was used for gunnery training and wartime requisition became permanent military occupation. Tyneham has never returned to civilian ownership, and the villagers' expectations of returning to their homes were never realized.

In 1943, a report on the condition of timber on the Tyneham estate was prepared for the War Department. The report was illustrated by a dyeline copy of part of a 1:10,560 Ordnance Survey sheet, coloured in ochre to indicate the areas of woodland mentioned. The black ink numbers added to the map are Ordnance Survey plot numbers taken from the 1:2500 map of Dorset. The report was also accompanied by a tracing of the 1:2500 map showing Tyneham House and gardens; the lettered areas on the tracing relate to that part of the report which describes the ornamental trees.

Reproductions of parts of Ordnance Survey maps were widely used in government departments and other organizations to provide base maps to which other information could be added. There are innumerable examples of such maps in the National Archives and other record offices.

The coloured boundary lines on the map are not contemporary with the report or with the ochre colouring and black ink annotations. They are crudely drawn in felt-tip pen, an implement which was not available in the 1940s, and no key appears either on the map or on any papers on the file. It is reasonable to assume that these lines were added to the map when the question of permanently closing public rights of way over the Lulworth Ranges was considered in the 1960s. War Office, Ministry of Defence, Treasury and Ministry of Transport records in the National Archives all contain bulky files about the ensuing debate, but none of the numerous maps on these files depict the same lines as this does. One is left to speculate who drew them and why.

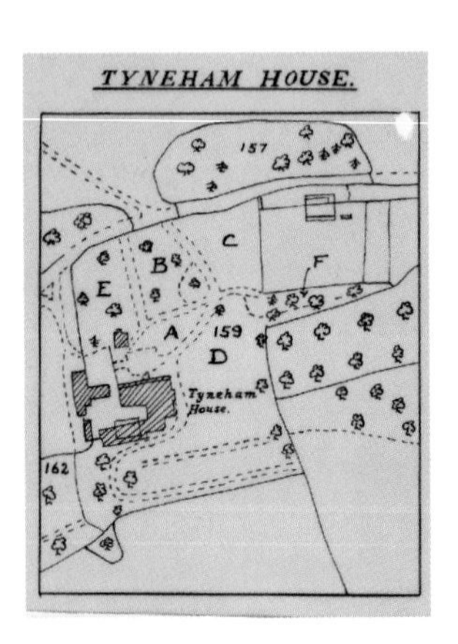

Right Ordnance Survey map showing Tyneham annotated in 1943 and again in about 1960. **Above** Tyneham House and its surroundings.

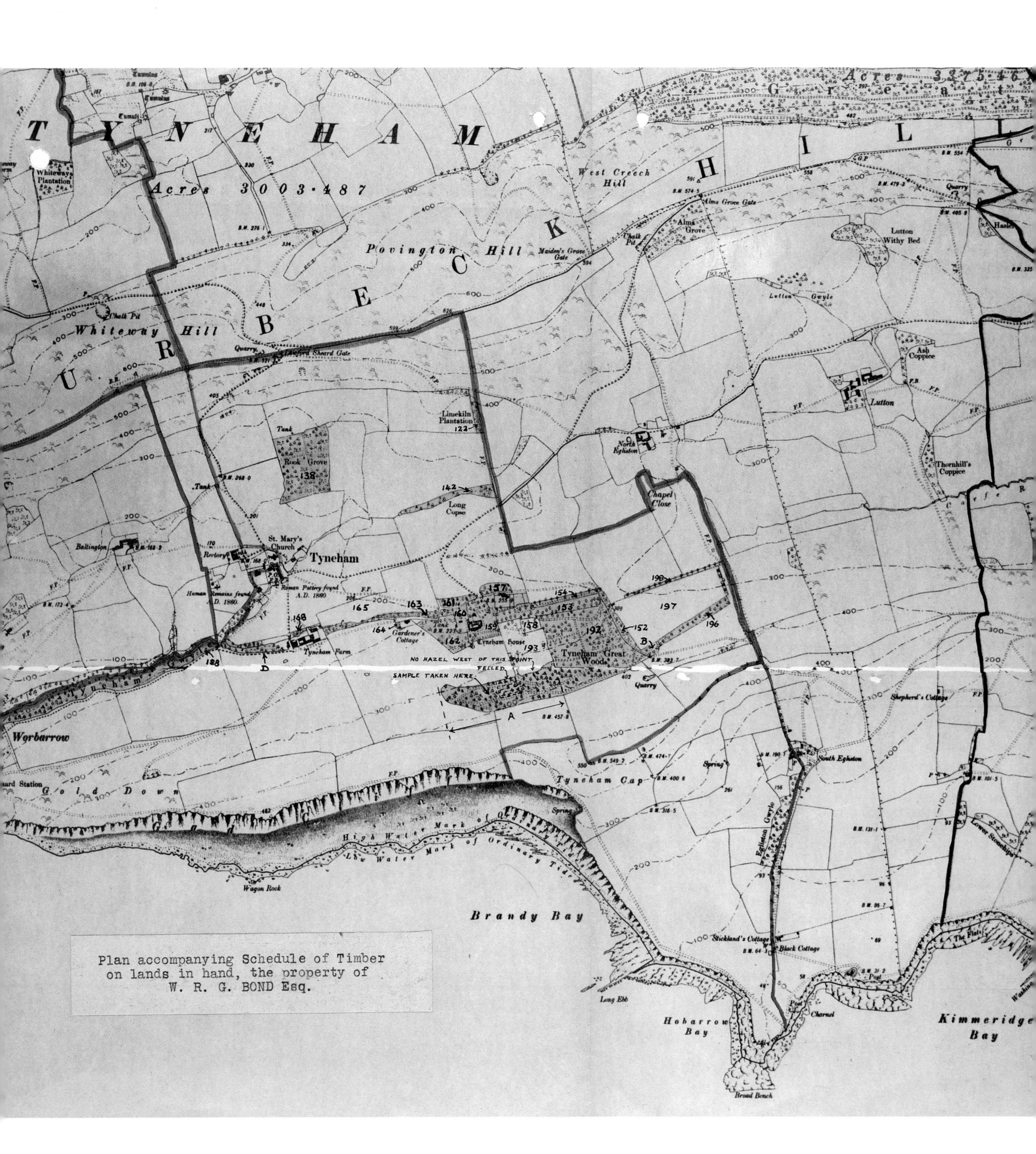

Plan accompanying Schedule of Timber
on lands in hand, the property of
W. R. G. BOND Esq.
TYNEHAM
Acres 3003·487
PURBECK HILLS
Acres
Whiteway Plantation
Whiteway Hill
Povington Hill
West Creech Hill
Alms Grove Gate
Alms Grove
Maiden's Grave Gate
Lutton Withy Bed
Lutton Gwyle
Ash Coppice
Lutton
Limekiln Plantation
Rook Grove
Long Copse
North Egliston
Thornhill's Coppice
Chapel Close
Baltington
St. Mary's Church
Rectory
Tyneham
Human Remains found A.D. 1860.
Roman Pottery found A.D. 1880
Gardener's Cottage
Tyneham House
Tyneham Farm
Tyneham Great Woods
NO HAZEL WEST OF THIS POINT
FELLED
SAMPLE TAKEN HERE
Quarry
Shepherd's Cottage
Worbarrow
Gold Down
Tyneham Cap
South Egliston
Egliston Gwyle
Spring
High Water Mark of Ordinary Tides
Low Water Mark of Ordinary Tides
Wagon Rock
Brandy Bay
Stickland's Cottage
Black Cottage
Long Ebb
Hobarrow Bay
Charnel
Broad Bench
Kimmeridge Bay
The Flats
Lower Stonehips

Allied mapmakers were unaware of the true nature of Nazi concentration camps when they created maps to help deal with the flood of displaced persons at the end of the war.

Consciousness and Cartography

Right US Office of Strategic Services, *Billeting and Control Facilities for Displaced Persons: Hannover and Bremen*, 26 February 1945.

IN THE CLOSING months of the Second World War refugees streamed into Germany to escape the advancing Allies, particularly ethnic Germans fleeing ahead of the Soviet armies, while the wretched, mainly Jewish, inmates of the concentration and extermination camps in the East were force-marched into areas still under German occupation. Amidst the chaos, hundreds of thousands died of starvation or disease.

Refugees are a feature of most wars and it was only to be expected that the Allies would make plans to deal with them. This American map attempts to give some idea of how many beds would be available for the displaced persons in the areas that they expected to be liberating in the coming months. The main focus is the cities, indicated by red circles. The small empty circles show towns where, possibly as a result of Allied bombing, fewer than 7,000 beds were likely to be available. The largest red-filled circles are for towns like Hamburg, Bremen, Braunschweig and Hanover where it was reckoned that more than 50,000 beds would be available.

Much less prominent, because they are marked in black, are the additional beds that the Allies expected would be available in the various camps. Probably assuming that the German concentration camps were no worse than those opened by the British for the South African Boers during the last two years of the Boer War, the Allied mapmakers used similar-looking black triangle symbols for prisoner-of-war camps (black triangle), for concentration camps (black triangles with a white strip) and for foreign workers' hostels (blank triangles). In the middle of April 1945, British forces, accompanied by journalists, entered the concentration camp of Bergen-Belsen (at the centre of the map) and discovered, with horror, just how different the three sorts of camp were. All thought of using the concentration camps for the displaced persons was abandoned as the plight of their pitiful former inmates became primary focuses of attention – on a new category of map.

PROVISIONAL EDITION

NO. L4203 - SECRET

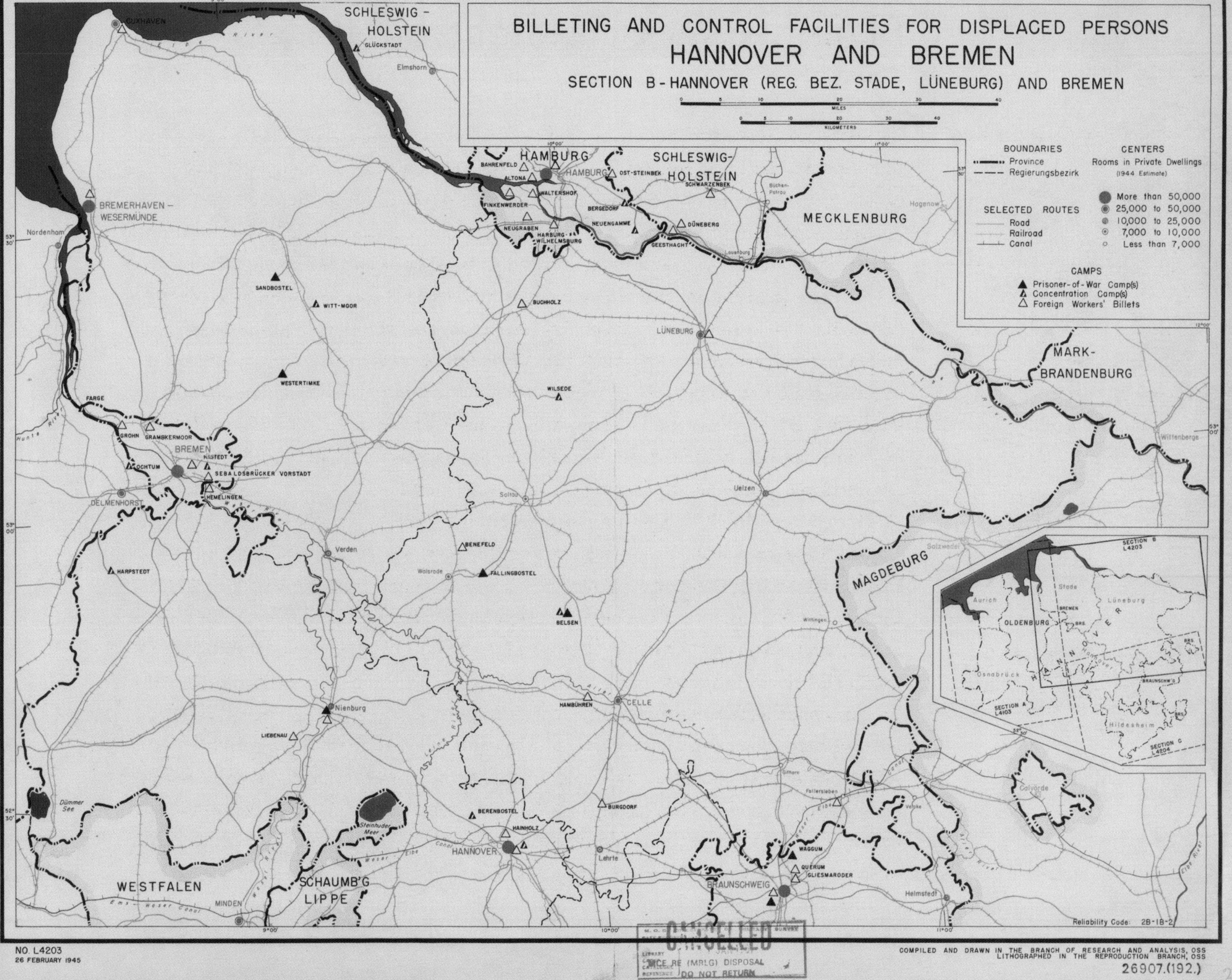

NO. L4203
26 FEBRUARY 1945

COMPILED AND DRAWN IN THE BRANCH OF RESEARCH AND ANALYSIS, OSS
LITHOGRAPHED IN THE REPRODUCTION BRANCH, OSS

26907.(192.)

The first Oblique Mercator Projection map was created to help pilots to fly straight from Chicago to Gander.

Flying Along a Straight Line

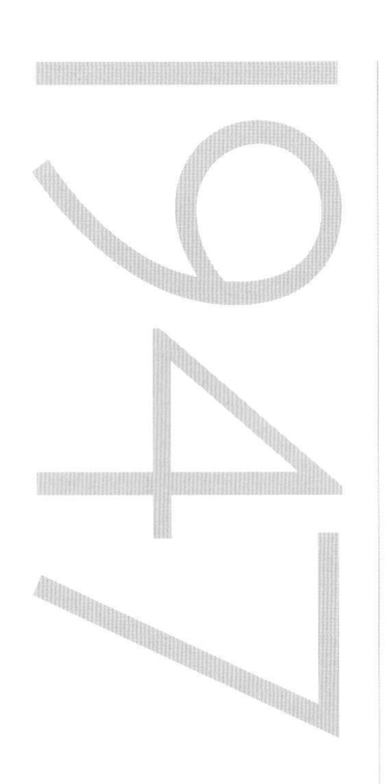

Right Detail of US Coast and Geodetic Survey, *Chicago-to-Gander Aeronautical Route Chart.* **Following pages** The complete map.

'MAPS ARE MADE to be looked at; charts are made to be worked on.' The Chicago-to-Gander route chart was the first international aviation chart designed by the US Coast and Geodetic Survey after the Second World War. The Coast and Geodetic Survey, the oldest scientific agency in the US government, had acquired responsibility for civilian aviation maps in 1926. Since the Survey had published nautical charts for sailors for a century, they devised aero-nautical charts for pilots, and the name subsequently became used universally. The original aeronautical charts were strip maps, oriented to the direction of the flight line between airports, which presented only those narrow swaths of land and water adjacent to the pilot's flight line. Colours and features evolved to better represent the terrain as it was seen from the air during flight, resulting in simplified but evocative maps in which 'the ground speaks for itself', as Survey cartographers described it.

After the war, the Survey turned to work on charts adapted to the needs of pilots in planes that flew faster and higher – but were still incapable of flying between the great cities of the eastern US and western Europe in one hop. The airport at Gander, in Newfoundland, was the critical midway point for trans-Atlantic travel. Accordingly, the Survey created a chart for planning flights from the population centres of eastern Canada and the United States to Gander. The chart was the first major map developed on the Oblique Mercator Projection published in the western hemisphere in the twentieth century. The projection is based on a cylinder tangent to the earth along the line from Chicago, Illinois (in the upper left-hand corner) to Gander, in the lower right-hand corner. Distances and angles are true only along the tangent line, but the distortions for adjacent areas are small enough that the chart could readily be used for plotting courses along the entire Atlantic seaboard. Every complex curve of the parallels of latitude was scribed by hand using the Survey's great projection ruling machine.

It was in 1970 that the Survey became part of the National Oceanic and Atmospheric Administration (NOAA), and in 2000 the aeronautical charting division was transferred to the Federal Aviation Administration (FAA). Modern aeronautical charts, like the Chicago-to-Gander chart, are still printed in the basement of the Commerce Department headquarters, in a press room in continuous operation since 1932.

SAINT LAWRENCE RIVER
GULF OF SAINT LAWRENCE
GASPE PENINSULA
NOTRE DAME MOUNTAINS
CHALEUR BAY
ANTICOSTI ISLAND
MINGAN PASSAGE
MINGAN ISLANDS
PRINCE EDWARD ISLAND
NORTHUMBERLAND STRAIT
CAPE BRETON ISLAND
CABOT STRAIT
BAY OF FUNDY
ST GEORGES BAY
Magdalen Is
Grand Manan Island
Miscou Island
Shippegan Island
HIGHEST ELEV NOT KNOWN
Halifax
St John
Moncton
Sydney
Cape Sable
Cape Gaspe
Cape Ray
Cape St George
Grande Miquelon (France)
Petite Miquelon
Seven Islands
Baie Comeau
Chatham
Newcastle
Campbellton
Mont Carleton
Fox Bay
Heath Point

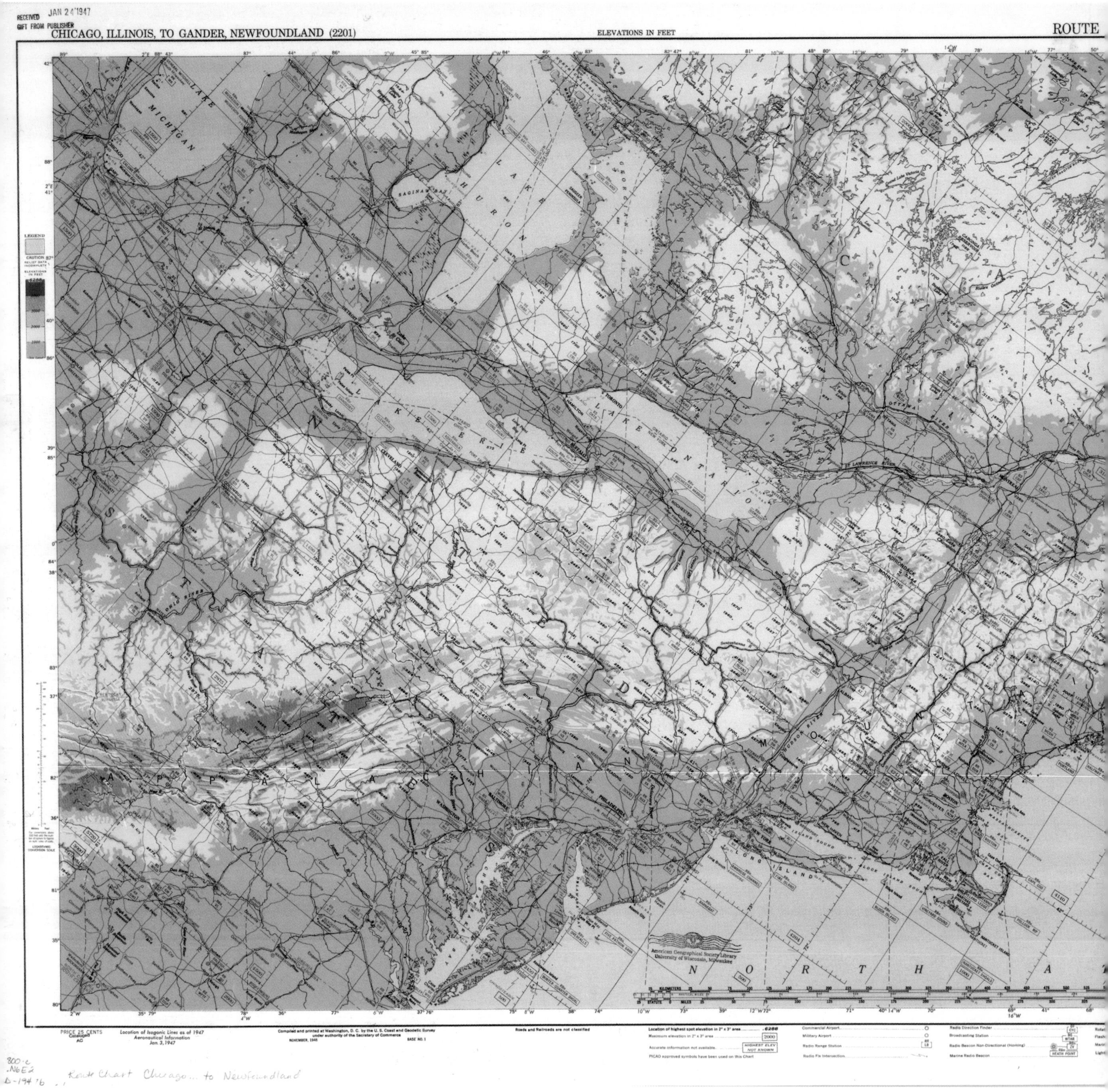

CHICAGO, ILLINOIS, TO GANDER, NEWFOUNDLAND (2201)

ELEVATIONS IN FEET

ROUTE

LAKE MICHIGAN

LAKE HURON

SAGINAW BAY

GEORGIAN BAY

LAKE ERIE

LAKE ONTARIO

ST LAWRENCE RIVER

OTTAWA RIVER

UNITED STATES

CANADA

APPALACHIAN MOUNTAINS

LONG ISLAND

NORTH ATLANTIC

LEGEND

CAUTION

ELEVATIONS IN FEET

KILOMETERS

STATUTE MILES

PRICE 25 CENTS

Location of Isogonic Lines as of 1947
Aeronautical Information
Jan. 3, 1947

Compiled and printed at Washington, D. C. by the U. S. Coast and Geodetic Survey under authority of the Secretary of Commerce

NOVEMBER, 1946

BASE NO. 1

Roads and Railroads are not classified

Location of highest spot elevation in 2° x 3° area .6288

Maximum elevation in 2° x 3° area 2000

Accurate information not available. HIGHEST ELEV NOT KNOWN

PICAO approved symbols have been used on this Chart

Commercial Airport

Military Airport

Radio Range Station

Radio Fix Intersection

Radio Direction Finder

Broadcasting Station

Radio Beacon Non-Directional (Homing)

Marine Radio Beacon

Route Chart Chicago... to Newfoundland

HART
Oblique Mercator Projection
NORTH ATLANTIC OCEAN
LABRADOR
SAINT LAWRENCE RIVER
GULF OF SAINT LAWRENCE
NEWFOUNDLAND
PRINCE EDWARD ISLAND
CAPE BRETON ISLAND
CABOT STRAIT
BAY OF FUNDY
ATLANTIC OCEAN
STRAIT OF BELLE ISLE
LAKE MELVILLE
NOTRE DAME BAY
ST GEORGES BAY
FORTUNE BAY
PLACENTIA BAY
CHICAGO, ILLINOIS, TO GANDER, NEWFOUNDLAND (2201)
SCALE 1:2,000,000
ADVANCE EDITION—JANUARY 3, 1947
OBLIQUE MERCATOR PROJECTION

Soviet-era maps give a distinctive and detailed but not totally up-to-date image of British towns and cities.

Brezhnev is Watching You!

1973

Right Soviet General Staff, *Oksford.*

THE TOWN PLAN is defined as: 'a comprehensive large-scale map of a city or town delineating streets, important buildings, and other urban features' [Wallis and Robinson, 1987]. For town plans with a difference, however, those produced in the Cold War by the General Staff of the Soviet military are worthy of further examination. During the course of the 1970s numerous such plans were made for many of the United Kingdom's principal settlements. The example selected in this case is a two-sheet 1973-published offering at a scale of 1:10,000 entitled ОКСФОРД, transliterated as 'Oksford', and lavishly printed in ten colours.

Compiled in 1972, the city is divided into West and East sheets, covering an area of 11 by 7 kilometres (7 by 4 miles) (West) and 11 by 6 kilometres (7 by 3¾ miles) (East). The West sheet extends from Yarnton through the city centre to Kennington, and from Cumnor to South Parks; the East sheet runs from Beckley to Toot Baldon, and from South Parks to Wheatley. All streets are included and named, as are principal buildings. There is an alphanumeric street gazetteer included on the East sheet. Especially noteworthy are the presence of metric contours at a time when Ordnance Survey contour information output was still displayed in imperial units. Thus this is much more than a conventional street map. There is a wealth of topographic information included in addition to the street layout. Forty-one key installations are identified and numbered, ranging from named locations, for example Morris Motors and Pressed Steel, to anonymous post and sub-post offices; there are gasworks, bus depots, warehouses and publishers, and also the more specific, such as the police station, the radio mast at Beckley, the town hall, and University College mistakenly identified as the 'University'.

In terms of the text, transliteration, rather than translation is employed, hence the suburb of Park Town is named as ПАРК ТАУН, instead of using the likely direct translation Parkovyi Gorod. Of some concern though, is the case of Marston Ferry Road, a major new route linking north Oxford with Marston. This road was opened on 12 November 1971, but does not appear on the map.

Such mapping was largely acquired in Britain after the break-up of the USSR, as the former Soviet Army withdrew from Latvia leaving behind a depot full of maps, which the Latvian authorities chose to make available for sale.

Ред-Барн
АППЕР-
ВУЛВЕРКОТ
ЛОУЭР-
Юниверсити-Филд
Уайтем
парк Уайтем
Натслоу
САННИМИД-ИСТЕЙТ
САННИМИД
САММЕРТАУН
МАРСТОН
ПАРК-ТАУН
НЬЮ-МАРСТОН
УОЛТОН-МАНОР
НОРЕМ-МАНОР
ДЖЕРИКО
Бинси
Медли-Манор
Тилбери
БОТЛИ
ОСНИ
СЕНТ-ЭББЕС
ГРАНДПОНТ
НЬЮ-ХИНКСИ
НОРТ-ХИНКСИ
Саут-Хинкси
Чоли
Темза

A realistic-looking composite map by a Japanese artist depicting an island containing Pearl Harbor and Hiroshima.

A Modern Utopia

JAPANESE ARTIST SATOMI Matoba has digitally manipulated sections of two topographic maps into an image of 'Utopia'. Like many contemporary artists, Matoba finds the map a powerful medium for exploring and expressing ideas and exploiting the creative opportunities of today's technology.

Utopia – 'non-place' and 'good-place' – dates to Thomas More's 1516 essay, written in the first flush of European oceanic 'discovery' and world mapping. A wood engraved map of More's imaginary island accompanied the first edition of his book, introducing a long tradition of 'mapping' imaginary, exotic islands. Matoba's composite image uses this island tradition to register personal geographical experience, alongside broader historical and political realities. She exploits the contemporary availability of digitized national map series to fuse two topographies into a seamless image. Her choice of locations is both ironic and powerful. Pearl Harbor on the Hawaiian island of Oahu, and Hiroshima, arguably 'book-end' the Japanese-American Pacific war. The 3 December 1941 bombing of Pearl Harbor dramatically exposed the United States' vulnerability to aerial attack (stimulating a revolution in American popular cartography) and plunged America into global war. The war's end was signalled by the 6 August 1945 bombing of Hiroshima that levelled the city, killed over 150,000 people and opened the nuclear age. The scale and artistry of twentieth-century topographic map series rendered them icons of nationhood, and Matoba's ability to exploit the graphic similarity of the US and Japanese topographic sheets reflects the enduring cultural influence of America's post-war occupation of Japan.

Satomi Matoba's *Map of Utopia* is one of her many cartographic artworks, including fused images of world city-regions and cartographic animations. Herself a Hiroshima resident, Matoba maps her personal 'place in the world': as a Japanese woman from a traditional family in a conservative city who travels and connects through her art diverse places and cultures.

Their ubiquity in modern society, as well as their silent authority and power as 'true' representations of the world, make maps and mapping a rich source for contemporary art. Contemporary art's critical engagement with cartography extends well beyond the traditional aesthetic concerns that have long connected the two practices, uniting them today around contested questions of culture, environment and politics.

Right Satomi Matoba, *Map of Utopia*, 1998. **Above** The island of Utopia as depicted for the first edition of Thomas More's 1516 essay. **See also** 1596.

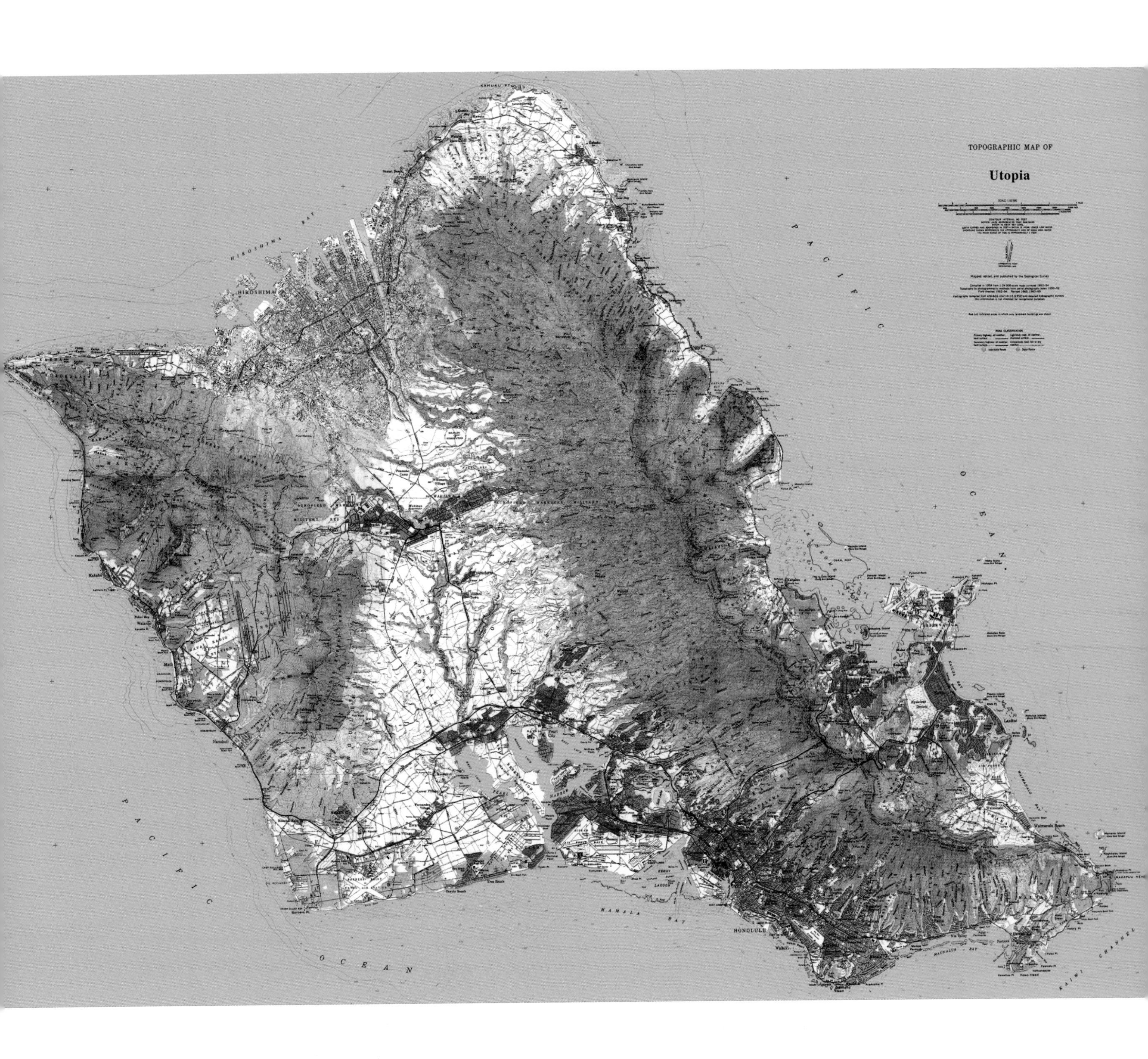
TOPOGRAPHIC MAP OF
Utopia
HIROSHIMA BAY
HIROSHIMA
PACIFIC
OCEAN
PACIFIC
OCEAN
MAMALA BAY
HONOLULU

A modern cycling tourist map combines clear road and tourist information with subtle depiction of relief in a handy but robust format.

On Your Bike?

ON THE SHELF above my desk, amongst the jumble of maps, is a copy of the *Cyclists Guide to the English Lake District*, published in 1899. It carries an advert for Philips Waterproof Maps for Cyclists, which boast that they are 'impervious to water', 'take up less room', are 'readily cleaned with a sponge' and, intriguingly, 'give greater satisfaction'.

Over a century later, these qualities continue to be relevant when producing maps for cyclists. Although the roads are now covered in tarmac, the hills are just as steep and the weather still affects the progress of cyclists.

In 2001 Stirling Surveys, publishers of Footprint Maps, joined forces with Scottish Borders Tourist Board to produce this map. Its development was part of the tourist board's Land of Creativity promotion, which celebrated the region's long and diverse literary tradition. The cycle tour evolved as an ideal way for active visitors to take a few days to soak up the ambience which inspired and informed the work of James Hogg, Sir Walter Scott and John Buchan, amongst others.

The Border Loop map is designed to be a distillation of all the essential information cyclists will need en route, and encourage further exploration of the area. Layer colours are used to give a subtle but vivid impression of terrain, while steep gradients, tricky junctions and busy roads are clearly marked to allow riders to anticipate what may lie ahead. Colour is used to group and structure similar types of information; dark brown symbols for tourist attractions, cafes, pubs and so on, while a pale orange is used for text points explaining the literary theme of the tour. The map won the British Cartographic Society Design Award, for outstanding work in map design, in 2001.

It was decided to keep the sheet size modest; handling large sheet maps outdoors can be a very tricky business. A traditional glued on cover was omitted; the map folds to a compact 210 by 119 millimetres (8 by 5 inches) and a cover would only make getting it in and out of pockets more difficult than it ought to be. The most important feature of the map is that it is printed on waterproof paper, which the ink stays on even when soaking wet. However, in this state the map can be dried out and reused repeatedly – treatment a standard paper map wouldn't survive.

Perhaps it should adopt the advertisement copy from over a century ago?

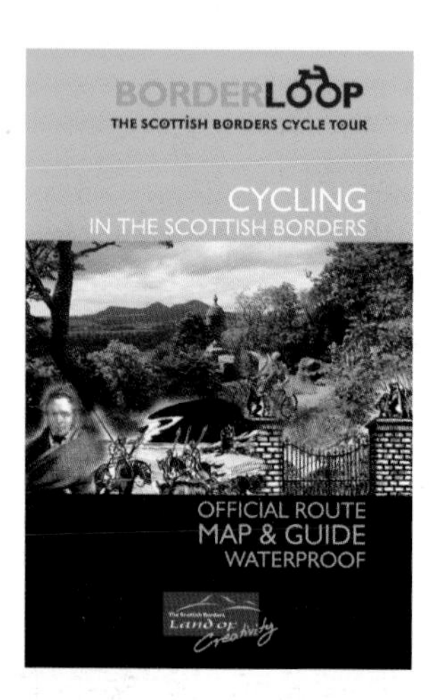

Right Two maps from the Border Loop guide. **Above** Stirling Surveys, *Border Loop. The Scottish Borders Cycle Map: Cycling in the Scottish Borders*. **See also** 1890.

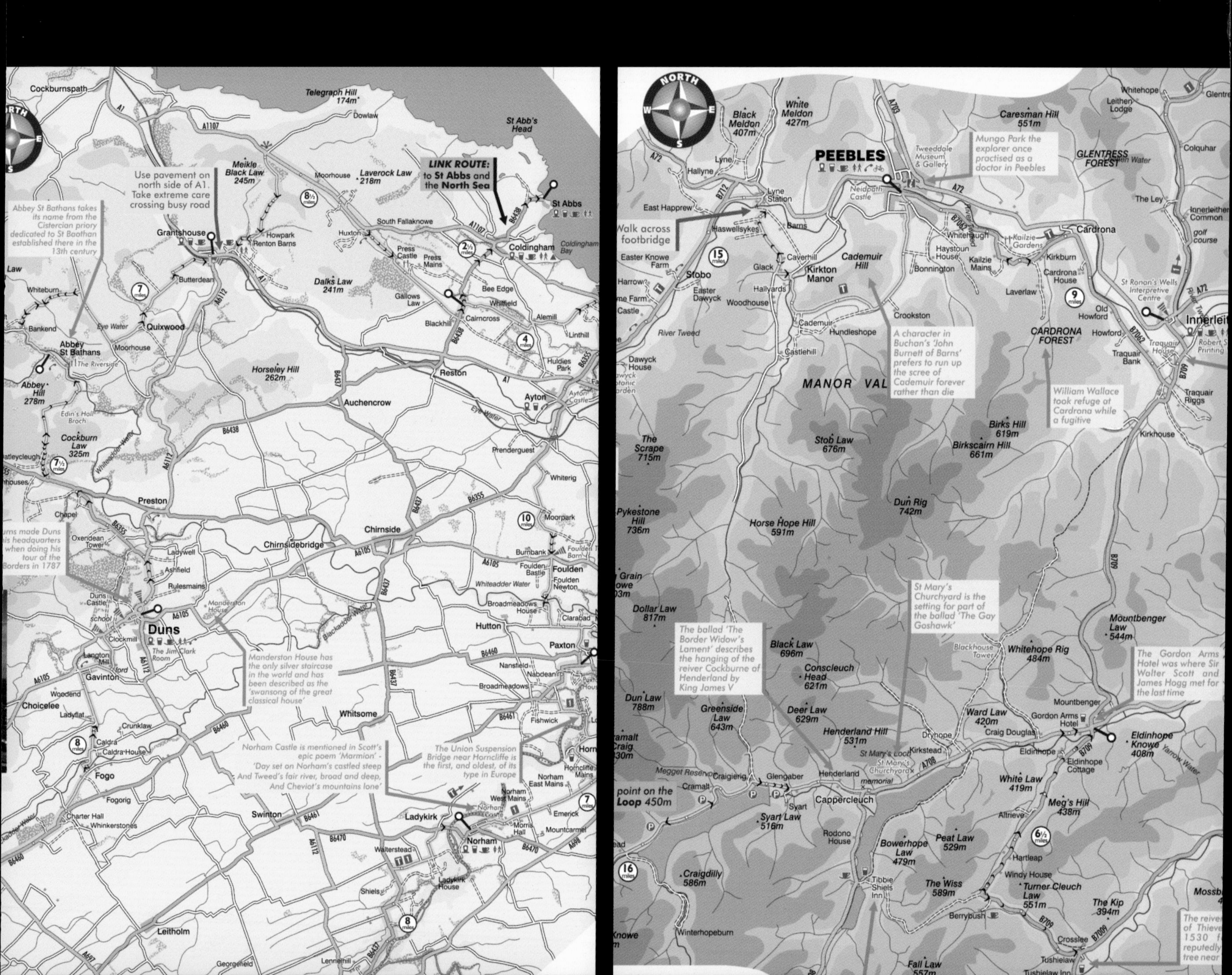

NORTH
Cockburnspath
Telegraph Hill 174m
Dowlaw
St Abb's Head
A1
A1107
Meikle Black Law 245m
Use pavement on north side of A1. Take extreme care crossing busy road
Moorhouse
Laverock Law 218m
LINK ROUTE: to St Abbs and the North Sea
St Abbs
8½ miles
Abbey St Bathans takes its name from the Cistercian priory dedicated to St Baothan established there in the 13th century
Grantshouse
Howpark
Renton Barns
South Fallaknowe
Huxton
Press Castle
Press Mains
2½ miles
Coldingham
Coldingham Bay
Law
Whiteburn
Butterdean
Dalks Law 241m
Gallows Law
Bee Edge
Whitfield
7 miles
A6112
Blackhill
Cairncross
Alemill
Linthill
Bankend
Eye Water
Quixwood
Abbey St Bathans
Moorhouse
The Riverside
4 miles
Horseley Hill 262m
B6438
B6437
Reston
Huldies Park
B6355
Abbey Hill 278m
Auchencrow
Ayton
Ayton Castle
Edin's Hall Broch
Eye Water
Cockburn Law 325m
B6438
Whiteadder Water
Prenderguest
7½ miles
Whiterig
Preston
B6355
Chapel
10 miles
Moorpark
Burns made Duns his headquarters when doing his tour of the Borders in 1787
Oxendean Tower
Chirnside
Chirnsidebridge
Ladywell
A6105
Burnbank
Foulden Barn
Ashfield
Foulden Bastle
Foulden
Rulesmains
Whiteadder Water
Foulden Newton
Duns Castle
Manderston House
Broadmeadows House
school
A6105
Blackadder Water
Clarabad
Clockmill
Duns
Hutton
Paxton
The Jim Clark Room
Langton Mill
Manderston House has the only silver staircase in the world and has been described as the 'swansong of the great classical house'
B6460
Nansfield
Nabdean
Gavinton
A6105
Broadmeadows
Woodend
Choicelee
Whitsome
Ladyflat
B6461
Fishwick
Crunklaw
B6460
8 miles
Caldra
Caldra House
Norham Castle is mentioned in Scott's epic poem 'Marmion' - 'Day set on Norham's castled steep And Tweed's fair river, broad and deep, And Cheviot's mountains lone'
The Union Suspension Bridge near Horncliffe is the first, and oldest, of its type in Europe
Horncliffe Mains
Fogo
Norham East Mains
Fogorig
Norham West Mains
Swinton
Charter Hall
Whinkerstones
Ladykirk
Emerick
B6461
Norham Castle
Morris Hall
Mountcarmel
B6470
Norham
A6112
Walterstead
B6470
A698
Ladykirk House
Shiels
8 miles
Leitholm
Lennelhill
Georgefield
A697
Black Meldon 407m
White Meldon 427m
A703
Caresman Hill 551m
Whitehope
Leithen Lodge
Glentress
A72
Lyne
Hallyne
PEEBLES
Tweeddale Museum & Gallery
Mungo Park the explorer once practised as a doctor in Peebles
GLENTRESS FOREST
Colquhar
Neidpath Castle
A72
The Ley
East Happrew
B712
Lyne Station
Innerleithen Common
Walk across footbridge
Haswellsykes
Barns
Whitehaugh
Kailzie Gardens
Cardrona
golf course
Easter Knowe Farm
15 miles
Caverhill
Haystoun House
Kailzie Mains
Kirkburn
Stobo
Cademuir Hill
Glack
Kirkton Manor
Bonnington
Cardrona House
Easter Dawyck
Hallyards
St Ronan's Wells Interpretive Centre
Woodhouse
Laverlaw
9 miles
A72
Castle
Crookston
Old Howford
Innerleithen
River Tweed
Cademuir
Hundleshope
CARDRONA FOREST
Howford
A character in Buchan's 'John Burnett of Barns' prefers to run up the scree of Cademuir forever rather than die
B7062
Traquair House
Dawyck House
Castlehill
Traquair Bank
B709
MANOR VAL
William Wallace took refuge at Cardrona while a fugitive
Traquair Riggs
Birks Hill 619m
The Scrape 715m
Stob Law 676m
Birkscairn Hill 661m
Kirkhouse
Dun Rig 742m
Pykestone Hill 736m
Horse Hope Hill 591m
B709
Dollar Law 817m
St Mary's Churchyard is the setting for part of the ballad 'The Gay Goshawk'
Mountbenger Law 544m
The ballad 'The Border Widow's Lament' describes the hanging of the reiver Cockburne of Henderland by King James V
Black Law 696m
Blackhouse Tower
Whitehope Rig 484m
The Gordon Arms Hotel was where Sir Walter Scott and James Hogg met for the last time
Conscleuch Head 621m
Dun Law 788m
Greenside Law 643m
Deer Law 629m
Mountbenger
Ward Law 420m
Gordon Arms Hotel
Henderland Hill 531m
Dryhope
Craig Douglas
Eldinhope Knowe 408m
Yarrow Water
St Mary's Loch
Kirkstead
Eldinhope
St Mary's Churchyard
A708
Eldinhope Cottage
Megget Reservoir
Craigierig
Glengaber
Henderland
memorial
White Law 419m
point on the Loop 450m
Cramalt
Cappercleuch
Syart
Syart Law 516m
Meg's Hill 438m
Altrieve
Rodono House
6½ miles
Bowerhope Law 479m
Peat Law 529m
Hartleap
16 miles
Craigdilly 586m
Tibbie Shiels Inn
Windy House
The Wiss 589m
Turner Cleuch Law 551m
The Kip 394m
Berrybush
B709
Mossburn
Winterhopeburn
Crosslee
B7009
Tushielaw
Tushielaw Inn
Fall Law 557m

Satellite mapping enables subsidence and uplift, in this case to do with water extraction and tunnel boring, to be charted over time.

Mapping Millimetric Ground Movement from Space

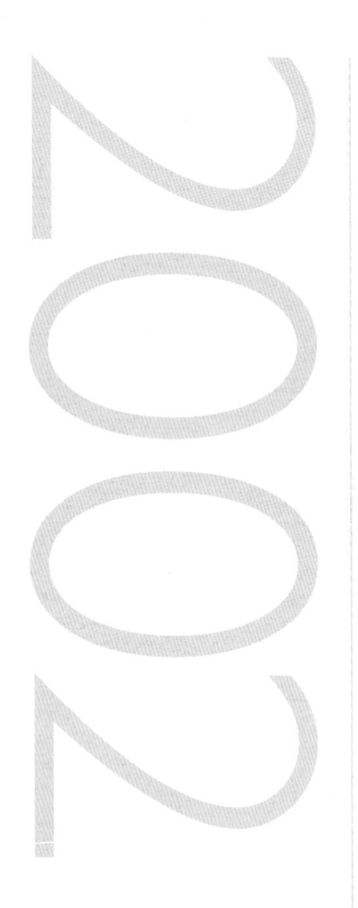

Right Analysis of multiple satellite radar images acquired by the European Radar Satellites between 1992 and 2003.

THE VALUE OF observations from vantage points was recognized in the most ancient writings. Today the ultimate vantage point, and the basis of much modern mapping, is aerial or satellite imagery. Satellite images with a ground resolution of one metre or better are now readily available (albeit at a cost).

The Landsat ETM image (15-metre/50-foot resolution) forming the backdrop to this map of London, with well-known features clearly visible, is no surprise. What is absolutely remarkable, however, is that the colour scale represents elevation changes of millimetres per year recorded from 780 kilometres (485 miles) away! Many will find this scarcely believable; such accurate measurements over a decade would be almost impossible to make on the ground, and certainly not for the thousands of points represented on this map.

So what are these phenomena, and are they really true? The uplift (blue) in the north-east around Docklands is the former industrial heart of London. Today factories no longer extract large volumes of water from boreholes and the water table is rising, lifting the area. Conversely, the large area of subsidence (red) in the south-west is where the water table is falling due to increased extraction. Even more striking are the linear areas of subsidence (red), one from St James's Park to London Bridge – clearly the Jubilee Line Underground extension – and the other running westwards from Battersea Park – a tunnel for electricity cables. A hearing of the All-Party Parliamentary Group for Earth Sciences in 2002 established that the design had predicted the Jubilee Extension would create this subsidence, and it was identical to that measured in-situ and by satellite.

This technique uses data from ERS1 & ERS2 (European Radar Satellites) acquired approximately monthly since 1991. The images (around 100) are 'stacked' in the computer and small changes in the phase component of the radar returns are computed and statistically analyzed over the complete period. This is analogous to a police radar speed camera – but used over a very long distance and time!

Not only is space imaging providing similar accuracy to ground survey, but it reveals subtle features over wide areas that only a time-travelling surveyor could achieve. The benefits in geotechnical engineering are immense, particularly since trends towards urbanization and mega-cities indicate that much future urban infrastructure and construction will need to be underground.

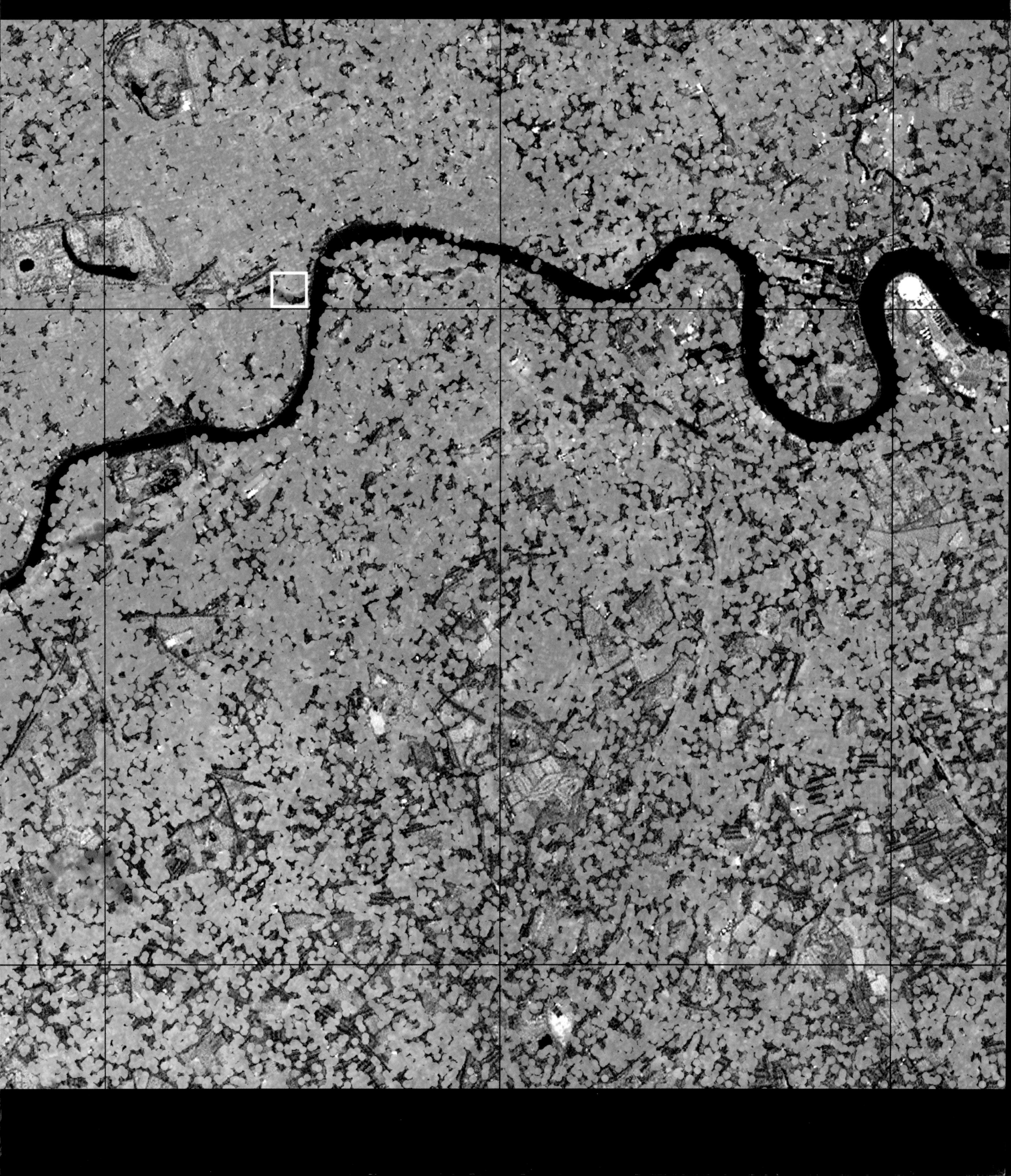

Geographical Information Systems enable databases of information and statistics and graphic images to be combined by anyone for almost any purpose.

Everyone a Mapmaker!

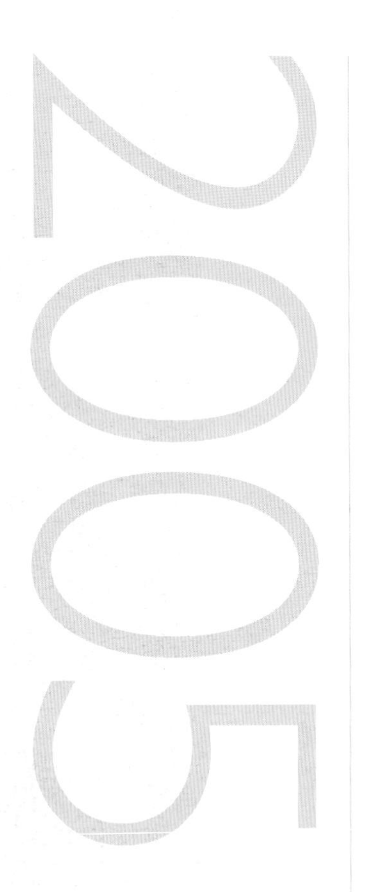

Right A computer screen showing a GIS relating to Detroit, USA.

THE AVAILABILITY OF digital cartographic data presents users with the ability to interact with maps and information in a way previously impossible. Whereas in the past, a single cartographer made decisions about the information to be included in a map and how it was to be portrayed, it is now possible for users to customize maps to satisfy specific interests using Geographic Information Systems (GIS). Cartographic data may be linked to databases of information and statistics to illustrate a phenomenon or make a point in an immediate and specific way.

The menagerie of images here represents what might be seen on a computer screen; it is a snapshot of the dynamic components – images, databases, and other tools – that make up a GIS. Digital database files (numerical statistics) and boundary files (graphic map boundaries) have been combined and linked together to allow a more complex interrogation and analysis of the relationships between these elements. The small map appearing in the upper portion of the page is a simple, choroplethic map, based upon a map of census tract boundaries and a geographically referenced database of inhabitants' income levels. The larger map depicts urban land use, upon which the same income data has been overlaid as an additional thematic layer. From merging these databases and displaying them cartographically, the relationships between the two variables are elucidated. For instance, it is apparent that lower economic conditions prevail in the residential blocks surrounding the airport facilities and industrial activity in these areas. The maps here present just two manifestations of these data, however, and further data integration, analysis and querying, as well as changes in display attributes, are possible.

GIS technologies are used widely among map and geographical data producers, particularly government agencies. The digital files used in these maps were produced variously by central and local government departments (United States Census 2000 and the City of Detroit) and made freely available on the web. Through online data download, maps and information may be quickly and easily distributed and applied towards any number of purposes. The increasing availability of GIS data complements the functionality and new accessibility of geographic technologies, expanding the way spatial information of all types is used and understood.

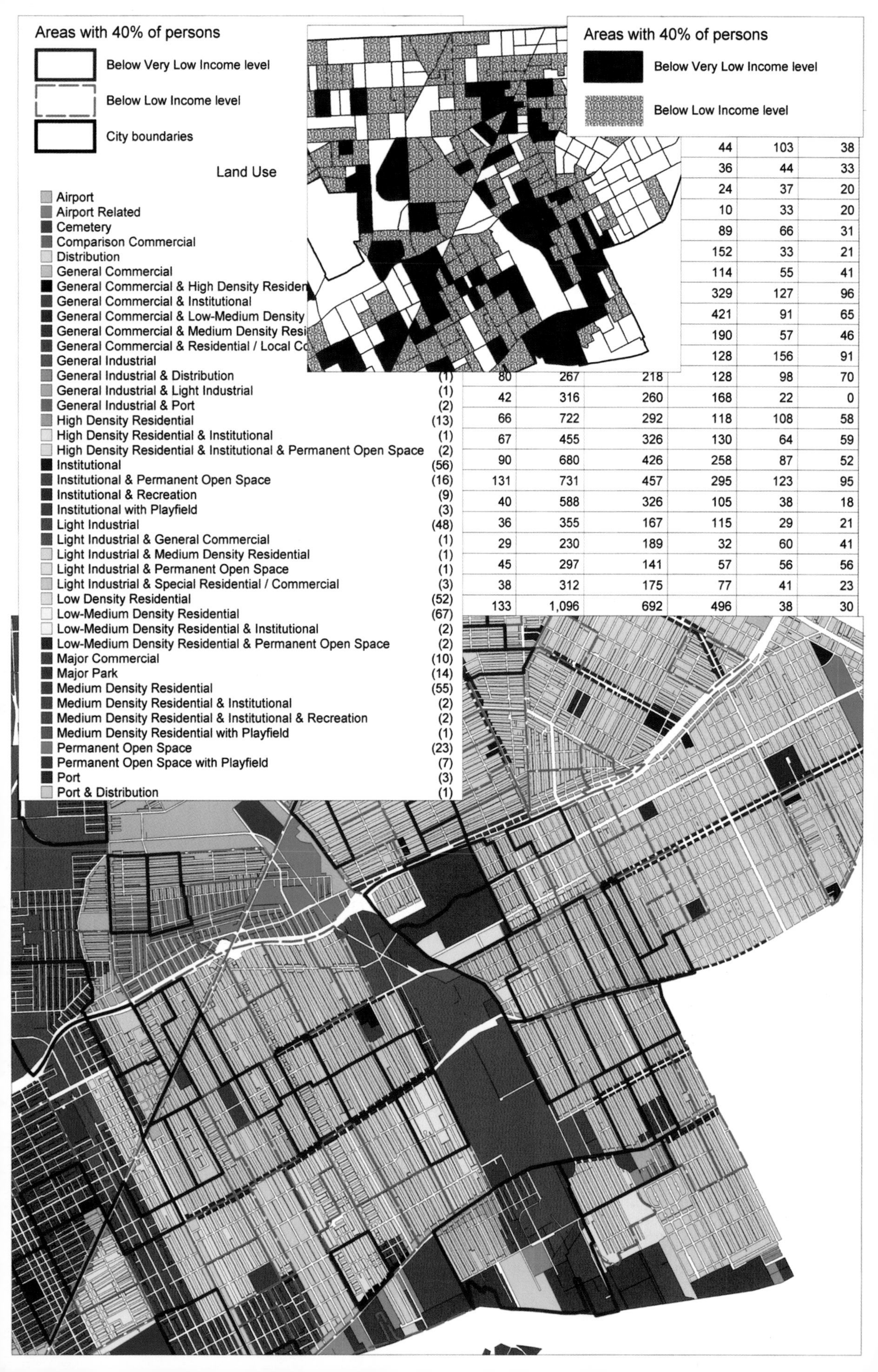
Areas with 40% of persons
Below Very Low Income level
Below Low Income level
City boundaries
Land Use
Airport
Airport Related
Cemetery
Comparison Commercial
Distribution
General Commercial
General Commercial & High Density Residen
General Commercial & Institutional
General Commercial & Low-Medium Density
General Commercial & Medium Density Resi
General Commercial & Residential / Local Co
General Industrial
General Industrial & Distribution (1)
General Industrial & Light Industrial (1)
General Industrial & Port (2)
High Density Residential (13)
High Density Residential & Institutional (1)
High Density Residential & Institutional & Permanent Open Space (2)
Institutional (56)
Institutional & Permanent Open Space (16)
Institutional & Recreation (9)
Institutional with Playfield (3)
Light Industrial (48)
Light Industrial & General Commercial (1)
Light Industrial & Medium Density Residential (1)
Light Industrial & Permanent Open Space (1)
Light Industrial & Special Residential / Commercial (3)
Low Density Residential (52)
Low-Medium Density Residential (67)
Low-Medium Density Residential & Institutional (2)
Low-Medium Density Residential & Permanent Open Space (2)
Major Commercial (10)
Major Park (14)
Medium Density Residential (55)
Medium Density Residential & Institutional (2)
Medium Density Residential & Institutional & Recreation (2)
Medium Density Residential with Playfield (1)
Permanent Open Space (23)
Permanent Open Space with Playfield (7)
Port (3)
Port & Distribution (1)
Areas with 40% of persons
Below Very Low Income level
Below Low Income level
44 103 38
36 44 33
24 37 20
10 33 20
89 66 31
152 33 21
114 55 41
329 127 96
421 91 65
190 57 46
128 156 91
80 267 218 128 98 70
42 316 260 168 22 0
66 722 292 118 108 58
67 455 326 130 64 59
90 680 426 258 87 52
131 731 457 295 123 95
40 588 326 105 38 18
36 355 167 115 29 21
29 230 189 32 60 41
45 297 141 57 56 56
38 312 175 77 41 23
133 1,096 692 496 38 30

This remote sensing image of Mount St Helens seems to be a view but has been artificially constructed by an individual.

Is This Really a Map?

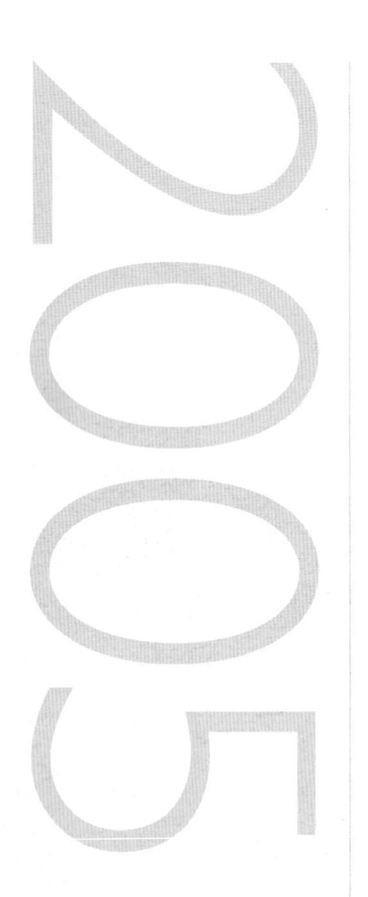

THE IMAGE TO the right is just that: a picture, derived from satellite-generated data, of Mount St Helens just after the volcanic activity of March 2005. It was created as part of a mission to record and monitor changes in the earth's environment and climate utilizing multi-spectral technology to display and map surface conditions. By providing imagery of the earth with great frequency – its orbit is repeated every sixteen days – and spatial resolution – each pixel represents 15 metres on the ground – conditions associated with earth processes, from volcanic activity and agricultural conditions to pollution detection and ocean dynamics, can be portrayed visually.

Unlike a paper map, where most of the information about a given area is accessible visually, the majority of data collected by this remote sensing instrument cannot be depicted in a single frame. But it is like mapmaking of the past in that it exhibits decisions made by an individual. Similar to the mathematical notations of a surveyor, the multi-spectral data gathered on the satellite are meaningless without additional interpretation and compilation to identify place and make decisions about what is depicted and how. Of the fourteen spectral bands gathered by the ASTER instrument, the individual compiling this image selected visible and infrared bands; of the swathes worldwide travelled by the Terra spacecraft, this area of Washington state in the USA at a scale making visible the lava in the volcano. Following image processing and analysis, wavelengths indicating vegetation were assigned the colour green; snow, light blue; bare rocks, tan; and lava, red.

Already cartography has moved into a digital realm, as seemingly 'traditional' maps are oftentimes no longer printed on paper by the publishers for distribution, but rather are designed to be accessed and compiled dynamically according to the needs of the user, who may then commit the digital information to paper. Many would say that this satellite image, lacking the requisite labels, title, scale bar and other hallmarks of a trained cartographer, is not a map. But conventional boundaries are continually being blurred, as maps have ever more in common with representations of data such as this one. As technology develops, geographic data is increasingly manipulatable, scalable and flexible, to be displayed in innumerable combinations and appearances for countless purposes.

So what makes a map a map?

Right Remote sensing image of Mount St Helens by NASA's Terra satellite.

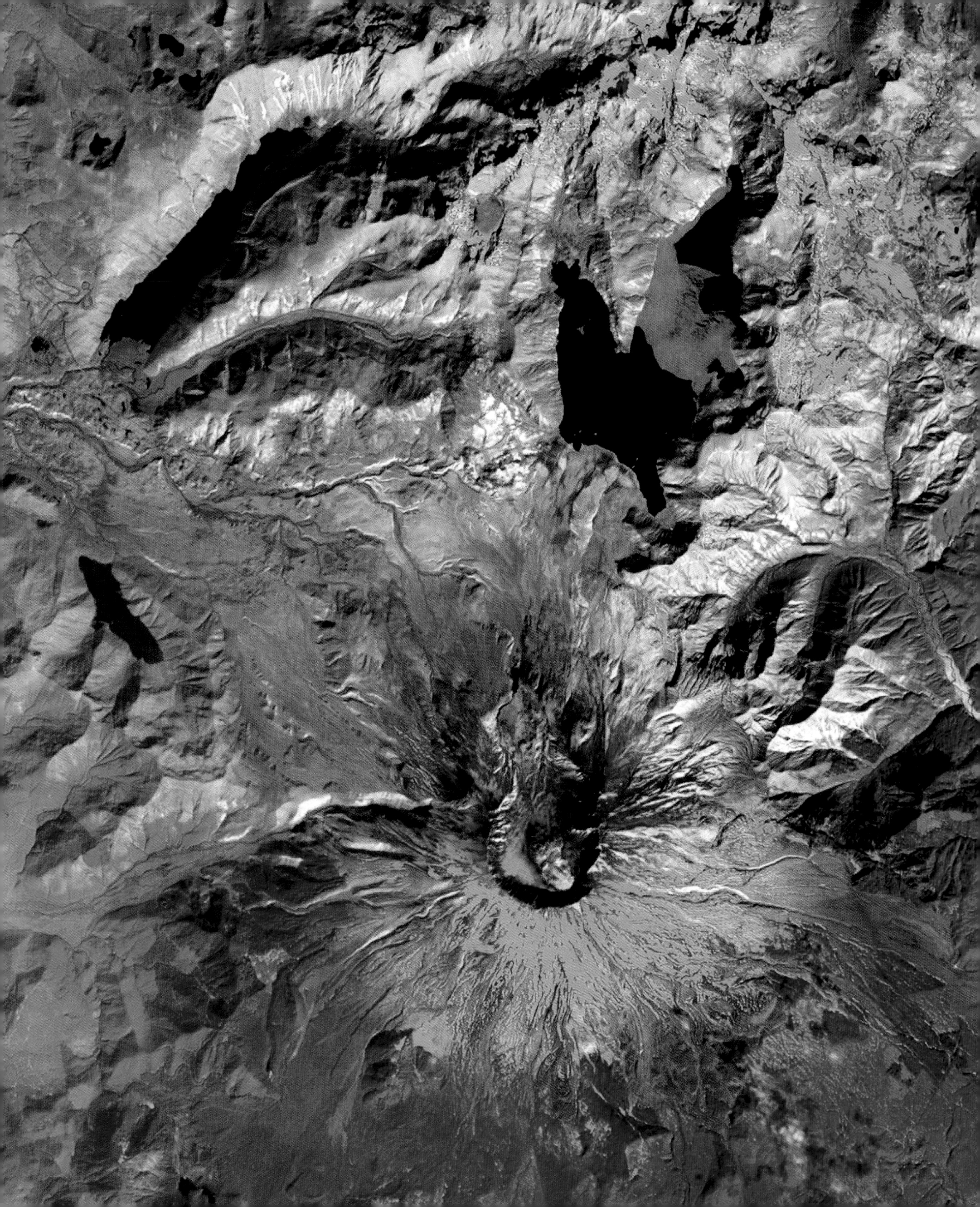

List of contributors

The publishers would like to thank all those who contributed to the book. Texts written by each author are indicated by date. All texts not listed are by Peter Barber.

Dr James Akerman is Director of the Herman Dunlap Center for the History of Cartography at the Newberry Library, Chicago. He has published widely, particularly on twentieth century American mapmaking. **1941 (p. 332)**

Geoff Armitage is a Curator and Maps Reading Room Manager in British Library Map Collections. Previous publications include *Shadow of the Moon: British Eclipse Mapping in the Eighteenth Century* (1997). He is currently undertaking research into British twin-hemisphere maps. **1764, 1785**

David Barbour, MA Dip. Cart., is a partner in Stirling Surveys, Publishers of Footprint maps. Stirling Surveys won the British Cartographic Society Design Award in 1996, 2001 and 2004. **2001**

Ashley Baynton-Williams is Editor of MapForum, a quarterly journal devoted to antique maps. **1715**

Geraldine Beech has worked at the National Archives (formerly the Public Record Office) since 1977, has specialized in maps and related records since 1980, and is now Senior Map Archivist. **1943 (p. 336)**

Dr Sarah Bendall is Fellow and Development Director of Emmanuel College, Cambridge. She works on large-scale local maps made in Great Britain and Ireland and is currently updating the *Dictionary of Land Surveyors and Local Map-Makers of Great Britain and Ireland 1530-1850*. **1720**

Daniel Birkholz is Assistant Professor of English at the University of Texas at Austin. He is the author of *The King's Two Maps: Cartography and Culture in Thirteenth-Century England* (2004). **1360**

Dr Chris Board OBE, formerly Senior Lecturer in Geography at the LSE and past president of the British Cartographical Society, is Chairman of the Charles Close Society for the study of Ordnance Survey maps. He has published extensively on modern and twentieth century cartography. **1943 (p. 334)**

Steve Brace is Head of Collections at the Royal Geographical Society. He has previously worked in a development charity and was Director of Education at the Commonwealth Institute. **1804, 1851, 1858, 1873, 1914 (p. 310)**

Marcel van den Broecke has been studying the works of Ortelius since 1982 and has published two books and about 25 articles about his maps. He is now writing a study on the translated text on the back of Ortelius atlas maps. **1596**

Dr Jerry Brotton is Senior Lecturer in Renaissance Studies at Queen Mary, University of London. He is the author of *Trading Territories: Mapping the Early Modern World* (1997). **1513**

David Buisseret has worked at Cambridge University, at the University of the West Indies in Jamaica, and at the Newberry Library in Chicago. He is at present Garrett Professor of History at the University of Texas in Arlington. **1600, 1730 (p. 198)**

Philip Burden is a map dealer and author of *The Mapping of North America: A List of Printed Maps 1511–1670* (1996), the standard work on the subject. **1670**

John Cloud is is presently the historian of the National Oceanic and Atmospheric Administration (NOAA), the oldest scientific agency in the US government. **1947**

Peter Collier lectures in geography in the University of Portsmouth. His main research interest is the history of topographic mapping in the 19th and 20th centuries. **1901, 1914 (p. 312)**

Karen Severud Cook, whose research interests are the history of geological maps, map reproduction and Australian exploration, is the Assistant Special Collections Librarian, Spencer Research Library, University of Kansas. **1823**

Denis Cosgrove is professor of Geography at UCLA. He writes extensively on maps, pictures and geographical knowledge. **1998**

Surekha Davies is a Curator at the British Library Map Library and a doctoral candidate at the Warburg Institute, University of London. She has published on sixteenth-century cartography and Hispanic-Amerindian encounters in the New World. **1597 (p. 134), 1599**

Dr Catherine Delano-Smith is currently a senior research fellow at the Institute of Historical Research, University of London. She is also editor of *Imago Mundi. The International Journal for the History of Cartography*. **1500 BC, 600 BC, 500, 565 Map Signs**

Michael Dover is a publisher in London. **1555, 1675, 1933, 1938**

Matthew H. Edney is associate professor and Osher Map Library faculty scholar, University of Southern Maine, Portland, and director of The History of Cartography Project, University of Wisconsin-Madison, USA. **1677**

Evelyn Edson is a Professor of History at Piedmont Virginia Community College. She is the author of *Mapping Time and Space: How Medieval Mapmakers Viewed Their World* (1997), and co-author of *Medieval Views of the Cosmos: Picturing the Universe in the Christian and Islamic Middle Ages* (2004). **425**

James Elliot is a former Curator at the British Library Map Library. He has written numerous articles on map storage, catalogue automation and urban mapping. **1891**

Dr Francesca Fiorani is assistant professor of art history at the University of Virginia. Her special interest in art and cartography in the Late Renaissance is reflected in her book *The Marvel of Maps Art, Cartography and Politics in Renaissance Italy* (2005). **1580**

Chris Fleet has worked at the National Library of Scotland from 1994, with particular interests in the Pont and Blaeu maps of Scotland, in digital mapping, and presenting map images over the Internet. **1590 (p. 126)**

Christopher Fletcher is a curator of Literary Manuscripts at the British Library. He has published on a number of literary topics and has a particular interest in Joseph Conrad. **1899 (p. 294)**

Dov Gavish is a Historian of aerial Photography and Cartography. From 1970–2003 he was Director of Aerial Photographs Archives, and lecturer at the Department of Geography, Hebrew University of Jerusalem. **1914 (p. 308)**

Susan Gole lived in India for many years, and has published several books on Indian maps. She was Chairman of the International Map Collectors' Society (1989–96), and is currently Editor of its quarterly journal. **1619, 1782**

Debbie Hall is a curator in the Map Library of the British Library and has participated in several exhibitions including 'The Lie of the Land' at the British Library in 2001–2. **1753, 1831 (p. 264), 1846, 1941 (p. 330)**

P. D. A. Harvey is Professor Emeritus of Medieval History at the University of Durham. His publications include works on medieval English social and economic history as well as on the history of cartography. **1250**

Yolande Hodson is a historian of military cartography and of the Ordnance Survey; currently preparing a catalogue of King George III's military maps in the Royal Collection at Windsor Castle. **1815, 1914 (p. 304)**

Crispin Jewitt has been Head of the British Library Sound Archive since 1991, having previously worked as a curator in the Library's map department. He published *Maps for Empire* in 1992 and is currently working on British military mapping during the early 19th century. **1818**

Kimberly C. Kowal is the Curator of Digital Mapping at the British Library. She has worked with GIS and electronic mapping in university libraries in the United States, and holds advanced degrees in Geography and Library & Information Science. **2005 (p. 352, p. 354)**

Roger Kain CBE is Montefiore Professor of Geography and Deputy Vice-Chancellor, University of Exeter. He is Vice-President and Treasurer of the British Academy. **1587**

Robert W. Karrow, Jr. is Curator of Maps and Curator of Special Collections at the Newberry Library in Chicago. His principal research interest is 16th-century cartography. **1540**

Robert Laurie is a Curator at the British Library Map Library and also editor of the Bulletin of the Marx Memorial Library. **1926**

Joseph Loh is completing his doctorate at Columbia University and is currently a research fellow at the British Library. His field is Asian art and culture and his research concerns the relationship between painted practice and the cartographic imagination. **1666, 1800 (p. 244)**

J. P. Losty has published widely on many aspects of the Indian manuscript and painting traditions. He recently retired from the Asia Pacific and Africa section of the British Library. **1750 (p. 206)**

Maria Luisa Martín-Merás has been Head of the Map Department at the Museo Naval in Madrid since 1970. **1500, 1525, 1544 (p. 102),**

Earl B. McElfresh is the author of *Maps and Mapmakers of the Civil War* (1999) as well as being a cartographer/map historian and partner of McElfresh Map Company. **1862**

Marica Milanesi is Professor of History of Geography at the University of Pavia, Italy. Her main field of research are geography and cartography in the Renaissance. **1561, 1688**

Nick Millea has been Map Librarian at the Bodleian Library, University of Oxford since 1992; previously Map Curator at the University of Sussex (1990–2). He is Bibliographer for *Imago Mundi* and Chairman of the Working Group for Education, LIBER Groupe des Cartothecaires. **1973**

Rose Mitchell is a map archivist at the National Archives in London. She specializes in medieval and 16th-century maps in he National Archives, and is also co-author of a book on aspects of English cadastral and agricultural mapping. **1430, 1608**

Barbara E. Mundy is an Associate Professor at Fordham University in New York City. Her book, *The Mapping of New Spain*, covers the cartography of Aztec and early colonial Mexico. **1524, 1579**

Dr Tim Nicholson, a retired publisher's editor, came to studying and writing about maps through his books on transport history. **1890, 1930**

Dr Thomas O'Loughlin was born in Dublin and read Philosophy & Medieval History at University College Dublin. He later specialized in Medieval Theology and manuscript studies. He is Reader in Historical Theology in the University of Wales, Lampeter. **750, 1210**

Ferjan Ormeling holds the cartography chair at Utrecht University and studies the thematic cartography of the Netherlands as well as the geographical naming process. **1596**

Dr Mary Sponberg Pedley is assistant map curator at the William Clements Library in Ann Arbor and an associate editor of *Imago Mundi*. She is the author of *The Commerce of Cartography: Making and Marketing Maps in Eighteenth-Century France and England* (2005). **1777**

Monique Pelletier was the head of the National Library Map Division in Paris 1977–99. She is carrying out research into 18th-century French mapping, both civilian and military. **1756**

Nigel Press is chairman of Nigel Press Associates Ltd., a remote imaging company specialising in the commercial application of earth observation satellite data. **2002**

Dr Franz Reitinger is an Austrian scholar with a special interest in cartographical curiosities, allegorical maps and their related texts on which he has published widely. **1660 (p. 162)**

Rear Admiral Steve Ritchie CB, DSC, FRICS was Hydrographer of the Navy 1966–71 and subsequently President of the International Hydrographic Organisation. **1539**

Neil Safier is a postdoctoral fellow in the Michigan Society of Fellows and assistant professor in the department of History at the University of Michigan. His forthcoming book on the European exploration of South America, *Itinerant Enlightenment*, will be published in 2007. **1780 (p. 230)**

Emilie Savage-Smith, Ph.D., is Senior Research Fellow at St Cross College, and Senior Research Associate of the Khalili Research Centre for the Art and Material Culture of the Middle East, University of Oxford. **1050**

Joseph E. Schwartzberg taught geography at the University of Minnesota for 36 years. He is the principal author and editor of *A Historical Atlas of South Asia* and an associate editor of and major contributor to volume 2, books 1 and 2 of *The History of Cartography*. **1690, 1770, 1800 (p. 242), 1889**

Zur Shalev (Ph.D. Princeton 2004) teaches early modern European history at the University of Haifa. His research interests include the history of science, religion, and travel. **150, 1646**

Rodney Shirley is a researcher and writer on historical cartography. His book *The Mapping of the World: Early Printed World Maps 1472-1700* is a standard reference work, as are his two books on the early maps of the British Isles. In 2004 he published a detailed work *Maps in the Atlases of the British Library c.800 AD–1800*. **1564, 1576, 1603, 1623, 1730 (p. 196), 1830**

David Smith has published extensively on the history of cartography with special reference to Victorian commercial mapping and early printed town plans. He is the author of *Maps & Plans for the Local Historian and Collector* (Batsford, 1988). **1790**

Sylvia Sumira specializes in the conservation of globes. After working at the National Maritime Museum in Greenwich for several years, she set up her own studio and has worked for many institutions, including The British Library, as well as for private clients. **1766**

Ilana Tahan, M.Phil. is Curator of the Hebraica collection at the British Library. Born in Romania, she was educated at the Hebrew University in Jerusalem and at the Aston University in Birmingham, UK. **1695**

Richard Talbert is Kenan Professor of History and Classics at the University of North Carolina, Chapel Hill. He edited the *Barrington Atlas of the Greek and Roman World* and is now preparing a major new edition of Peutinger's Roman map. **200, 250, 300**

Captain Albert E. Theberge spent 27 years as an ocean surveyor for the U.S. National Oceanic and Atmospheric Administration. He is a historian on the reference staff of the NOAA Central Library and a member of both the United States and international committees responsible for the nomenclature and terminology of undersea features. **1899 (p. 296)**

Hilary L. Turner has worked as a freelance researcher and translator in England and Greece, with publications on architectural history as well as maps. **1590 (p. 128)**

Sarah Tyacke, CB, Chief Executive of the National Archives (formerly Keeper of the Public Record Office) worked for many years in the Map Library of the British Library and is an authority on the history of cartography, with particular emphasis on early English chart-making on which she has published extensively. **1621**

Diana Webster has worked in the National Library of Scotland since 1988, becoming Head of the Map Library in 1999. With a degree in geography, her interests include Scottish sea charts, cartouches and surveying methods. **1746, 1750 (p. 208)**

Frances Wood is curator of the Chinese collections in the British Library. She is author of the *Blue Guide to China* and other books. **1751**

Laurence Worms is an antiquarian bookseller in London (Ash Rare Books), former Hon. Sec. of the Antiquarian Booksellers' Association (International) and Fellow of the Rare Book Society. Contributor to *The Cambridge History of the Book in Britain* and the *University of Chicago Press History of Cartography*. **1748, 1793**

Bülent Yilmazer is a Mechanical Engineer. He researches the history of Ottoman/Turkish aviation as a private interest and lectures on the History of Aviation as a part-time instructor at the Aerospace Engineering Department of Middle East Technical University, Ankara-Turkey. **1914 (p. 208)**

Further reading

J. B. Harley and David Woodward (eds.), *The History of Cartography Vol. 1 :* Cartography in Prehistoric, Ancient, and Medieval Europe and the Mediterranean (1987); vol. 2 : Cartography in the Traditional Islamic and South Asian Societies (2 books: 1992, 1994), Volume 2, Book 3 (1998): *Cartography in the Traditional African, American, Arctic, Australian, and Pacific Societies.* ; Volume 3 (in press, 2005): *Cartography in the European Renaissance* (Chicago: University of Chicago Press).

Imago Mundi, the principal scholarly journal on the history of cartography, contains a detailed annual bibliography of recent publications, articles and essays.

List of illustrations

BL = British Library

Front jacket: see pages 53, 73, 107, 135, 197, 261, 335, 333, 321, 355.
Back jacket: see page 105.
Endpapers: *Atlas, sive Cosmographicae Meditationes de Fabrica Mundi et Fabricati Figura. De nuò auctus excusam sub cane vigilanti Henrici Hondii Amsterodami An.D.1623*. BL Maps C.3.c.9.

1 Shan Map Relating to a Border Dispute Between (British) Burma and China, circa 1889. By permission of the Syndics of Cambridge University Library (Scott LR. 13.34). **2** Sheldon tapestry maps of Shropshire and Worcestershire, (details). Produced for Ralph Sheldon, 1590 (Bodleian Library, University of Oxford / V&A Images, Victoria and Albert Museum. **3** 16th-century compass surrounded by bearings, Lambeth Palace Library Ms 463 f. 18 / The Bridgeman Art Library. **9** 'The Map describing the situations of the several Nations of the Indians between South Carolina and the Mississippi'. Redrawing of *c.*1723 The National Archives: Colonial Office Library 700 North American Colonies General No. 6 (2). **10** Map of a hut and fields, Roccia di Napoleone, Val Fontanalba, Monte Bego. From Clarence Bicknell, *A Guide to the Prehistoric Rock Engravings in the Italian Maritime Alps* (Bordighera, 1902), Tavola XIX (1). BL 7709.d.25. **11** Line-drawing of whole Bedolina assemblage, adapted from Walter Blumer, 'The oldest known plan of an inhabited site dating from the Bronze Age, about the middle of the 2nd millennium BC' in *Imago Mundi. The International Journal for the History of Cartography*, 18 (1964), 9-11, Fig. 2. **13** Plan of Nippur, *c.*1500 BC. By permission of the Hilprecht Collection, Friedrich-Schiller University, Jena, Germany. **15** Papyrus plan of an Egyptian tomb, probably that of Rameses IV. Soprintendenza per le Antichita Egizie, Turin. By permission of the Museo Egiziano, Turin. **17** Map of the World, *c.*600 BC (stone), Babylonian. British Museum, London/The Bridgeman Art Library. **19** Nero. Bronze sestertius, AD 64-68 showing Ostia harbour. BMCRE 132. Copyright of the trustees of the British Museum. **21** World map in Claudius Ptolemy's *Geographia* (Constantinople (?), ca. 1400). BL Burney MS 111, fols. 105verso-106. **23** Forma Urbis Romae, [Severan Marble Plan of Rome]. Capitoline Museums: Antiquarium Comunale, Rome. Fragment 10opqr. Section of the Subura neighborhood (*Subura*) including the Porticus of Livia (*porticus Liviae*) and the Baths of Trajan (*thermae Traiani*). By permission of the Stanford Digital Forma Urbis Romae project and the Musei Capitolini, Rome. **25** Black Sea Map from Dura-Europos. Bibliothèque Nationale, Paris, Suppl. Gr. 1354.2-V. **27** *Tabula Peutingeriana*. Österreichische Nationalbibliothek, Vienna, Codex Vindobonensis 324. **29** *Corpus Agrimensorum*. Germany (?) *c.*800. Vatican Library Pal. Lat. 1564 fol. 89r from *Vedere i Classici: l'illustrazione libraria dei testi antichi dall' eta romana al tardo medioevo* (Italy, Rose Srl, 1996). 31 Macrobius world map. 9th century. BL, MS Harl. 2772, fol. 70v. **33** (top) Classical T-O diagram from Isidore of Seville, *Etymologiae* (compiled 622-633), (9th or 10th century copy), at the end of chapter II of Book 14. Stiftsbibliothek, St Gallen Switzerland, MS 237, folio 219. (bottom) From Sallust's *de Bello Jugurtha* (written before 34 BC) from a 10th century copy now in Leipzig (MS no. XL, kl. Folio). From Konrad Miller, *Mappaemundi. Die älteste Weltkarten* (Stugttgart, J. Roth, 1895-98, 6 vols.) iii, p. 11. **34** Byzantine gold coin (Histamenon) of Romanus IV Ca. 1070. Private Collection. **35** 'Cosmas Indicopleustes' (*fl. c.* AD 540), world map in *Topographia Christiana*, Mount Athos., 11th century. Florence, Biblioteca Medicea Laurenziana, Plut.9.28 fol. 92v **36** Mosaic,'Old Church', Madaba, Jordan, *c.*565. Detail showing Jerusalem. From M. Piccirillo and E. Alliatta (eds.) *The Madaba Map Centenary 1897–1997: Travelling through the Byzantine Umayyad period* (Jerusalem Stdium Biblicum Francscanum, 1998). **37** Mosaic, 'Old Church', Madaba, Jordan. *c.*565. Extract from main fragment. From M. Piccirillo and E. Alliatta (eds.) *ibid*. **39** (top) Anonymous, so-called 'Isidore' world map; Provence , 776. Vatican Library Ms lat. 6018 fols. 63v-64. From *Segni e Sogni Della Terra* (Novara: De Agostini, 2001). (bottom) Explanatory diagram in Brigitte Englisch, *Ordo orbis terrae: Die Weltsicht in den Mappae Mundi des frühen end hohen Mittelalters* [Orbis mediaevalis. Vorstellungswelten des Mittelalters, Band 3] (Berlin: Akademie Verlag, 2002). **41** *Figura terrae repromissionis* from the *Pauca problesmata de enigmatibus ex tomis canonicis*. France (?) c.750. Paris, Bibliothèque Nationale de France, lat. 11561 fol. 43v. **43** The Albi world Map, South-western France or Spain. Albi: Bibliothèque Municipale, MS 29, fol. 57v. From *Segni e Sogni Della Terra* (Novara: De Agostini, 2001). **45** Commentaries on the Apocalypse, Silos. BL Add. MS 11695, fols. 39v-40. **47** The Anglo-Saxon or Cottonian world map. England: Christ Church Canterbury, *c.*1025–50. BL Cotton MS Tiberius B. V. fol. 56v. **48** World map from *The Book of Curiosities*, Egypt, *c.*1020–50. Bodleian Library, University of Oxford, MS Arab. c. 90, fols. 23b-24a. From E. Edson and E. Savage-Smith, *Medieval Views of the Cosmos* (Oxford: Bodleian Library, 2004). **49** (top) Map of the Mediterranean from *The Book of Curiosities*. Egypt, ca. 1020-50, Bodleian Library, University of Oxford, MS Arab. *c.*90, fols. 30b-31a 49; (bottom) Key to the map from E. Edson and E. Savage-Smith, *Medieval views of the Cosmos* (Oxford: Bodleian Library, 2004). **51** The Sawley world map, (Durham?) late 12th century. Parker Library, Cambridge MS 66, p.2. By permission of the Masters and Fellows of Corpus Christi College, Cambridge. **53** A Map of Europe drawn under the direction of Giraldus Cambrensis (*c.*1146–*c.*1223). Diocese of Lincoln, 1210–23. The National Library of Ireland, Dublin, 700, fol. 48r. **55** Map of Palestine in the hand of Matthew Paris. *c.*1250. Oxford, Corpus Christi College, MS. 2*, fols. 1r, 2v. By permission of the Masters and Fellows of Corpus Christi College, Oxford/The Bridgeman Art Library. **56** Psalter World Map BL Add. MS 28684 fol.9 verso. 9 cm diameter. **57** Psalter World Map, London, *c.*1265. London, BL Add. MS 28681 fol. 9. **59** Ebstorf World Map; Ebstorf, Lower Saxony, *c.*1240–90. Reproduced with permission of the Ebskart project, University of Lüneburg (http://kulturinformatik.uni-lueneburg.de/projekte/ebskart/content/start.html). **61** Richard of Holdingham or Sleaford, The Hereford World Map. Lincoln and/or Hereford, *c.*1300. By kind permission of the Dean and Chapter of Hereford and the Hereford Mappa Mundi Trust. **63** Portolan chart of the Atlantic coasts of Europe and Africa attributed to Pietro Vesconte, *c.*1325. From an atlas accompanying Pietro Sanudo's *Liber Secretorum Fidelium Crucis*. BL Add. MS 27376*, fols. 180v-181. **65** The Gough Map of Britain. England, prototype *c.*1280, extant copy *c.*1360. Bodleian Library, University of Oxford, Gough Gen.Top.16. **67** Evesham world map (verso of sixth membrane of genealogy of Ralph Boteler, Lord Sudeley, created 1447–52); Evesham circa 1390/1415 under the patronage of Nicholas Herford, Prior of Evesham Abbey 1352–92 and Roger Yatton, Abbot of Evesham 1372–1418. College of Arms, London: Muniment Room 18/19. **69** Map of Pinchbeck Fen *c.*1430: Duchy of Lancaster records. The National Archives, Kew: MPCC 1/7. **71** Fra Mauro and Andrea Bianco, Fra Mauro world map; Venice, 1450 (?). Venice: Biblioteca Marciana / Photo SCALA, Florence. **73** *La Fleur des Histoires*. Valenciennes, 1459–63 by Jean Mansel. Map attributed to Simon Marmion (active, Amiens and Valenciennes, 1449–89). Bruxelles: Bibliothèque Royale Albert Ier, MS 9231, fol. 281v. **75** [Chart of Europe and the Mediterranean] 'Hanc cartam composuit Petrus Roselli in civitate maioricana (?) anno domini mcccclxv'. BL Egerton MS 2712. **77** The top 2/3 of chart is 4 x 4 matrix of 16 town signs from *c.*1500 BC–1800; bottom 1/3 of chart is a 2 x 2 matrix showing 4 groups of hill, vegetation and settlement signs: M. Paris (c. 1259); Wm. Smith 1601/2; Ordnance Survey 1801 & 1805; Ordnance Survey 2003. **79** World map of Juan de la Cosa, 1500. Museo Naval, Madrid P-257. **81-3** Martin Waldseemueller with Matthias Ringmann, *Univeralis Cosmographiae Secundum Ptholomaei Traditionem Et Americi Vespucii Aliorumque Lustrationes*, (St Dié or Strasbourg, 1507). Library of Congress, Washington. **85** Piri Reis. Fragment of World Map, Turkey, *c.*1513. Topkapi Saray Museum, Istanbul. **87-9** Jörg Seld and Hans Weiditz, *Sacri Romani Imperii Civitas Augusta Vindelicorum Augsburg*, 1521. BL Maps *30415 (6). **91** From Cortés, Hernando. *Praeclara Ferdinandi...* (Nuremberg, 1524), Rare Books and Manuscripts Division, The New York Public Library, Astor, Lenox and Tilden Foundation. **93-5** Nuño Garcia de Torreno (attrib,), The Salviatti World Map. Florence: Biblioteca Laurenziana Med. Pal. 249. **96** Detail showing Spouter from facsimile of original map produced in Malmo in 1949. BL Maps 184.g.3. **97** Olaus Magnus, *Carta Marina et Descriptio Septentrionalium Terrarum*, Rome, 1539. Colour reproduction produced by Generalstabens Litografiska Anstalt 1967. **98** View of the eruption of Monte Nuovo from Simone Porzio, *De conflagratione Agri Puteolani* (Naples[?]: Giovanni Sultzbach[?], 1538). Rare Books Division, New

York Public Library. **99** 'The master GA with the coltrop.', *Il vero disegnio in svi propio lvogho retratto del infelice paese di Pozvolo . . . et del monte nvovo nato in mare et in terra . . . ali 29 de setenbre 1538* (Rome [?], ca. 1540 [?]) BL K. Top. 83.71. **101** Richard Cavendish, [mouths of the Thames and Medway from Ipswich to Sandwich and Maldon and Rochester to the Sea]. BL Cotton MS Augustus I.i.53. **103** Sebastian Cabot, World map. Antwerp, 1544. Paris, Bibiliothèque Nationale de France, Départment de Cartes et Plans, Rès. Ge. AA 582. **105** One of the sheets from Le Testu's *Cosmographie Universelle*, Chateau de Vincennes, Bibliotheque du service Historique de l'Amee de Terre. Min. Defense - Service Historique de l'Armée de Terre, France, Giraudon/The Bridgeman Art Library. **107** [Chart of the Atlantic] 'Bastiam lopez a fez 1558 nouembro 15'. BL Additional MS 27303. **109-11** Giacomo Gastaldi and others, *Cosmographia universalis et exactissima iuxta postremam neotericorum traditionem / A Iacobo Castaldio nonnullisque aliis huius disciplinae peritissimis nunc [pr]imum revisa ac infinitis fere in locis correcta et locupletata* (Venice, *c.*1561). BL Maps. C. 18. n. 1. **113** Laurence Nowell, *A general description of England and Ireland with the costes adioyning*. BL Add MS 62540. **114** *Londinum feracissimi Angliae regni metropolis* from Georg Braun and Frans Hogenberg, *Civitates Orbis Terrarum* (Cologne, 1572), Detail. BL G.3603. **115** *Londinum feracissimi Angliae regni metropolis* from Georg Braun and Frans Hogenberg, *Civitates Orbis Terrarum* (Cologne, 1574), edition of 1623, volume 1. BL Maps C.29.e.1. **116-7** Frontispiece from Saxton's Atlas of the counties of England and Wales (1579) and Christopher Saxton, *Promontorium hoc in mare projectum Cornubia dicitur*... [map of Cornwall] Engraved by Lenaert Terwoort, 1576. Private Collection. **119** Map of Cholula, Mexico. Nettie Lee Benson Latin American Collection, University of Texas at Austin, USA. **121** Danti, Egnazio (1536–86): Territory of Bologna. Galleria delle Carte Geografiche, Vatican Palace, Rome. Photo SCALA, Florence. **122-3** Galleria delle Carte Geografiche viewed towards the Galleria degli Arazzi., Vatican Palace, Rome. Photo SCALA, Florence. **125** John Hooker, *Isca Damnoniorum . . . anglice . . . Exeter* engraved by Remigius Hogenberg *c.*1587. BL, Maps C.5.a.3. **127** Pont's map of the Loch Tay environs, Scotland, *c.*1583-1600. National Library of Scotland, Adv.MS.70.2.9, Pont 18 (detail). **128-9** Sheldon tapestry maps of Shropshire and Worcestershire, (details). Produced for Ralph Sheldon, 1590 (Bodleian Library, University of Oxford / V&A Images, Victoria and Albert Museum. **130** Portrait of William Cecil, 1st Baron Burghley by an unknown artist *c.*1585. National Portrait Gallery. **131** Anonymous, ['Lord Burghley's Map of Lancashire'], BL Royal MS 18.D.III, fols 82v-83. **133** *Abraham Ortelius, Utopiae Typus, Ex Narratione Raphaelis Hythlodaei, Descriptione D. Thomae Mor, Delineatione Abrahami Ortelii* (Antwerp: Ortelius, 1595/6). Private collection. **135** Jodocus Hondius, *Typus totius orbis terrarum, in quo & Christiani militis certamen super terram (in pietatis studiosi gratiam) graphicè designatur* (Amsterdam, *c.*1597). BL Maps 188.k.1.(5.). **137** Anonymous, [Survey of the manor of Chilton, Suffolk, 1597]. BL Add. MS 70953. **139** Petrus Plancius, engraved by Jan van Duetecum, *Orbis terrarum typus de integro multis in locis emendatus auctore Petro Plancio 1594* (Amsterdam, 1594), in Linschoten. *Navigatio ac Itinerarium*... (Amsterdam, 1599), BL 569.g.10(1). **141** Jean de Beins, 'Paisage de Grenoble', From the Sully Atlas BL Add. MS 21117, fols. 49v-50. **143** *Hans Woutneel, A Descripsion of The Kingdoms of England, Scotland & Ireland...1603*. (Georg-August-Universität Göttingen). **145-7** Map of Aldbourne Chase, Wiltshire 1608: Duchy of Lancaster records. The National Archives, Kew: MPCC 1/8 (ex DL 4/54/51). **149** Peter's Projection World Map. From a leaflet published by UNESCO *c.*1995. **151** William Baffin/ Thomas Roe, *A Description of East India, conteyninge th'Empire of the great Mogoll*. 1619. BL: Maps K. Top. 115.22. **153** Map of the Celebes from the East Indies Atlas by Gabriel Tatton, 1621. Royal Naval Museum, Admiralty Library. MSS 352. By permission of the Royal Naval Museum VZ 7/39. **154** Title page of Abraham Ortelius, *Theatrum Orbis Terrarum* (Antwerp, 1570). Private collection. **155** Title page of *Atlas, sive Cosmographicae Meditationes de Fabrica Mundi et Fabricati Figura.. De nuò auctus excusam sub cane vigilanti Henrici Hondii Amsterodami An.D.1623*. BL Maps C.3.c.9. **156** Bochart's engraved portrait: from Samuel Bochart, *Opera Omnia*, 4ed., 3 vols. (Leiden, 1712), BL, vol.1, 7.f.6-8. **157** *Tabula universalis locorum quae Phoenicum navigationibus maxime frequentata sunt a Taprobana Thulen usque*, opposite p. 361 ('Liber primus de Phoenicum coloniis'), from Bochart, *Geographia sacra* (Caen, 1646), BL 3105.f.4. **158** Portrait of Sir William Petty: Frontispiece from W. Petty, Hiberniae Delineatio (London, 1685) BL Maps 18.e.10. **159** *The greatest part contiguous of the Barony of Philiipstown in the Kings County [Leinster] by Patricke Ragett*. 'E. Lucas deliniavit'. BL Add. MS 72869, ff.72v-75. **160** Detail from Balthasar and Frans Florentsz. van Berkenrode, *Nova Et Accurata Totivs Hollandiae Westfrisiaeq. Topographia* (Amsterdam: Willem Jansz. Blaeu, 1621). Reproduced in Guenter Schilder, *Monument Cartographica Neerlandica* (Aalphen-an-den-Rijn: Canaletto), v (1996). **161** Johannes Vermeer, *Soldier and Laughing Girl*, Oil on canvas. Frick Collection, New York. **163** [Nicolas Cochin], *Jansenia* in Louis de Fontaines, Sieur de St Marcel, known as Père Zacharie de Lisieux, *Relation du pays de Jansenie où est traitté des singularités qui s'y trouvent, des coustumes, moeurs et religion des habitans* (Paris, 1660). BL 4051.a.13. **164** Silver medal by Thomas Simon commemorating Charles II's coronation in 1661. Private Collection. **165** *Soli Britannico Reduci Carolo Secundo regum augustissimo hoc Orbis Terrae Compendium humill. off. I. Klencke* [A collection of forty-two maps of all parts of the world, published by J. and W. Blaeu, H. Allard, N.J. Visscher and others.] 1613–60. 175 x 99 cm. BL K.A.R. **167** Anonymous [after Jacques Cortleyou] *A Description of The Towne Of Mannados: Or New Amsterdam: as it was in September 1661...Anno: Domini.1664*. BL K. Top. 121.35. **169** *Nihon Bunkei-zu* [Atlas of Japan] (Kyoto: Yoshida Tarobe, 1666). BL Maps c. 21 b. 28. **171** Richard Blome, *A Generall Mapp of the Empire of Germany with its severall Estates. Designed by Monsieur Sanson. Geographer to the French King*. (London, 1669). BL Maps 188.b.3. **173** *Virginia and Maryland as it is Planted Inhabited this Present Year 1670 Surveyed and Exactly Drawne by the Only Labour and Endeavour of Augustine Herman Bohemiensis*. 1673 (1674). Engraved by W. Faithorne. BL G. 7434. **175** John Ogilby 'The Road from London to the City of Bristol' from John Ogilby, *Britannia, volume the first; or, an illustration of the Kingdom of England and Dominion of Wales; by a geographical and historical description of the principal Roads thereof* (London: John Ogilby, 1675). Private Collection/The Bridgeman Art Library. **177** William Hubbard, *A Map of New England*, cut by John Foster, in Hubbard's *Narrative of the Troubles with the Indians* (Boston, 1677). BL G.7146. **179** Romeyn de Hooghe, *Luxemburgum* (Amsterdam: N. Visscher, 1684). From the Beudeker Collection BL Maps C.9.d.1 (27). **181** Coronelli, Vincenzo, *MARE DEL SUD,/ detto altrimenti/ MARE PACIFICO/Auttore il P.M. Coronelli M.C. Cosmografo della/SERENISSIMA REPUBLICA DI VENETIA/dedicato/ALL'ILL.MO ET ECC.MO SIGNOR IL SIGNORE/CAVALIERE GIULIO GIUSTINIAN/Savio Grande*, Venezia, [1688]. In Vincenzo Coronelli, *Atlante Veneto*... , (Venice: Albrizzi, 1691). BL Maps 44.f.6(1), tav.28. **183** Diwali Celebrations at the Royal Palace at Kotah, Rajasthan, *c.*1690. By permission of the National Gallery of Victoria, Melbourne (cat. No. 52), Felton Bequest, 1980. **185** Avraham bar Ya'akov, Map of the Holy Land: the Route of the Exodus. From *Haggadah shel Pessach* (Amsterdam, 1695). BL 1974.f.5. **187** Frederik de Wit, *'t Koninckleyck Huys te Ryswyck. Amsterdam*, 1697. From the Beudeker Collection, BL Maps C.9.e.8 (68). **189** John Harris, *A New Map of Europe Done from the most Accurate Observations Communicated by the Royal Societies at London & Paris Illustrated with Plans &c Views of the Battles, Seiges and other Advantages obtained by her Majesties Forces and those of Her Allies over the French* (London: Thomas Bowles, *c.*1720). BL Maps CC.6.a.2. **191** Hermann Moll *A New and Exact MAP of the DOMINIONS of the KING of GREAT BRITAIN on ye Continent of NORTH AMERICA*, London, 1715. **192-3** Isaac Causton, *A map of Madam Margaret Bonnell's land lying in the parishes of Brent-Eligh and Preston in the County of Suffolk*, 1719. BL Add. MS 69459. **195** Thomas Hogben, *The Topography and Mensuration of all the Flow'd Lands lying in the Valley between Ashford Bridge and Long Bridge Bidges [sic] Measured by order of the Honble Commissioners of Sewers. Containing one Hundred Thirty Six Acres; Three Roods & Thirty five Perches*. 1720. Centre for Kentish Studies, Maidstone, S/EK P1. **197** Homann Heirs, *Accurater Prospect und Grundriss der Königl: Gros-Britan[n]isch: Haupt- und Residentz-Stadt. London*. Nuremberg, *c.*1730. Private collection. **199** Francisco Alvarez Barreiro, *Plano corographico é hydrographico de las provincias de el Nuevo Mexico*. Reduced and copied by Luis de Surville, 1770. BL Add.MS17650.B. **201** John Finlayson, *A plan of the battle of Culloden and the adjacent country, shewing the incampment of the English Army at Nairn, and the march of the Highlanders in order to attack them by night*. *c.*1746. BL Maps*9115(3). **203** Map screen with maps by William Berry, Herman Moll, Henry Popple and Emanuel Bowen. Assembled and attached to a pine frame after 1746. BL Maps Screen 1. **204** *An exact View of the Ruins of the Houses burned down by the fire in Cornhill, the 25th March, 1748. Taken from Sam's Coffee House in Exchange Alley*. BL Maps K.Top.27.34. **205** Thomas Jeffreys, *A Plan of all the Houses Destroyed and Damages by the Great Fire which begun in Exchange Alley, Cornhill, on Friday March 25th 1748*. BL Maps K. Top.21.9. **207** Jain Cosmological Map, Western India, mid-18th century, Adhai-dvipa. BL Or.MS. 13937. **208** Plate II in *A Treatise of maritim [sic] surveying*... by Murdoch Mackenzie, Senior. London: Edward and Charles Dilly, 1774. BL 60.d.6. **209** *The north east coast of Orkney with the rocks tides soundings &c.*, surveyed and navigated by Murdoch Mackenzie: Map IV in: *Orcades: or, a geographic and hydrographic survey of the Orkney and Lewis Islands, in eight maps*... London: (the author), 1750. BL Maps 13.Tab.5. **211** Maps of Hangzhou and the imperial 'travelling palaces' prepared for the Qianlong Emperor *c.*1751. BL Or. 13994. **213** John Speed, *Cambridgeshire Described with the devision of the hundreds, the Townes situation, with the Armes of the Colleges ... and also the Armes of all such Princes and noble-men as have ... borne the ... tytles ... of the Earldome of Cambridg. Performed by John Speede ... Anno. 1610 ... from John Speed, The Theatre of the Empire of Great Britain*. Private collection. **215** *Plan stolichnago goroda Sankt Peterburga... (Plan de la Ville de St. Pétersbourg...)* by I. Truscott. St Petersburg: Academie Imperiale des Sciences & des Arts, 1753. BL Maps K.Top.112.76.5 TAB.END. **217** Detail showing Paris and Versailles, Cassini Map of France, *Addition aux tables astronomiques de M. Cassini. Par M. Cassini de Thury.*, Paris, 1756. BL T.C.6.a.24. Feuille 1. **218** James Gillray, *A New Map of England: The French Invasion 1793*. Courtesy of the Warden and Scholars of New College, Oxford/The Bridgeman Art Library. **219** *La Manche*. From a volume of plans for a proposed invasion of Britain, *c.*1762. Depot de la Marine. BL Add. MS 74751. **221** Madame Le Pauté Dagelet *Passage de l'Ombre de la lune au travers de l'Europe*... (Paris, engraved by Mme, Lattré and by Elisabeth Claire Tardieu, 1764) BL Maps CC.5.a.400. **223** George Adams. Terrestrial and Celestial Globes. *c.*1766. Copyright Erik de Goederen from Elly Dekker and Peter van der Krogt, *Globes from the Western World*, (London: Zwemmer, 1993). **225** Eclectic World Map, middle to late eighteenth century. Museum für Islamische Kunst, Staaatliche Museen Preussischer Kulturbesitz, Berlin (inv. No. I. 39/68). **226** James Gillray, *The Plum Pudding in Danger; -or- State Epicures taking un Petit Souper*. (London: H. Humphrey, 1805). Private Collection/The Bridgeman Art Library. *227 The Troelfth Cake/Le Gateau des Roys* (London: Robert Sayer, 1772). BL Maps cc.5.a.115. **229** Le Chevalier Jean de Beaurain, *Carte du Port et Havre de Boston*, Paris, 1777. William L. Clements Library, Ann Arbor Michigan; Atlas W-A-B. **231** Anonymous, *El Collao*. William L. Clements Library, Ann Arbor, Michigan. **233** I. W. Green, *Plan of the Encampment on Finchley Common under the command of Major Genl Faucit from 10th Augt to 20th Octr. 1780* BL Add. MS 15533, fol. 51. **235** *Hindoostan*, by James Rennell. 1782. BL Maps C.25.b.8. **237** Richard Marshall *A New and Correct Map of the World*... 1785. BL Maps M.T.11.g.1.(9.). **239** *Cary's Survey of the High Roads from London...On a Scale of one Inch to a Mile...with...the different turnpike gates, shewing the connection which one trust has with another*. (London, J. Cary, 1790). BL 577.e.11. **240** Fitzroy Square. From Thomas Malton: *A Picturesque Tour through the Cities of London and Westminster, illustrated with the most interesting views, accurately delineated and executed in Aquatinta. London 1792-[1801]* BL 190.f.7. **241** Extract from Richard Horwood, *Plan of the cities of London and Westminster, the Borough of Southwark, and parts adjoining shewing every house*. BL Maps 148.e.7. **243** Gujarati Chart of the Red Sea and Gulf of Aden, late eighteenth or early nineteenth century. Royal Geographical Society (Asia S.4). **245** *Ch'onhado*, 'Map of All Under Heaven', Korean woodcut, 1800. BL Maps c.27.f.14. **246** W. Edy, *View from Alderman Macauley's garden at Blackheath*, 1790: BL K. Top. 18.5-d. **247** Thomas Milne, *Milne's Plan of the Cities of London and Westminster, circumjacent Towns and Parishes &c, laid down from a Trigonometrical Survey taken in the Years 1795-99* (London: Thomas Milne, 1800) BL K.Top, 6.95. **249** *To the King's most excellent Majesty, this Map of the Island of Jamaica, constructed from actual surveys, under the Authority of the Hon. House of Assembly ... is inscribed by ... James Robertson, A.M. ... Engraved by S.J. Neele* (London: J. Robertson, 1804), Royal Geographical Society. **251** *The Waterloo Map*. Reconniassance survey by Royal Engineer Officers, 1814–15. By permission of the Royal Engineers Museum, Chatham. **253** *The South End of Hounslow Heath*. Quarter Master Gens Office, Horse Guards, May 20th 1818. BL Maps C.18.m.1.(10b.). **255** *Map of 24 miles round the city of Bath... by... C. Harcourt Masters. Coloured geologically by the Revd. W.D. Conybeare F.R.S. &c. and H.T. De la Beche Esq. F.R.S. &c.* (Bath, J. Barratt, 1823). Earth Sciences Library, Natural History Museum. **257** *The Court Game of Geography* (London: William and Henry Rock, 1839 but originally 1827). BL. **259** *Map of Munster prepared for the purpose of presenting ... the state of this disturbed and distressed province by ... N.P.Leader* (1827-8). BL Add. MS 63362. **260** Detail from Osborne's map of the Grand Junction Railway (Birmingham: E. C. & W. Osborne, 1840). BL Maps 1233 (15). **261** *A Map of the Inland Navigation, Canals and Rail Roads with the situations of the various Mineral Productions throughout Great Britain...by J Walker, Land and Mineral Surveyor, Wakefield...1830*. Private collection. **263** Auguste-Henri Dufour and Leonard Chodzko, *Carte Routiere, Historique et Statistique des Etats de l'ancienne Pologne*, Paris 1831. BL Maps C.44.b.52. **265** Joseph James Forrester, *O Douro Portuguez e Paiz adjacent, com tanto do Rio quanto se pode tornar navegavel em Espanha = The Portuguese Douro and the adjacent country, and so much of the river as can be made navigable in Spain*... (London: J.Weale, 1848). BL Maps 19935. (4.) **267** Anon., *O'Connorville, the first Estate purchased by the Chartist Co-operative Land Company, Situate in the Parish of Rickmansworth, Hertfordshire. Founded by Feargus O'Connor Esq[ui]r[e] 1846*. BL. Maps 162.s.1. **269-71** Smith Evans, Emigration map of the world, (London: Letts, Printed by Son and Steer, 1851). Royal Geographical Society. **273** Karl von Czoernig, *Ethnographische Karte der oesterreichischen Monarchie* (Vienna: Militärgeographisches Institut, 1855). BL Maps 6.b.53. **275** [Map of part of the Himalayas] in *Proceedings of the Royal Geographical Society* (1858). Royal Geographical Society. **276** Washington A. Roebling in Civil War Uniform. Roebling Collection, Institute Archives and Special Collections, Rensselaer Polytechnic Institute, Troy, NY. **277** *Map of the Battle-Field of Bull Run*. Manuscript. Cartographic and Architectural Branch, National Archives, Washington, D.C., U.S.A RG 77:G 122 1/2. **279** Victor Marlet, *Carte Utopique de l'Europe Pacifiée*. Paris, (1867?). BL Maps 1035 (36). **281** Map of South Africa before and after Livingstone's explorations. Royal Geographical Society. **283** Detail from Peter Schenk, *Le Royaume de Bohème, divisée en ses douze cercles; Carte réduite sur celle de 25 feuilles par J.C. Müller*. (Amsterdam: Peter Schenk, 1745). Private collection. **285** F. W. Rose, *Serio-Comic War Map for the year 1877. Revised Edition*. (London: G.W. Bacon & Co., 1877) BL Maps *1078.(46.). **287-9** Shan Map Relating to a Border Dispute Between (British) Burma and China, circa 1889. By permission of the Syndics of Cambridge University Library (Scott LR. 13.34). **291** *The Cycle Road-Map of 50 Miles Round London*. (London: G.W. Bacon & Co., 1890). Private Collection. **293** Descriptive map of London poverty. From: Charles Booth, *Labour and Life of the People*. London: Williams & Norgate, 1891. BL Maps C21.a.18. **295** (top) *Stanley in Africa 1867 to 1889* (London and Liverpool: George Philip and Son, *c.*1890). BL Maps 66110 (7). (bottom) *A New Map of Africa Exhibiting its Natural and Political Divisions* (C. F. Cruchley, Map-seller, 81 Fleet Street, 1865). BL Maps 63510 (96). **297** Alexander Supan, *Tiefenkarte des Weltmeeres (Gotha: Justus Perthes, 1899) in Dr A. Petermanns Mitteilungen 45* (1899) 45. BL PP. 3946/2. **299** Claude O. Lucas under the supervision of Captain A.J. O'Meara. R.E., *Siege of Kimberley*. 1900. BL Add. MS 71232.A. **301** Eastern Turkey in Asia: 1:250,000 scale map of Van-Bitlis, Sheet 19 of Intelligence Division, War Office No. 1522. Published 1901. BL Maps 152.d.2. **303** Admiralty Chart. *Port Mudros*. 1903. BL Maps Sec.5.(1661.) **305** Ordnance Survey of England and Wales. Sheet 144. Revised 1912–13. Printed 1914. Private Collection. **307** Macdonald Gill, *The Wonderground Map of London Town*. London, 1914. BL Maps 3485.(199.) **308-9** Aviators in Constantinople before take-off and Plan of air crash of 27th February 1914. Ottoman military archive of the Chief of General Staff, Military History and Strategic Studies (Askeri Tarih ve Stratejik Etut Baskanligi), Ankara. **311** Sketch Map of Antarctica drawn by Sir Ernest Shackleton. Royal Geographical Society. **313** Propaganda poster from India, ca. 1914. By permission of the Imperial War Museum, Duxford. **314** Detail of the right side of the library taken from Mary Sterwar-Wilson, *Queen Mary's Doll's House* (London: Bodley Head. 1988). **315** *Atlas of the British Empire. Reproduced from the original made for Her Majesty Queen Mary's Dolls' House.* (London: E. Stanford, 1924). 3.5 x 4.2 cm. BL Maps C.7.a.26. **317** *The New Map of Europe* from J. F. Horrabin, *The Plebs Atlas* (London: The Plebs League, 1926). BL Maps 47.b.9. **319** *Paris-Sud: Carte Officielle du Service Géographique de l'Armée* (Paris: Société EPOC, 1930). Private Collection. **320** Copyright Ken Garland from *Mr Beck's Underground Map: A History*, Ken Garland (Middlesex, Capital Transport publishing, 1994). **321** (top) 1906 Map of London's Electric Railways. (bottom) Harry Beck's original design, 1931. Both with permission of the London Transport Museum. **323** *One People, One Empire, One Leader*, 1938 (colour litho), German School (20th century)/Private Collection/ Archives Charmet/The Bridgeman Art Library. **325** J. I. Lenke, *Aufmarsch der Politischen Leiter Reichsparteitag 1939 Zeppelinwiese*, overprinted on Leonhard Amersdorfer, *Plan von Nürnberg-Fürth*. BL Add. MS 63747 A. **327** G. Hulbe, *Einwanderer erster und zweiter Generation aus Mittel- und Westeuropa*, Stuttgart and Hamburg, 1940-1. BL Maps X.3817. **329** Germany, Luftwaffe, Generalstab. [Great Britain bombing maps.] Sheet showing Manchester. 30 June 1941. BL Maps Y.415. **331** Letter from Captain Clayton Hutton of M.I.9 to Norman Watson of Waddington, 26th March 1941. Escape map of Italy, printed on silk. S.l., M.I.9, 1941. BL Maps C.49.e.55. (1.). **332-3** *Standard Oil Interstate Route Map, Southern Routes, US 60, 66, 70, 80* (Chicago and San Jose: H.M. Gousha for Standard Oil Company of California, 1941). Private collection of the author. **335** Ordnance Survey (Cartography by Phyllis M. Boyd). *Great Britain. Land Utilisation. Sheet 2* Scale 1:625,000. (Southampton, 1943). National Library of Scotland. **337** Copy of an Ordnance Survey map annotated 1943 (and circa 1960) on Ministry of Defence (Defence Lands Service) file. The National Archives, Kew: DEFE 51/9. **339** United States Office of Strategic Services (Branch of Research and Analysis), *Billeting and control facilities for displaced persons Hannover and Bremen*, 26 February 1945. BL Maps 26907. (192). **341-3** *Chicago-to-Gander Aeronautical Route Chart (No. 2201)* (United States Coast and Geodetic SurveyAdvanced Edition, January 3, 1947), 66cm x 137cm. 1:2,000,000 scale, Oblique Mercator Projection. American Geographical Society Library, at the University of Wisconsin-Milwaukee. **345** *Oksford*. 1:10,000. Moscow: Soviet General Staff, 1973. Bodleian Library, University of Oxford – C17:70 Oxford (154). **347** Satomi Matoba, *Map of Utopia*, 1998 utilising the Japan Geographical Survey Institute, 1:50000 topographic sheet N1-53-33-10 HIROSHIMA (1994) and the U.S.Geological Survey, 1:62500 topographic sheet N2115-W15739/28x38 ISLAND OF OAHU (1954, limited revision in 1970). By permission of Satomi Matoba. **348-9** Stirling Surveys, *Border Loop – The Scottish Borders Cycle Tour* (Footprint Maps and Scottish Borders Tourist Board, 2001). By permission of Stirling Surveys. **351** Analysis of multiple satellite radar images of London acquired by the ERS satellite between 1991 and 2002. (Copyright Nigel Press Associates Ltd.). **353** City of Detroit, Planning Information & Resource Center. Future General Land Use. City boundaries. US Census. Census 2000 Low and Moderate Income Summary Data. Block Groups. Maps and graphics created using MapInfo Professional Version 7.8 software. **355** Data gathered by Advanced Spaceborne Thermal Emission and Reflection Radiometer (ASTER) image, on board NASA's Terra satellite. Earth Observing System mission. Captured 15 March, 2005. NASA.

Index

First published in Great Britain in 2005
by Weidenfeld & Nicolson
10 9 8 7 6 5 4 3 2 1

A CIP catalogue record for this book is available from the British Library.

ISBN 0 297 84372 9

Design director: David Rowley
Designed by Clive Hayball
Picture research by Brónagh Woods
Editorial by Jennie Condell, Nicky Granville, Jo Murray and Brónagh Woods

Printed and bound in Italy

Weidenfeld & Nicolson
The Orion Publishing Group Ltd
Wellington House
125 Strand
London, WC2R 0BB

Acknowledgements

This book would not have been possible without the enthusiasm and commitment of Michael Dover, who approached me to edit it and who contributed several texts. I am most grateful to all the other contributors (including many of my British Library colleagues) who provided their pieces, ideas for other pieces and often illustrations at short notice and without complaint. No fewer thanks go to Jennie Condell, whose enthusiasm and hard work ensured that the book came together and got printed, and to her predecessor Nicky Granville. I owe great debts of gratitude to the inspired designer, Clive Hayball, to Brónagh Woods who co-ordinated the work on the pictures, to my colleagues at the British Library and particularly to Nicola Beech, Judith Cooper, Neil Fitzgerald, Andy Ogilvie and the other BL photographers, and to David Way who suggested to Michael Dover that I might be interested in the project.

Lastly, and above all, I thank my family and particularly my wife Christiane for the good humour with which she has run the house and kept me fed and clothed at what was a stressful time for her despite my absences and erratic hours, even if she says that she is getting used to them. I dedicate this book, with all my love, to her, and to Philip and Cecily.

VER
TERRA
SEPTENTRIO
AMERICA
Anno Domini 1492 a Christophoro Columbo nomine Regis Castellæ primum detecta, et ab Americo Vesputio nomen sortita 1499
AMERICA SEPTEN TRI ONALIS
Nova Albion
Hic ad latitud. gr. 72 delatus erit anno 1586 Ioannes Davisius cum præ nimia glacie coactus fuit reverti.
Magni huius Oceani primus Obiector Ille fuit Mr. Hudson.
Frisland
TERRA DE LABRA DOR
NOVA FRANCIA
Asores Ins.
al. Flandricæ
Virginia
Barmuda
Sinus Mexicanus
Hispaniola
Cuba
MAR DEL NORT
Tropicus Cancri
MAR DEL ZUR
Circulus Æquinoctialis
Tropicus Capricorni
Cariban a
Guiana
AMERICA MERIDIONALIS sive PERUVIA
CHILI
Rio de la Plata
Fretum Magellanicum
TERRA AUST RALIS INCOGNITA
Circulus Antarcticus
MERIDIES
AUTUMNUS
L.S.
Multa priscis incognita, hodie deteguntur: nec veterum modò errores, at novæ etiam terræ demonstrantur. De iisdem non pauci tabulas ediderunt; sed (liceat dicere) nullus adhuc prodijt Mundi Typus qui tam concinnè, in exiguo spatio singula contineret: videbis in America Septent. plurima mutata esse, sunt etiam in Oceano Tartarico et circa illum nova permulta; sunt et alibi. Amice Lector, utere hac Tabula novissima et perfectissimà; sic enim est: nam ut cæteri taceant, Pindarus testis est: dies inquit sequentes testes sunt sapientissimi.